Issues in Environmental Geology: a British Perspective

Issues in Environmental Geology: a British Perspective

Edited by
Matthew R. Bennett & Peter Doyle
School of Earth Sciences, University of Greenwich

With editorial assistance from
Linda Murr, Stephen J. Edwards & Jonathan G. Larwood
University of Greenwich & English Nature

1998
Published by The Geological Society

THE GEOLOGICAL SOCIETY

The Society was founded in 1807 as The Geological Society of London and is the oldest geological society in the world. It received its Royal Charter in 1825 for the purpose of 'investigating the mineral structure of the Earth'. The Society is Britain's national society for geology with a membership of around 8500. It has countrywide coverage and approximately 1500 members reside overseas. The Society is responsible for all aspects of the geological sciences including professional matters. The Society has its own publishing house, which produces the Society's international journals, books and maps, and which acts as the European distributor for publications of the American Association of Petroleum Geologists, SEPM and the Geological Society of America.

Fellowship is open to those holding a recognized honours degree in geology or cognate subject and who have at least two years' relevant postgraduate experience, or who have not less than six years' relevant experience in geology or a cognate subject. A Fellow who has not less than five years' relevant postgraduate experience in the practice of geology may apply for validation and, subject to approval, may be able to use the designatory letters C Geol (Chartered Geologist).

Further information about the Society is available from the Membership Manager, The Geological Society, Burlington House, Piccadilly, London W1V 0JU, UK. The Society is a Registered Charity, No. 210161.

Published by The Geological Society from:
The Geological Society Publishing House
Unit 7, Brassmill Enterprise Centre
Brassmill Lane
Bath BA1 3JN
UK
(*Orders:* Tel. 01225 445046
Fax 01225 442836)

First published 1998

British Library Cataloguing in Publication Data
A catalogue record for this book is available from the British Library.

ISBN 1-86239-014-2

Typeset by Bath Typesetting, Bath, UK
Printed by The Alden Press, Osney Mead, Oxford, UK

Distributors

USA
AAPG Bookstore
PO Box 979
Tulsa
OK 74101-0979
USA
(*Orders:* Tel. (918) 584-2555 Fax (918) 584-2652)

Australia
Australian Mineral Foundation
63 Conyngham Street
Glenside
South Australia 5065
Australia
(*Orders:* Tel. (08) 379-0444 Fax (08) 379-4634)

India
Affiliated East-West Press PVT Ltd
G-1/16 Ansari Road
New Delhi 110 002
India
(*Orders:* Tel. (11) 327-9113 Fax (11) 326-0538)

Japan
Kanda Book Trading Co.
Cityhouse Tama 204
Tsurumaki 1-3-10
Tama-shi
Tokyo 206-0034
Japan
(*Orders*: Tel. (0423) 57-7650
Fax (0423) 57-7651)

Contents

List of Contributors

Jane Anderton
School of Earth and Environmental Sciences, University of Greenwich, Medway Campus, Chatham Maritime, Kent ME4 4TB

Jan Barron
Earth Resources Centre, University of Exeter, North Park Road, Exeter EX4 4QE

Andrew F. Bennett, MP
Chair of the House of Commons Environment Select Committee, House of Commons, Westminster, London SW1A OAA

Matthew R. Bennett
School of Earth and Environmental Sciences, University of Greenwich, Medway Campus, Chatham Maritime, Kent ME4 4TB

Anthony R. P. Cosgrove
School of Earth and Environmental Sciences, University of Greenwich, Medway Campus, Chatham Maritime, Kent ME4 4TB

Gary D. Couples
Department of Geology and Applied Geology, University of Glasgow G12 8QQ

David Cuckson
Environmental Law Group, Stephenson Harwood, One St Pauls Churchyard, London EC4M 8SH

Peter Doyle
School of Earth and Environmental Sciences, University of Greenwich, Medway Campus, Chatham Maritime, Kent ME4 4TB

Stephen J. Edwards
School of Earth and Environmental Sciences, University of Greenwich, Medway Campus, Chatham Maritime, Kent ME4 4TB

J. Murray Gray
Environmental Science Unit, Department of Geography, Queen Mary & Westfield College, University of London, Mile End Road, London E1 4NS

R. S. Haszeldine
Department of Geology and Applied Geology, University of Glasgow G12 8QQ

Elizabeth Y. Haworth
Institute of Freshwater Ecology, The Ferry House, Ambleside, Cumbria LA22 0LP

David J. Horne
School of Earth and Environmental Sciences, University of Greenwich, Medway Campus, Chatham Maritime, Kent ME4 4TB

Jeff Jones
Earth Resources Centre, University of Exeter, North Park Road, Exeter EX4 4QE

John Lamont-Black
Department of Civil Engineering, University of Newcastle, Castle Buildings, Claremont Rd, Newcastle-upon-Tyne NE1 7RU

Jonathan G. Larwood
English Nature, Northminster House, Peterborough PE1 1UA

Chris Lee
Minerals Division, School of Applied Sciences, University of Glamorgan, Pontypridd, Mid Glamorgan, South Wales CF37 1DL
Jonanthan McCue
Sir William Halrow and Partners Ltd, Burdrop Park, Swindon, Wiltshire SN4 0QO
C. McKeown
Department of Petroleum Engineering, Heriot-Watt University, Edinburgh EH14 4AS
Alan P. McKirdy
Scottish Natural Heritage, Anderson Place, Edinburgh EH6 5NP
John Merefield
Earth Resources Centre, University of Exeter, North Park Road, Exeter EX4 4QE
Tom Moat
English Nature, Northminster House, Peterborough PE1 1UA
Siegbert Otto
Kimberley Service, Reading RG1 6LA
Jane Poole
Symonds Travers Morgan, Symonds House, Wood Street, East Grinstead, West Sussex RH19 1UU
Fergal Quinn
School of Earth and Environmental Sciences, University of Greenwich, Medway Campus, Chatham Maritime, Kent ME4 4TB
Jo Roberts
Earth Resources Centre, University of Exeter, North Park Road, Exeter EX4 4QE
Michael Rosenbaum
Faculty of Environmental Studies, Nottingham Trent University, Burton Street, Nottingham NG1 4BU
Edward P. F. Rose
Department of Geology, Royal Holloway, University of London, Egham, Surrey TW20 OEX
D. K. Smythe
Department of Geology and Applied Geology, University of Glasgow G12 8QQ
Ian Stone
Earth Resources Centre, University of Exeter, North Park Road, Exeter EX4 4QE
Alan Thompson
Symonds Travers Morgan, Symonds House, Wood Street, East Grinstead, West Sussex RH19 1UU
Nick R. G. Walton
Department of Geology, University of Portsmouth, Burnaby Road, Portsmouth PO1 3QL
David S. Wray
School of Earth and Environmental Sciences, University of Greenwich, Medway Campus, Chatham Maritime, Kent ME4 4TB

Preface

Environmental geology is currently in vogue. Popular with students because it presents an 'acceptable face' to what has often been perceived as an exploitative science, most earth science departments offer undergraduate or postgraduate degrees in the subject. Yet there is debate about its significance and relevance: is it a passing phase; an adjunct to engineering geology/applied geomorphology; or a separate discipline in its own right?

It is our belief that environmental geology is a logical and much needed development in the earth sciences; a meld of applied geomorphology, economic and engineering geology – a truly integrative applied science. Effectively, environmental geology is managing the interaction of humans in all its diverse forms with the geological environment. This involves the provision of economic resources, mitigation of natural hazards and management of the built environment. Within this framework one can identify four major issues which are currently important in Britain and northern Europe: the conservation and management of natural areas; the disposal of waste and the challenges of engineering geology; the mitigation of coastal hazards; and the provision of geological information and training of future practitioners. The need to debate these issues stimulated the conference and led to this volume. Its aim was to bring together practitioners from academic, professional, legal and political spheres in order to debate, inform and educate a wider audience in the vast potential of this burgeoning subject. We hope this volume provides a collection of case studies and ideas of interest to the earth science community at large.

Matthew R. Bennett & Peter Doyle
Chatham Maritime
September 1997

Acknowledgements

This publication arose from a conference held at the University of Greenwich on 17 January 1997. We are grateful for the support and financial aid provided by English Nature who co-convened the conference with the School of Earth & Environmental Sciences. The effectiveness of the meeting in putting across many of the current issues in environmental geology to a wide audience was greatly enhanced through the administrative ability of Linda Murr, who has also been responsible for gathering together and formatting the papers in this volume. Our colleagues, Jonathan Larwood and Stephen Edwards, aided us in planning both the conference and the volume, and we are grateful for their input and interest. Finally we would like to thank those postgraduates who were dragooned into assisting with all those difficult jobs which a conference demands: Jane Anderton, Kez Baggley, Nick Bergquist, Tony Cosgrove, Richard Pole, Jo Rosso, Kate Spencer, Alistair Wells and Jason Wood.

1 Issues in environmental geology: a British perspective

Matthew R. Bennett & Peter Doyle

Summary

- The definition of environmental geology and the evolution of the discipline is explored.
- We argue that environmental geology is about environmental management or 'sustainable development'.
- Key themes which define the environmental geology agenda in Britain are:
  (1) the conservation and management of natural areas;
  (2) problems of waste disposal and engineering geology;
  (3) coastal hazards; and
  (4) the need to educate future practitioners in environmental geology.

Environmental geology can be defined as the interaction of humans with the geological environment (Bennett & Doyle 1997). It attempts to place human activity within the broader context of the Earth (Woodcock 1994). The geological environment includes not only the physical constituents of the Earth, its rocks, sediments, soils and fluids, but also the Earth's surface, its landforms, and in particular the processes which operate to change it through time. This environment is both a resource and a hazard to human development, and is essential to life. It provides us with water, industrial minerals, building materials and fuel. It constrains the location and architecture of our urban settlements and transport infrastructure. It provides us with the means for the effective disposal of our waste products. However, although the geological environment sustains us with these essential elements, it also generates some of the most potent hazards to our existence in the form of earthquakes, volcanoes and floods.

Environmental geology and sustainable development

To many, environmental geology is simply the study of human impacts on the environment (e.g. Nathanail 1995). In this context the main role for environmental geology is the documentation and mitigation of the environmental impacts of mineral exploration and exploitation, of the disposal of wastes and of construction and engineering projects. A subsidiary branch of environmental geology, concerned with natural hazards, is focused introspectively; what impact will these hazards have on us? This approach, although valid, tends to cast all geological exploitation as potentially damaging, leading to conflict with the needs of society for resources and defences against natural hazards. An alternative view is that environmental geology is about management of a natural system, a concept which has recently been subsumed into the popular term 'sustainable development'. While this term is widely applied, often meaning different things to different people, it is here defined as:

> the management of natural resources to support continued social and economic development in such a way that renewable resources are not depleted, and that the impact of the extraction and use of non-renewable resources is minimized (Bennett & Doyle 1997).

Put simply, this provides for development, but not at an unacceptable cost to the natural environment. The key to sustainable development is effective environmental management in, for example, balancing the need for the exploitation of geological resources with any environmental impacts that this exploitation may have.

Environmental geology should have as its central philosophy the goals of effective environmental management within the context of sustainable development. It has four main components (Coates 1981; Keller 1992; Woodcock 1994; Bennett & Doyle 1997).

1. Managing geological resources, such as fossil fuels, industrial minerals and water. This is a process which involves not only mineral exploration and exploitation, but also the limitation and mitigation of the environmental damage caused by this

resource use.

2. Understanding and adapting to the constraints on engineering and construction imposed by the geological environment; a subject of particular importance in regions of climatic extreme.
3. Appropriate use of the geological environment for waste disposal in order to minimize problems of contamination and pollution.
4. Recognition of natural hazards and the mitigation of their impacts on human infrastructure.

The history of environmental geology

Environmental geology is not a new subject, but a meld of the three traditional applied earth sciences – applied geomorphology, economic geology and engineering geology (Fig. 1.1; Bennett & Doyle 1997). This meld is the result of two main trends. First, the rise in interest in the environment which was popularized under the banner of sustainable development during the 1990s and has placed environmental management firmly on the political agenda. Second, there has been a gradual shift over the last three decades from a reactive to proactive stance within applied earth science. Traditionally, the role of the applied earth scientist was a reactive one; a hazard or resource requirement was identified, and an engineering solution or resource frontier sought. However, in dealing with geomorphological hazards, for example, prevention is often better than cure. By promoting development away from areas vulnerable to natural hazards, costly future mitigation work can be avoided. As a consequence, there has been a progressive shift towards proactive work within applied earth science as the importance of its contribution to environmental and resource management has become better appreciated by planners, politicians and decision-makers in general. Increasingly, the end-user of applied earth science data requires a breadth of information not only, for example, on resource distribution, but also on hazards and ground conditions. These types of integrated applied earth science data are becoming the basic raw material for local and regional structure plans (Culshaw *et al.* 1987; McCall & Marker 1989). The challenge

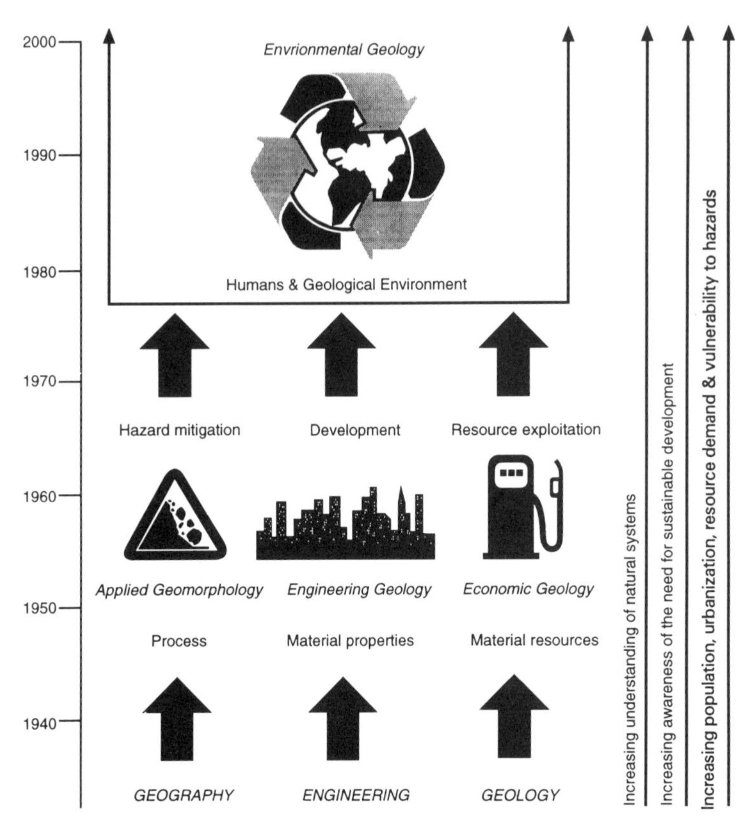

Fig. 1.1. Environmental geology as the product of applied geomorphology, economic geology and engineering geology (modified from Bennett & Doyle 1997)

of supplying these types of data in an integrated and user-friendly format has led to the first real and effective integration between applied geomorphology, and the twin fields of engineering and economic geology.

Issues in environmental geology in Britain

As one of the most densely populated and urbanized nations in Europe, Britain has a wide range of environmental geology problems (Lumsden 1992). The nature of these problems to some extent reflects the political will and current economic constraints of the day. Four themes, applicable to Britain and the whole of northern Europe, can be identified: (1) conservation and management of natural areas; (2) waste disposal and engineering geology; (3) coastal hazards; and (4) education.

Conservation and management of natural areas is of increasing importance as land-use pressure on them builds. Managing these areas and achieving a balance between the ever-increasing demand for green-field development sites and the undiminished and increasing need for recreation is one of the most important challenges. Within the heart and hinterland of the built environment, the challenges are even greater, as ever more sophisticated civil engineering projects stretch the envelope of knowledge surrounding the design of safe geological foundations. Equally, our urban areas produce vast quantities of waste which must be effectively managed. A major issue is that of contaminated land produced by poor waste management in the past – this must be cleansed to reduce land-use pressure on undeveloped areas and is a key area for future study. Natural hazards are often perceived in Britain as problems restricted to foreign soil, since the few earthquakes experienced are limited in magnitude and we are hardly in a position to be influenced directly by active volcanoes. Yet coastal erosion and flooding pose major hazards for a crowded island such as ours and these issues are highly emotive. Finally, education is a vital part of modern environmental geology, both in training future environmental geologists and in educating decision-makers of the need to incorporate geological data within the decision-making process (Culshaw *et al.* 1987; Woodcock 1995).

The chapters within this volume explore these four themes and in so doing provide a series of case studies which illustrate some of the major challenges faced by environmental geologists, be they academics, politicians or decision-makers in general. Together they help shape the frontiers of environmental geology in Britain as we approach the new millennium.

References

Bennett, M. R. & Doyle, P. 1997. *Environmental Geology: geology*

and the human environment. John Wiley & Sons, Chichester.
Coates, D. R. 1981. *Environmental Geology*. John Wiley & Sons, New York.
Culshaw, M. G., Bell, F. G., Cripps, J. C. & O'Hara, M. (eds) 1987. *Planning and Engineering Geology*. Geological Society, London, Engineering Special Publication, **4**.
Keller, E. A. 1992. *Environmental Geology* (6th edn). Macmillan, New York.
Lumsden, G. I. (ed.) 1992. *Geology and the Environment in Western Europe*. Oxford University Press, Oxford.
McCall, J. & Marker, B. (eds) 1989. *Earth Science Mapping for Planning, Development and Conservation*. Graham & Trotman Ltd, London.
Nathanail, P. 1995. What is environmental geology? *Geoscientist*, **5**, 14–15.
Woodcock, N. 1994. *Geology and Environment in Britain and Ireland*. UCL Press, London.
Woodcock, N. 1995. Environmental geology: educational threat or opportunity? *Geoscientist*, **5**, 11–13.

2 Environmental geology: achieving sustainability

Jonathan G. Larwood & Tom Moat

Summary

- Environmental geology is concerned with 'geology as it relates to human activities' which requires an understanding of earth processes in order to develop and use the Earth's resources, and to manage waste and pollution.
- Sustainable development aims to reconcile the seemingly opposed goals of economic development and environmental protection.
- Environmental geologists provide a key link between understanding geological processes and the use of geological resources and are therefore uniquely placed to influence decision-makers into making appropriate decisions when managing resources or conserving the environment.

The inevitable outcome of the Earth's expanding population is the increased consumption of natural resources and the consequent increased production of wastes which, if poorly managed, may lead to contamination and pollution. Most people now accept that human action is a significant cause of environmental change today, and unless this issue is addressed, there may be adverse physical and socio-economic impacts on a global scale. International approaches to solving these problems are being co-ordinated under the banner of 'sustainable development.' This term was popularized by the 1992 Rio de Janeiro UN Conference on Environment and Development, UNCED or 'the Earth Summit'. It is a term which means different things to different people, but is used here as the essential requirement for economic development to take full account not only of the costs of resource consumption, waste production and pollution, but also the wider impact on the environment.

A sustainable approach to development is critical in all sectors of society, not least for geologists where the study of the planet and its systems is clearly relevant for the exploration and exploitation of the Earth's resources. Environmental geologists, in particular, have a critical role to play in this arena. They should have an appreciation of the twin value of the Earth as a resource, and as the provider of the wider environment in which we live, and contribute vigorously to the campaign for sustainable development.

What is sustainable development?

Modern society is driven by two ideals, both of which look to the future and aim to improve the quality of life. The first is the desire to achieve economic development to maintain and enhance the standard of living. The second is the desire to protect and enhance the quality of the natural and human environment around us. They appear, however, to conflict. In order to provide the basic resources to maintain and enhance the quality of life, economic development is essential. To achieve this, however, we must place demands on natural resources which, more often than not, degrade, rather than improve, the wider environment.

The concept of 'sustainable development' aims to reconcile this problem, achieving both goals. Sustainable development is most commonly quoted as

> 'development that meets the needs of the present without compromising the ability of future generations to meet their own needs'
>
> (Brundtland Commission 1987).

Importantly, the Rio Earth Summit placed the emphasis on the wider environment; not only achieving sustainable management of economic resources, but also emphasizing the impact that such development has on the natural environment. Within the sustainable development debate, environmental issues have moved to the fore. The reasons for this shift are complex, but the main areas are as follows (Department of Environment 1994*b*).

1. The desire to maintain public health and reduce the risk from harmful pollutants.
2. The conservation of finite natural resources; not only minerals and the land itself, but also the diversity and quantity of species which provide food stocks and the opportunities for research and development. This concept is commonly referred to as biodiversity.
3. Recognition of the value of the landscape, wildlife and habitats in their own right, and for present as well as future generations to enjoy.

Environmental concern has, in recent years, focused on the concept of 'environmental sustainability' (e.g. English Nature 1993). This means maintaining the environment's natural qualities and characteristics as well as its capacity to function naturally, thereby maintaining biodiversity. Fundamentally, this approach involves defining environmental capacities and establishing the environmental limits necessary to maintain them. All development must therefore be planned within the natural limits of the environment. In this concept, those aspects of the environment that cannot readily be replaced are termed 'critical natural capital'. Other, non-critical elements may, nonetheless, be vital for maintaining a constant quality of environment and should not be allowed (in total) to fall below minimum levels. These are referred to as 'constant natural assets'. For successful environmental sustainability, environmental concerns must be integrated into all levels of policy formulation, development and land-use planning (Fig. 2.1). In theory, by establishing environmental limits, unacceptable environmental impacts can be avoided.

What is being done?

The Earth Summit in Rio de Janeiro in 1992 established the principles of sustainability by initiating a global programme of action to achieve sustainable development for the next century (commonly referred to as Agenda 21). As one of four main documents signed up to by most of the world's nations (the others

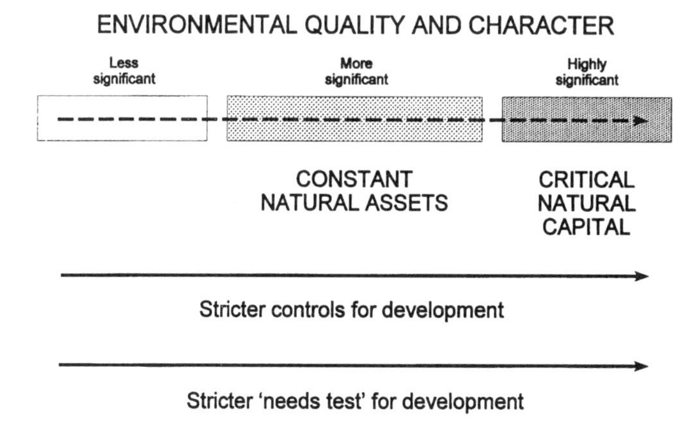

Fig. 2.1. Setting environmental quality and character against development control and requirements.

being agreements on climate change, biodiversity and forests), Agenda 21 is an agreement not just for governments, but for everyone. Notably, it stresses the need for contributions from business and industry. The Fifth Environmental Action Programme: 'Towards Sustainability' sets the framework for achieving Agenda 21 within the European Union, whilst the UK government's contribution to this initiative is set out in *Sustainable Development: the UK Strategy* (Department of the Environment 1994b). The UK strategy stresses the need for international and national policy decisions to be governed by sustainable concepts. It outlines the need to link development with environmental appraisal and the importance of the adoption of Agenda 21 at a local or community level (Local Agenda 21). Central to its implementation, therefore, is the response and action of local government authorities. In the UK, land-use planning, which is normally managed at a local level, is probably the strongest tool with which to establish a sustainable approach. Since 1990, all government planning legislation and guidance has been reviewed in the light of environmental priorities. For example, planning policy guidance and mineral policy guidance notes are currently being revised within a sustainable context.

Environmental geology and sustainability

Environmental geology is the interaction of humans with the physical or geological environment (Montgomery 1989; Bennett & Doyle 1997) and can be divided into three main themes.

1. Resource management, including exploration and exploitation, and its environmental consequences.
2. The geology of construction, of urban development, and of waste/pollution management.
3. The geology of natural hazards.

Sustainable development has a relevance to all three areas of environmental geology, in particular in the management of natural resources and its consequent impact on the environment. All have an impact on the environment which, unless managed from the point of view of sustainable development, may become increasingly detrimental and may ultimately lead to environmental degradation and economic loss.

Exploration and exploitation of natural resources and sustainability

Natural resources include not only minerals for use in industry, construction and as fuels, but also water and air. In this context, the concept of sustainability is perhaps most familiar to the environmental geologist. Due to the length of most geological timescales, most surface and sub-surface minerals are ultimately non-renewable on a human timescale; their use must therefore be considered as non-sustainable. In this context, however, air and water which are naturally replenished, are sustainable resources and should, therefore, be managed as such.

Minerals extraction

There is an ever-increasing demand for minerals. This can, however, have a direct and negative impact on landscape, wildlife and its habitats. This is further exacerbated by the increased production of waste which leads to pollution when poorly managed. Clearly, mineral extraction must look to minimize its impact on the environment.

There are two opposing views of long-term mineral availability. Optimistically, mineral resources will always be available and increased demand will be met by lower grade resources whose exploitation will be achieved through technological advance, though potentially at much greater financial cost – the Ricardian perspective. Pessimistically, the Malthusian principle, suggests that mineral resources are finite and will eventually run out (Woodcock 1994; Bennett & Doyle 1997).

In practice, the Malthusian perspective probably applies to most mineral resources and ultimately alternatives must be sought. In the long term, therefore, 'pure' sustainability is an unrealistic concept. However, through the application of the principles of sustainable development it is possible to ensure that this resource depletion does not have a detrimental impact on other environmental resources and, through the application of resource conservation strategies, the life of critical resources may be prolonged. It is important to note that placing environmental limits on minerals extraction may make previously economic extraction operations environmentally unacceptable. In turn, this may affect assessment of the total resource availability.

Environmental geologists have a key role to play in a number of areas in ensuring effective resource management, in keeping with the sustainable principles outlined above.

1. Seeking alternatives. A recent report for the Council for the Protection of Rural England (Owens & Cowell 1996) recognizes that the 'current dependency on virgin minerals for development, construction and other uses' is not sustainable. There is clearly a need to focus on seeking alternative sources for minerals, as well as improving efficiency of current use and increasing mineral recycling in order to reduce the pressure on future resources. Currently, only high grade minerals are sought (i.e. those where the cost of exploitation is low). In the long term, a shift to lower grade resources may increasingly be required and as such, increased research into these areas is desirable.
2. Global resources. Mineral resources should be considered on a local, national and global scale. In seeking sustainable

development in Britain, it is essential to recognize global needs and not export unsustainable demands. This is increasingly important with the development of wider European markets. For example, the development of superquarries in remote and environmentally sensitive locations in Scotland can be viewed as an export of unsustainability, since demand is not just driven by the UK market, but by a European demand for construction materials (Owens & Cowell 1996).

3. Restoration. Former mineral workings and associated spoil heaps represent the most extensive form of land dereliction in England, accounting for about 38% of derelict or contaminated land (Department of Environment 1996). Effective restoration can offer a number of environmental gains. Return to agricultural land may reduce agricultural land use pressure on natural habitats and landscapes. Similarly, development of recreational after-uses and housing/industrial development on damaged or contaminated land ('brownfield') can reduce environmental pressure on virgin ('greenfield') sites. Restoration for nature conservation provides valuable environmental enhancement whether through natural colonization and habitat creation or through the provision of long-term geological exposures. This can contribute to constant natural assets, and promote greater environmental awareness. Both professional ecologists and earth scientists have a critical role in such projects through survey and evaluation, strategic planning of reclamation after-use and through the design and management of reclamation schemes.
4. Environmental assessment. The EC Directive on environmental assessment was implemented in 1985 and formally adopted through the land planning system in the UK in 1988 (Department of Environment 1988). The principles of the directive are first, to ensure that the environmental consequences of new development are taken into account prior to planning permission being granted. Second, to encourage developers to consider environmental issues from the earliest stage of project planning and design where the development is likely to have potentially adverse environmental effects. Mineral extraction is now widely subjected to environmental

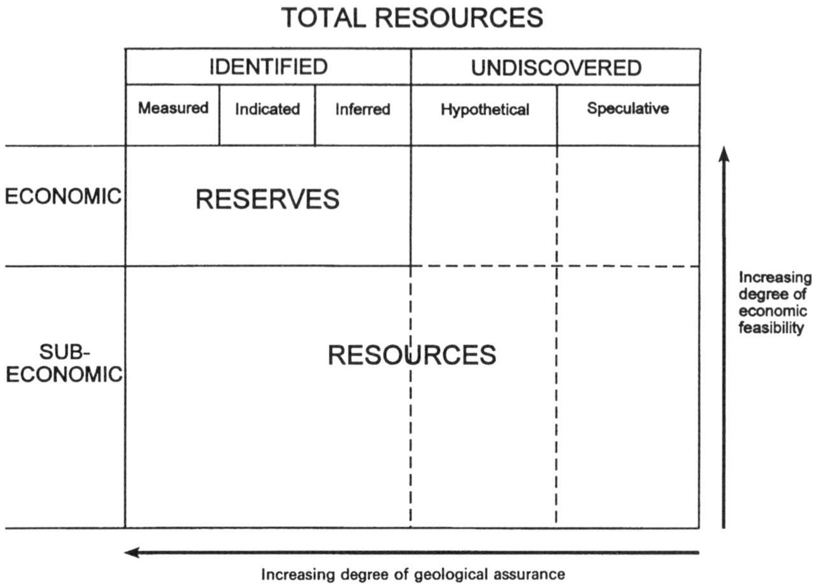

Fig. 2.2. The McKelvey Box: geological certainty measured against economic feasibility (modified from Fernie & Pitkethly 1985).

assessment. The potential impact that extraction has on the environment during and post-extraction is emphasized. The environmental geologist has a key role to play in assessing the impact of minerals extraction on the environment and ensuring environmentally sustainable development.

A standard approach to appreciating the economic viability of a reserve is to assess geological certainty against economic feasibility, an approach known as the 'McKelvey Box' (Fig. 2.2; Fernie & Pitkethly 1985). As both these factors increase, so does the viability of the mineral reserve. This traditional approach presents a common problem; there is no clear place for environmental variables. This can be overcome by the introduction of a third dimension to the McKelvey Box making environmental considerations integral; the McKelvey Box then becomes a cube (Fig. 2.3; Owens & Cowell 1996). Environmental unacceptability may now

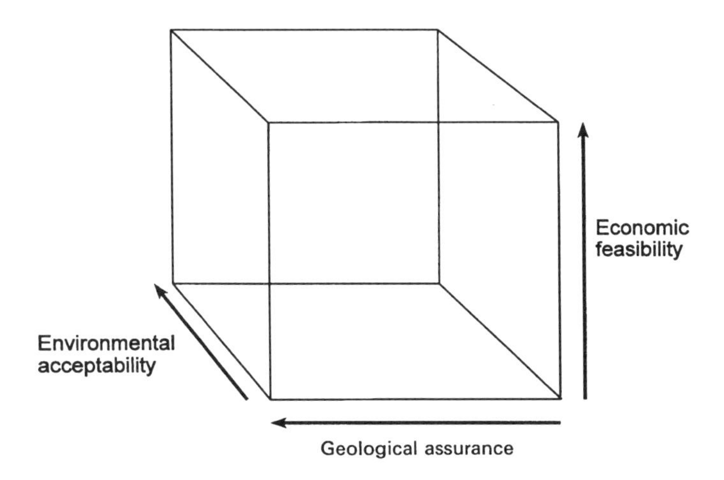

Fig. 2.3. The McKelvey Cube: environmental acceptability measured against geological certainty and economic feasibility (modified from Owens & Cowell 1996).

constrain an otherwise acceptable reserve. Such a shift in philosophy, if adopted by the environmental geologist, will further establish the principles of sustainable development.

Energy

The sustainable development of energy resources requires a balance between the economic need for an assured supply at competitive prices, and the environmental need to minimize adverse environmental impacts associated with energy use. Traditionally, the discovery of new fossil fuel reserves has resulted from the increasing demands of energy use and this will certainly continue into the immediate future. As these resources diminish, more sophisticated extraction techniques will be needed in increasingly more inaccessible and hostile regions. Burning of fossil fuels is also associated with atmospheric pollution, promoting climate warming, and also having a major role in the acidification and contamination of freshwater.

The UN Framework Convention on Climate Change, opened for signature at the Rio Earth Summit in 1992, aims to stabilize greenhouse gases at levels that will prevent human activities from interfering dangerously with the global climate system. Essential to this is the reduction of emissions and the enhancement of carbon sinks and reservoirs in order to achieve a globally sustainable environment.

Given this commitment, the drive for efficiency and search for better alternatives is an important current and future role for the environmental geologist. Natural gases offer a cleaner alternative to other fossil fuels and constitute a potentially larger resource. Nuclear power is relevant but presents long-term problems of waste disposal. Alternatives to fossil and nuclear fuels which are renewable and cleaner will, therefore, become increasingly necessary. Renewable energy covers a diverse range of energy sources and technologies, however, they may be broadly categorized (English Nature 1994).

1. Harnessing direct environmental energy resources: wind, wave, hydroelectric and solar technologies produced by the interaction of the Sun, Earth and Moon and geothermal energy source from sub-surface hot rocks and water.
2. Biomass technologies: solar energy recently trapped by photosynthesis, energy forestry and biofuel.
3. Urban or industrial waste technologies: utilizing energy from urban and industrial wastes including household waste, scrap tyres and landfill gas.

In 1988 the UK Government released Energy Paper 55, *Renewable Energy in the UK: the Way Forward* which suggested a 25% target for UK energy derived from renewable resources by 2025. The environmental geologist has a role in assessing the environmental impact of the renewable energy sources and in evaluating their engineering feasibility. As with the extractive industry, formal environmental assessment is required as part of the planning process.

Freshwater

Freshwater represents the most important of the life-sustaining natural resources (Bennett & Doyle 1997). It is an essential economic resource for industry, agriculture and waste disposal and is vital for maintaining ecological biodiversity. The problem facing a sustainable future for the water resources is to balance the demand for water from households, agriculture and industry against the maintenance of the aquatic environment and its associated fauna and flora.

English Nature's freshwater agenda (English Nature 1997) underlines the importance of managing freshwater resources sustainably to meet the full range of social, economic and environmental uses, including protection of wildlife. This approach is driven by Agenda 21 and also forms an integral element in achieving the UK Biodiversity Action Plan (Department of Environment 1994*a*).

If managed correctly, water is a renewable and therefore a sustainable resource. Supply is dependent not only on climate, but also on topography, geology and land use, which can be altered by human intervention. Careful modelling of the effects of groundwater and surface abstraction is essential to ensure that water depletion does not exceed recharge and therefore deplete the sustainable resource. Water depletion may have serious consequences for freshwater ecology and habitats and therefore for biodiversity, as well as giving rise to structural problems such as ground subsidence (Bennett & Doyle 1997).

To achieve this in England, water Catchment Management Plans (English Nature 1995) are being produced by the Environment Agency. Catchment Management Plans (now known as Local Environment Agency Plans) examine the link between the water environment (including coastal waters, estuaries, rivers, lakes and groundwater), land uses and activities that relate to it. Catchment management planning brings together a range of expertise in order to understand and manage the environmental effects of water usage. An essential link is provided by the environmental geologist, through the provision of detailed knowledge of the hydrological system which underpins the scientific basis for sustainable solutions to water use.

Waste and pollution disposal and sustainability

As population increases, so do waste products and pollution levels, particularly focused in urban and industrial areas (Bennett & Doyle

1997). There is a hierarchy of options for waste and pollution management which include reduction, re-use, recovery and disposal (Department of Environment 1994b). Environmental geologists are involved through the design of landfill sites and analysis of their potential interaction with the surrounding environment.

Currently, England and Wales generate enough waste to fill Lake Windermere every nine months (Department of Environment 1995). The document *Making Waste Work* (Department of Environment 1995) builds on the UK sustainable development strategy, emphasizes the need to (i) reduce the amount of waste, (ii) make the best use of the waste produced and (iii) to choose waste management practices which minimize immediate and future environmental pollution and threat to human health.

Irrespective of the impact of waste minimization strategies, there will always be a demand for landfill or land raise sites. Disposal through landfill or land raise requires a geological input in the selection of a suitable site and a clear assessment of the likely environmental impacts associated with the long-term degradation of the waste. To achieve environmental sustainability these impacts must be minimized.

Natural hazards and sustainability

Climate change and coastal erosion

Global warming, exacerbated by environmental pollution, has the potential to induce major shifts in global climate and change sea-level. The speed at which global warming occurs can be kerbed if the principles of sustainable development are applied to reducing world pollution levels. The consequences of sea-level rise, however, still require immediate management. Sea-level rise is likely to lead to land loss, increased coastal erosion and flooding, contamination of coastal aquifers and a resultant shift of ecosystems (Bennett & Doyle 1997).

As approximately 75% of the world's population live within 1.5 km of the coast (Bennett & Doyle 1997), the potential impact of increased coastal erosion and flooding are high through loss, not only of buildings and infrastructure, but also of agricultural land. This leads to demand for protective measures with which to slow or halt the rate of coastal recession. This, however, can have detrimental effects in two ways. First, environmental impacts may be high; many coastal habitats are reliant on a naturally evolving coastal system which, if interrupted by coastal defence are effectively destroyed. Equally, the very nature of an eroding coastline makes it a vital geological resource. Second, interrupting natural coastal processes can have significant knock-on effects, exacerbating coastal erosion by starvation of sediment and enhancement of erosive energy.

Recent research has shown that the coastline of England can be divided into 11 major sediment cells in which movement of sediment is broadly contained. These discrete units have now formed the basis for the development of Shoreline Management Plans (MAFF 1995) which take into account natural coastal processes, coastal defence needs, environmental considerations, planning issues and current and future land use (McCue 1998). The aim of a Shoreline Management Plan is 'to provide the basis for sustainable coastal defence policies within a sediment cell and to set objectives for the future management of the shoreline' (MAFF 1995).

Understanding how the sediment system works is essential for developing sustainable Shoreline Management Plans that cater for a wide range of needs, including environmental. Once more, it is the environmental geologist who can provide the detailed knowledge and understanding to inform such sustainable decisions.

Conclusions

The primary reason for needing to develop sustainable approaches is the world's ever-increasing population, and the need to satisfy its ever-rising expectations of a reasonable standard of life. There are three ways sustainability can be approached.

1. Severe population decrease. Short of natural disaster or something worse, this is not attainable.
2. Improving technology in order to find new and more resourceful ways of supplying our needs.
3. Changing life style so that we use resources more wisely and

have greater care for the natural environment.

Agenda 21 concentrates on these last two. Neither alone will solve the problems of resource scarcity and degradation of the environment. They call on the expertise of the scientist to help develop more efficient processes to 'farm' limited natural resources and to look for ways to re-use them and develop alternatives; the sustainable development approach. They also call for a recognition of concepts such as environmental limits in order to reduce or remove the environmental impacts of resource use.

The environmental geologist has a primary role to play in applying these principles to achieve sustainable development. Key areas of relevance include minerals extraction, energy development, freshwater use and waste and pollution disposal. The understanding of the interaction between humans and geology is of key importance for achieving sustainability in these areas. President Kennedy in his inaugural speech in 1961, referred to conservation as 'the wise use of natural resources'; this mirrors the current debate – for 'wise' read 'sustainable'.

References

Bennett, M. R. & Doyle, P. 1997. *Environmental Geology: geology and the human environment.* John Wiley, Chichester.

Brundtland Commission. 1987. *Our Common Future: the report of the World Commission on Environment and Development.* Oxford University Press, Oxford.

Department of the Environment 1988. *Environmental Assessment: A Guide to the Procedures.* HMSO, London.

——— 1994a. *Biodiversity: The UK Action Plan.* HMSO, London.

——— 1994b. *Sustainable Development: The UK Strategy.* HMSO, London.

——— 1995. *Making Waste Work: a strategy for sustainable waste management in England and Wales.* HMSO, London.

——— 1996. *Reclamation of Damaged Land for Nature Conservation.* HMSO, London.

English Nature 1993. *Position Statement on Sustainable Development.* English Nature, Peterborough.

——— 1994. *Nature Conservation Guidelines for Renewable Energy Projects*. English Nature, Peterborough.
——— 1995. *Conservation in Catchment Management Planning – a handbook*. English Nature, Peterborough.
——— 1997. *Wildlife and Fresh Water: an agenda for sustainable management*. English Nature, Peterborough.
Fernie, J. & Pitkethly, A. S. 1985. *Resources, Environment and Policy*. Harper & Row, London.
MAFF 1995. *Shoreline Management Plans: a guide for coastal defence authorities*. HMSO, London.
McCue, J. 1998. 'Sense and sustainability' – achieving geological conservation objectives as part of the present shoreline management plan process. *This volume*.
Montgomery, C. W. 1989. *Environmental Geology* (2nd edn). Wm. C. Brown Publishers, New York.
Owens, S. & Cowell, R. 1996. *Rocks and Hard Places: minerals resource planning and sustainability*. Council for the Protection of Rural England, London
Woodcock, N. H. 1994. *Geology and the Environment in Britain and Ireland*. UCL Press, London.

3 Dissemination of information on the earth sciences to planners and other decision-makers

Alan P. McKirdy, Alan Thompson & Jane Poole

Summary

- While the 'environment' currently has its highest profile, an understanding of the physical components of the environment amongst people involved in the development control process remains weak.
- The general public readily appreciates environmental issues, such as habitat loss or species extinction, but does not register a similar concern for environmental geology issues, such as the loss of a key geological reference section.
- A key task for the geological profession in the new millennium is to ensure that the general public gain an adequate understanding of the physical forces of nature.

Professional planners, the elected councillors who serve on local authority planning and coastal protection committees, developers, the minerals industry, strategists and policy makers in general would all benefit from a deeper and more complete understanding of what lies beneath their feet. Decisions are often taken that are meant to be in the public interest but demonstrate at best, ignorance of, and at worst, a disregard for, the natural processes that continue to shape the countryside. An over-emphasis on short-term consequences and a lack of appreciation of the dynamic nature of the Earth and its surface processes have all played their part in arriving at poor planning decisions, with often dramatic consequences. This chapter outlines the importance of the dissemination of earth science information to a wider audience.

Introducing the problem

Examples of poor practice

Hengistbury Head, Dorset

Hengistbury Head lies along a stretch of the Dorset coastline that has been much affected by engineering works over many years. The aerial photograph in Fig. 3.1 indicates the direction of movement of sediment along the coastline and also the degree to which a decision taken by one local council can affect the interests of its neighbour. The long groyne has interrupted the passage of sediment from west to east, causing a build up of sand and gravel in front of the structure and erosion problems in the lee of the groyne. This structure has undoubtedly helped to maintain the spit, but no regard was taken of its effects downdrift. Over the years since its construction after the Second World War, erosion of the cliffs has accelerated to the east of the long groyne and beach levels have dropped dramatically. The effects of this structure were exacerbated

Fig. 3.1. Hengistbury Head, Dorset. Interruption of sediment supply is clearly evident, causing problems of cliff instability downdrift. This slide is reproduced with the permission of Hydraulics Research Ltd.

by the removal of ironstone concretions (doggers) from the foreshore which acted as a natural defence against wave attack. The net result was a greatly increased rate of cliff retreat and a genuine fear that Hengistbury Head would be breached.

This example clearly indicates the 'interconnectedness' of natural systems and the need to enact a coastal protection strategy that will deliver effective management of the whole coastline. Coastal protection is the responsibility of district councils in England and Wales. Many councils have, in the past, acted independently without an appreciation of the way in which their individual actions may affect the sediment dynamics of the coastline as a whole. However, this picture has changed considerably in recent years and there is now a greater awareness of the need to understand the effects of natural processes on the coastal zone as a whole. This greater awareness has led to a more holistic view of coastal management which is to be greatly welcomed.

Durlston Bay, Dorset

Durlston Bay is the type locality for the Upper Jurassic Purbeck Beds and is internationally famous for its invertebrate and vertebrate fossil faunas. As reported in *Earth Science Conservation:*

> In 1972, it was proposed to build a block of flats adjacent to the rapidly eroding coastline at Durlston Bay. In spite of these unfavourable ground conditions, planning permission was granted and the flats were constructed. Some fifteen years on, successive cliff falls have pushed the cliffline back to within a metre or so of the building's foundations to the understandable alarm of the Council and more particularly the owners and occupiers of the flats. (McKirdy 1988, p. 36).

It was considered that the planning authority were unable to take account of the fact that the site for the flats lay immediately adjacent to a rapidly eroding coastline, as the cliff edge lay outwith the application area. In law, it was therefore not considered to be a 'material planning consideration' and the local authority were unable to take this matter into account when determining the planning application.

Dungeness, Kent
A nuclear power station has been built immediately adjacent to an eroding coastline at Dungeness in Kent. For obvious reasons, considerable effort has to be made to ensure that high water mark does not migrate landward and flood this installation. Shingle is dumped updrift of the power station and the process repeated on a regular basis. This site has experienced an interruption of sediment supply because of the construction of coast protection works and other engineering structures to the west of Dungeness. Although the siting of this facility was the subject of a planning inquiry prior to its construction in the 1960s, its precarious position next to the sea was apparently endorsed by the inquiry decision, despite the cost implications of maintaining the site.

River Tay, Perth
The Muirton Housing Estate in Perth was severely flooded in January 1993, causing much damage to property and distress to the local residents. This large council housing estate is built on the flood plain of the River Tay, separated from the flowing water by only a few metres of flat ground and a low flood bank. The River Tay drains much of central Scotland and carries more water than any other river in Britain. This river has a long history of flooding and it may seem surprising, therefore, that this area was selected as the site for a major housing development. Apparently, the residents of this public sector housing estate were ill-prepared for this flooding event, as few had taken out house contents insurance against such an eventuality.

Summarizing the problem
These case histories clearly illustrate what can happen if information on natural processes, such as the behaviour of coasts or rivers, is not taken into account. These planning decisions were taken at a time when most engineers had a Canute-like belief in the effectiveness of engineering works such as reinforced concrete walls, groynes and revetments and planners were largely unaware of the importance of earth science information. Thankfully, the situation has improved in recent years and both environmental impacts and the behaviour of natural systems are usually

considered to be important considerations at the design stage of any engineering works. Aspects of environmental geology have also been recognized increasingly in planning guidance issued by the Department of the Environment, Scottish Office and Welsh Office. However, a great deal more needs to be done to raise and maintain awareness amongst all professionals involved in the planning process.

Current initiatives in the dissemination of information

If the profession is to achieve a wider appreciation of the need to use earth science information, we must redouble our efforts to communicate effectively with those who must heed our concerns. It is a universal responsibility that must be shouldered by the earth science profession as a whole. The following initiatives have been helpful in this regard, although the process must continue over a broader front if it is to be successful in the long term.

Environmental Impact Assessment

The Environmental Assessment process was introduced formally in 1988, following EC Directive 85/337 *Assessment of the Effects of Certain Public and Private Projects on the Environment* (European Community 1988). Despite some lack of enthusiasm at the time of its introduction, there is now a general acceptance that Environmental Assessment greatly enhances the planning decision-making process and allows for a full exposition of all the relevant geological factors. However, only a limited number of major developments, such as radioactive waste repositories and motorway construction require Environmental Assessments by law, although that list is under constant review. There is a second list of developments including such diverse activities as opencast mining, reprocessing of nuclear fuels and pig-rearing, which only require an environmental assessment if the planning authority so decides.

An important part of the Environment Assessment process is the establishment of an environmental baseline which describes the present state of the environment and the way it is likely to change, assuming that the project, be it motorway construction or pig-rearing does not go ahead. This provides a yardstick against which any changes likely to be brought about by the development can be estimated. Many factors, including soil, geology, geomorphology, groundwater and surface water conditions are recommended for consideration in describing this baseline condition. The Environment Assessment process is relatively new and, from a reasonable cross-section of assessments studied by the authors, unless geology or surface processes are very obviously an important factor, this aspect is normally given fairly cursory treatment. This may be appropriate in many instances where the soil and rocks beneath simply provide firm foundations for the development above. Nevertheless, in the situations cited earlier, and in many other cases around the country, insufficient consideration is given to physical processes, such as coastal erosion and river flooding and the planning process is therefore ill-informed and incomplete. However, the Environmental Assessment process is an important mechanism that permits all relevant matters pertaining to environmental geology to be considered during the planning process. Although this important mechanism is now in place and backed by legislation, it requires a degree of vigilance by planners and others to ensure that geological and geomorphological issues receive the appropriate attention whenever required. This, in turn, requires planners to be more aware of the importance of 'environmental geology issues', so that they can make more effective use of the information provided to them by statutory consultees, such as environmental agencies and conservation bodies.

Encouraging planners to gain more awareness of such issues will also enable them to become more confident about asking the right questions in the first place. 'Is this area likely to flood and what effect would a fifty year event have on the viability of this proposal?', or 'what value can be placed on this international geological stratotype that would be irreparably damaged by this development proposal?' are examples of questions that should be asked in future under the appropriate circumstances.

Those who are most particularly affected by proposed development are the residents who live adjacent to development sites. As the planners and engineers debate the finer points of technical issues, local residents, who are most likely to be affected by the increased noise, vibration or pollution that the new development may bring, may not have a full appreciation of all the issues. However, where an Environmental Assessment is deemed necessary, there is a requirement for a non-technical summary of the issues to be prepared. This section of the report may be suitable for wider circulation to those with a direct interest in the proposed development, summarizing technical issues in plain English.

The Environmental Assessment process also allows for the development to be monitored over time and the predicted effects on the environment can be measured against the actual impacts that arise. However, the limited capability of most councils to monitor planning conditions and take the appropriate enforcement action means that monitoring predicted effects against actual impacts does not always take place as effectively as it should.

It is also important to note that the EC is considering a proposed directive on Strategic Environmental Assessment. If adopted, this might have the effect of bringing in Environment Assessment comparisons early in the site selection procedure rather than (as at present) trying to make the Environment Assessment fit the site which has already been chosen.

It is our view that the Environmental Assessment process is too often used by developers simply as a mechanism to gain planning permission. Information on environmental impact, such as noise or likely pollution effects, is often interpreted to put the best possible gloss on the proposed development. Environmental Assessments or reports derived therefrom, are normally presented at Public Inquiries as the most detailed account of likely environmental impacts and, in the adversarial nature of these proceedings, the case for the proponent has to be strongly argued. In such circumstances, it is unusual to concede any points of debate or uncertainty to one's opponents.

The rapidly increasing number of environmental consultancy companies in the UK, which has risen from around ten in the 1950s to approaching 400 today, is testament to the enhanced importance given to environmental considerations in the planning process (Glasson *et al.* 1994). There is therefore a reasonable prospect that this army of consultants, many with earth science backgrounds, will ensure that geological and geomorphological issues will be given adequate consideration in the future.

Initiatives from Department of the Environment (DoE)

The DoE has also been concerned with the need for a fuller understanding of earth science issues within the planning system and for many years the department has sought to promote an awareness of these issues through its Geological and Minerals Planning Research Programme. The aim of the programme has been to provide geological and other earth science information that is needed for the development of policy with respect to land-use planning, in a form that is suitable for use by planners and others without an earth science background. The programme has encompassed:

- national reviews of land instability (landslides, natural underground cavities, mining instability, erosion, deposition and flooding);
- more detailed investigations of specific earth science problems and the appropriate planning responses (e.g. subsidence hazard due to gypsum dissolution in Ripon, North Yorkshire; Thompson *et al.* 1996);
- national reviews of the availability of, and demand for, certain strategic minerals (e.g. high specification aggregates for road surfacing materials);
- regional assessments of the availability of mineral resources (e.g. sand and gravel, high purity limestone and sandstone);
- collation of information on mineral resources for use in development planning;
- area-specific applied geological mapping studies (e.g. Bathgate and Livingston, Stirling, and Glasgow; British Geological Survey 1991);
- subject-specific assessments of the environmental impact of surface mineral workings (e.g. noise, dust, and the effects on ground- and surface water).

The majority of these research projects have been specific to particular earth science themes or to particular geographical areas. It has not been practical to cover systematically all parts of the UK, and neither has there been any perception, amongst planners, of the need to do so. Within the last few years, however, the DoE has recognized the need for planners and developers in all areas to appreciate the importance and cost-effectiveness of taking earth science information into account. To address this need it has commissioned a series of four new research projects designed to promote the use of such information in different planning situations, these comprise:

- the Use of Earth Science Information in Coastal Planning (project completed by Rendel Geotechnics in 1996);
- the Use of Earth Science Information in Support of Major Development Initiatives (currently being undertaken by the British Geological Survey);
- the Use of Earth Science Information in Urban Planning (currently being undertaken by engineering and environmental consultants: Symonds Travers Morgan);
- the Use of Earth Science Information in Rural and Upland Planning (also currently being undertaken by Symonds Travers Morgan).

The four projects have broadly similar aims and objectives. The aims of the 'Rural' project, for example, are to provide a framework of advice for planners and developers on the need for, and cost-effective use of, earth science information to support decision-making in rural and upland areas, and on the best means for using and presenting such information.

In carrying out this work, Symonds Travers Morgan has investigated a total of 35 case studies (15 on the 'Urban' project and 20 on the more diverse 'Rural and Upland' project), in different parts of the UK. Each case study has focused on the particular earth science issues (natural hazards, natural resources and earth heritage conservation features) which are perceived to be of importance in each area, and on the ways in which earth science information has been, or could be, used by planners to deal with those issues.

Considerable thought has been given to the most effective way of promoting the various elements of good practice from these case studies amongst planners, developers and other potential users. As part of this process, Symonds Travers Morgan produced a dissemination strategy for consideration by the DoE and the

Table 3.1. Options for the Dissemination of Environmental Geology.

Output / Target Audience	Planners in central & regional Govt.Offices	Local Authority Planners	Local Authority Env. Health Officers	Local Authority Engineers	Building Inspectors	Major Developers	Other Planning Applicants	Engineering/ Environmental Consultants	Selected Statutory Consultees	Land & Property Professionals	Planning Students & CPD trainees	Earth Science Students	Earth Science Professionals
* Environmental Geology in Land Use Planning:- Advice for Planners and Developers	Main	Main	Possible	Possible	Possible	Possible	Unlikely	Main	Main	Possible	Main	Unlikely	Main
* Environmental Geology in Land Use Planning:- A Guide to Good Practice for Planners	Main	Main	Possible	Possible	Possible	Possible	Unlikely	Possible	Main	Possible	Main	Unlikely	Unlikely
Environmental Geology in Land Use Planning:- A Guide to Good Practice for Developers	Main	Possible	Possible	Possible	Possible	Main	Unlikely	Possible	Main	Main	Possible	Unlikely	Unlikely
Environmental Geology for Land Use Planners:- A Guide to Good Practice for Earth Scientists	Main	Unlikely	Unlikely	Unlikely	Unlikely	Unlikely	Unlikely	Possible	Possible	Unlikely	Unlikely	Main	Main
* Environmental Geology in Landuse Planning: Implications for current and emerging planning issues Planning Issues Report	Main	Main	Possible	Possible	Possible	Possible	Unlikely	Possible	Main	Possible	Main	Unlikely	Unlikely
* Environmental Geology in Land Use Planning Case study leaflet pack	Main	Main	Main	Possible	Possible	Main	Possible	Main	Main	Main	Main	Main	Main
Awareness of Environmental Geology leaflet	Main	Main	Main	Main	Main	Main	Main	Main	Main	Main	Main	Possible	Unlikely
Sources of Environmental Geology Information guide	Main	Main	Main	Main	Main	Main	Possible	Main	Main	Main	Main	Possible	Possible
* Combined project dissemination leaflet	Main	Main	Possible	Possible	Possible	Possible	Unlikely	Possible	Main	Possible	Unlikely	Unlikely	Possible
* Combined project dissemination seminars	Main	Main	Possible	Possible	Possible	Possible	Unlikely	Possible	Main	Possible	Unlikely	Unlikely	Possible
Articles in planning journals	Main	Main	Possible	Unlikely	Possible	Unlikely	Unlikely	Unlikely	Possible	Possible	Main	Possible	Possible
Articles in earth science journals	Possible	Unlikely	Unlikely	Unlikely	Unlikely	Possible	Unlikely	Possible	Possible	Possible	Unlikely	Main	Main
* RTPI Summer School Seminar	Possible	Main	Possible	Possible	Possible	Unlikely	Unlikely	Possible	Main	Unlikely	Possible	Possible	Possible
Curricula for training courses (incl. CPD)	Possible	Main	Possible	Unlikely	Unlikely	Possible	Unlikely	Main	Possible	Possible	Main	Main	Possible

Key

* Actual Project Output

Main target audience

Possible additional audience

Unlikely to be of interest

As put forward by Symonds Travers Morgan in relation to the possible output from DoE research contracts on the Use of Earth Science Information in Urban, and in Rural and Upland Planning.

project's steering group (Table 3.1). Some parts of the strategy relate to all four of the projects listed above, whilst others relate specifically to just the 'Rural' or 'Urban' projects. The actual output to be produced for the Department of the Environment is asterisked. The table serves to illustrate the range of different options available for 'getting the message across', and also the diversity of the potential target audiences.

In the past, the output from many of the DoE-funded research projects, whilst intended for use by planners, has been most beneficial to those who already have a background in earth sciences. With a small number of exceptions, the reports and maps that have been produced have often been too technical for use by those without such training, and the findings have not been sufficiently well linked to appropriate planning or building control responses. The work by Geomorphological Services Limited (and subsequently by Rendel Geotechnics) on the planning response to landslip problems in the Isle of Wight provides a good example of the way in which earth science information can be used to enable planners to deal directly with natural hazards, both in forward planning and development control.

More recently, Symonds Travers Morgan's two-year investigation of subsidence hazard caused by gypsum dissolution in Ripon (Thompson *et al.* 1996) has taken this a stage further by providing the local planning authority with a checklist-based development control procedure linked to a clearly defined development guidance map. This enables planners to be aware of the need to take earth science problems into account, and to exercise the necessary control over new development, whilst leaving the onus of responsibility for investigating the stability of each site in the hands of developers and their professional advisors (Thompson *et al.* 1996).

Promoting the awareness and use of earth science information is not only about solving problems, however, it is also about making the best use of natural resources and recognizing opportunities for conservation. This is well illustrated in Symonds Travers Morgan's case study of the Cairngorms area, as part of the 'Rural and Upland Planning' project. This is undoubtedly one of the 'classic' areas of British geomorphology, exhibiting some of the finest examples of glacial and periglacial landforms anywhere in the

Fig. 3.2. The Cairngorms provides some of the finest examples of glacial landforms in Great Britain. The physical landscape of the Cairngorms forms the basis for land-use planning and management of the area.

country (Fig. 3.2). Increasingly, however, growing development pressures and other land management interests are in conflict with the interests of conservation. To resolve these difficulties and to maintain proper balance between the various interests, the Cairngorms Partnership was established. Working closely with Scottish Natural Heritage, the Scottish Environmental Protection Agency, Highland Council planning authority and others, the partnership helps to ensure that a fully integrated approach to land-use planning, management and environmental protection is adopted. One of its fundamental principles is that the geological and geomorphological features of the area are the basis for all other aspects of its character and economy, from the rugged grandeur of the mountains and the fragile ecosystems thereon, to the modern interests of forestry and tourism.

Initiatives from Scottish Natural Heritage

Scottish Natural Heritage has made a considerable effort to present facets of the earth sciences in an exciting and challenging way during the five years of its existence. Many of these initiatives have not been specifically concerned with land management issues, but have a more general audience in mind. The Natural Heritage (Scotland) Act 1991, Scottish Natural Heritage's founding legislation, requires it to 'foster an understanding of the natural heritage of Scotland.' This requirement has been interpreted as a need for general information on the natural heritage in a form that will reach the widest possible target audience. Geology is perceived by many as a dull, bookish subject of limited relevance and interest, so considerable effort has been made to present the subject in an interesting way in the *Landscape Fashioned by Geology* series, which is published by Scottish Natural Heritage in conjunction with the British Geological Survey (Scottish Natural Heritage and British Geological Survey 1993–1997). Six titles have already been produced in this series covering areas such as the Cairngorms and Skye and the intention is to publish twelve titles in all, describing the relationship between landscape and the underlying geology and landforms in all the locations of Scotland where there is a particularly interesting story to tell. The teaching of geology has been so much neglected that awareness of the role of geology in shaping the landscape is not widely appreciated. The *Landscape* series has the long-term aim of raising awareness of the general public on the physical component of the natural heritage and the need for its conservation and active management. There appears to be an almost intuitive understanding of the need to conserve important ecosystems such as the last fragments of the Caledonian pinewoods of Scotland or the chalk downland of southern England, but few appreciate the importance and intrinsic value of rocks and landforms. Rocks are as old as the hills and have changed little on a human timescale, so are viewed as a backdrop, rather than a feature of interest in their own right. It is the intention of Scottish Natural Heritage and their counterparts in England and Wales that, with the *Landscape* initiative and other interpretative projects in a similar vein, that these long-held and deeply engrained views are changed.

Many of our most important geological and landform sites are in fact just as vulnerable to change as fragile habitats such as native pinewood or chalk grassland. Most of the developments that affect earth heritage sites require planning permission, so it is of vital importance that elected local councillors and their planning advisors, as well as other land managers, are aware of this aspect of the natural heritage and the need for its conservation. Scottish Natural Heritage is currently preparing a publication entitled *Legacy of the Land* with the subtitle *The Stewardship of Our Earth Heritage – A Practical Guide for Local Authorities*. The purpose of *Legacy of the Land* is to inform planners, landowners and developers of the vital role they can play in conserving our heritage of rocks, fossils and landforms. Practical guidance will be provided on the detailed management requirements of active process systems, such as rivers and sand dunes. Based on experience from throughout the UK, suggestions will be made on model planning policies which address issues relating to earth heritage conservation in a practical way. This publication also provides the ideal vehicle to describe geological and geomorphological Sites of Special Scientific Interest (SSSI) and their selection during the recently completed Geological Conservation Review.

Coastal Zone Management is a fairly recent phenomenon in the UK. To work effectively, coastal management must reconcile a large number of competing pressures, such as recreation, housing development and conservation. It also requires a detailed knowledge of the active processes, such as longshore drift and coastal erosion, acting on the coastline. In the past, planning decisions were made with little regard to natural processes in the mistaken belief that pouring a large enough amount of concrete on any problem area that emerged would solve the problem. The Ministry of Agriculture, Fisheries and Food (MAFF) pioneered the identification of coastal cells in the UK, and this marked a sea-change in the general attitude to coastal engineering. Natural processes are now regarded as one of the key determinants of coastal land use and hence of coastal management. Working with nature, rather than trying to contain its forces, was recognized as a sensible and cost-effective method of managing the coast. Shoreline Management Plans, which have recently been prepared for many stretches of coast in England and Wales, reflect the need to understand nature. The case for coastal engineering works or any other inter-tidal development can then be seen in its proper context.

The situation in England and Wales is fast reaching the point where elected councillors and their planning advisors, engineers, geomorphologists and many other interested parties are now collaborating and sharing information to develop a common vision for the coast. Unfortunately, the situation in Scotland is less advanced, and a fragmentary approach is still evident. However, Scottish Natural Heritage has taken the initiative in bringing the main parties together with the common purpose of managing the coastline in a more integrated fashion than hitherto. A coastal cells study, which will define and describe each cell in detail, will be published during 1997 and this series of twelve reports will form the basis of the new approach. The Scottish Office have provided financial support for the project over the last three years and are enthusiastic about this new approach. Fife Council have agreed to develop a pilot Shoreline Management Plan over the next two years and within a reasonably short space of time, the Scottish coastal authorities will have caught up with their English and Welsh counterparts.

Initiatives from English Nature

English Nature have done some excellent work with the minerals industry and mineral planners. Many SSSI throughout the UK are active or disused mineral workings and vigilance is required to ensure that this important resource is not lost through inappropriate restoration work or infill with domestic or industrial wastes. High profile seminars and other means have been used by English Nature to stress the importance of this resource to decision-makers within the industry and mineral planners. Research has been undertaken to demonstrate ways in which disused pits or worked parts of active quarries can be restored whilst retaining key features of geological or geomorphological importance.

More recently English Nature has been developing the Natural Areas approach to nature conservation (Prosser 1995). The characteristic distribution pattern of the natural features, species and habitats of England is determined by the physical factors of geology, soils, topography and climate and modified by land use. These physical characteristics of the landscape are used as the basis for the division of the country into Natural Areas. The Natural

Areas approach is a mechanism by which the landscape as a whole can be considered rather than focusing solely on protected sites. The idea behind the concept is to encourage people in their understanding and conservation of the natural environment.

Conclusions

However praiseworthy these initiatives are, there is a great deal more to be done and we all have our part to play in enhancing understanding of human interactions with the physical environment. As in many fields, effective communication is the key. Presenting the information in a comprehensible form to the wide variety of audiences that must be reached is a our main objective; another is to encourage those who receive this information to use it to best advantage. It is a slow process that is not susceptible to instant solutions, but those initial and most difficult steps have been taken. The legal requirement for Environmental Assessment, albeit for a limited number of activities, is perhaps one of the most positive developments in recent years as it provides an ideal vehicle for geological and geomorphological information to be assembled and to inform the planning process. However, perhaps more needs to be done to ensure that this process is objective, impartial and not self-serving.

The list of initiatives described above is not exhaustive; rather it is illustrative of what is being done by a cross-section of earth scientists who want a wider recognition of the importance of their subject in planning our developments and using our natural resources. However, we will not succeed in this important enterprise unless we all put our shoulders to the wheel.

Acknowledgement

The authors gratefully acknowledge Department of the Environment's permission to publish information relating to its on-going geological and mineral planning research programme.

References

British Geological Survey. 1991. *Geology For Land Use Planning: Stirling*. Technical Report WA/91/25.

European Community, 1988. *Assessment of the Effects of Certain Public and Private Projects on the Environment*. EC Directive 85/337. E.C., Brussels.

Glasson, J., Therivel, R. & Chadwick, A. 1994. *Introduction To Environmental Impact Assessment*. UCL Press, London.

McKirdy, A.P. 1988. Cliff Collapse Endangers Apartment Block. *Earth Science Conservation*, **24**, 35–36.

Prosser, C. D. 1995. Conserving our earth heritage through natural areas. *Geoscientist*, **5**, 19–20.

Scottish Natural Heritage/British Geological Survey. 1993–1997. *Edinburgh – A Landscape Fashioned by Geology*, 1993; *Skye – A Landscape Fashioned by Geology*, 1993; *Cairngorms – A Landscape Fashioned by Geology*, 1994; *Loch Lomond to Stirling – A Landscape Fashioned by Geology*, 1995; *Orkney and Shetland – A Landscape Fashioned by Geology*, 1996; *East Lothian and the Borders – A Landscape Fashioned by Geology*, 1997. Scottish Natural Heritage/British Geological Survey, Edinburgh.

Thompson, A., Hine, P. D., Greig, J. R. & Peach, D. W. 1996. *Assessment of Subsidence Arising from Gypsum Dissolution*. Technical Report for the Department of the Environment. Symonds Group Ltd., East Grinstead.

Part One

Conservation and recreational pressure

This part of the volume examines current issues of conservation and management of natural areas in Britain. Since the late 1940s landscape areas and nature reserves have been actively protected, and issues of over-use of resources and the relevance of the earth sciences to the general public are considered in this section. Earth heritage conservation has been dominated until recently by the concept of boundary delimited areas selected on purely scientific grounds. The current agenda recognizes the need for a great sense of context. Access to the countryside brings its own issues; a fear that over-use is detrimental, but more particularly an identifiable conflict between different user groups with their own environmental agendas.

4 Earth heritage conservation: past, present and future agendas

Peter Doyle & Matthew R. Bennett

Summary

- Conservation of Britain's earth heritage was formally initiated in the late 1940s and developed the philosophy of the protection of natural areas through what may be called the 'boundary concept' – the recognition of areas of scientific or scenic merit and the representation of these as a geographic entity on a map surrounded by a boundary.
- Current protection of Britain's earth heritage follows the boundary pattern, with the recognition by statutory and non-statutory organizations alike of 'networks' of sites each with a boundary. These exclude much which is of value in Britain's earth heritage, which is then sacrificed to save the precious 'jewels in the crown'.
- Today, English Nature is promoting a new initiative – the recognition of Natural Areas, each of which draws together geology, geomorphology and wildlife to define special areas of interest.
- Each Natural Area has the input of specialist interest groups of the community, both amateur and professional, but lacks the wider appreciation of the majority of the British population, (*c.* 80%), who live in our urban areas.
- The future agenda must address the fundamental promotion of a wider appreciation of geology in these urban areas.

Britain is usually seen as the birthplace of modern geology. Every year, scientists from across the world visit Britain to unlock the secrets of its unique rock record: 2900 million years of earth history conveniently concentrated into one small island. Yet the coasts and uplands, the quarries and cuttings which allow this story to be told are under threat from developers – with over 400 planning applications a year directly affecting earth heritage sites under statutory protection (Nature Conservancy Council 1990).

The protection of Britain's earth heritage – rock exposures and landforms – is at least a century old and commenced with the recognition and protection of small features deemed to be of exceptional merit on scientific grounds, and therefore worthy of protection. Typical of these is the preservation in 1887 of Fossil Grove in Victoria Park, Glasgow – a cluster of *in situ* Carboniferous tree trunks which is a relict of the exploitation of coal, and a monument to the wider Victorian interest in natural history. The protection of Fossil Grove is a microcosm of conservation. For over a century now, such geological features have been cordoned off – figuratively in most cases – and isolated from both human interference and the intrusion of the natural environment. The recognition of boundaries which enclose reserves for nature or landscape, and which can be denoted on a map, has been a defining concept in Britain since the end of the Second World War, when much of our current conservation legislation was first enacted. Yet despite protection afforded to Sites of Special Scientific Interest (SSSIs) – reserves in all senses of the word – by the National Parks and Access to the Countryside Act of 1949 and the Wildlife and Countryside Act of 1981, each year the precious threads of our earth heritage are threatened or lost to development. Few seem to object, and even fewer seem to care. The plain fact is that geology must be valued not just by a handful of geologists, but by all, if it is to be conserved.

Britain is one of the most heavily urbanized areas in the world. Over 80% of its people live in towns and cities. Fewer than 10% of the area of Britain is protected as designated country parks, national parks or nature reserves, and with a total population of 56 million, only around 7% of this population is actively involved in conservation (Evans 1997). Few of the rest are fully aware of the denigration of their natural environment, and of this number, even fewer are concerned about the demise of Britain's geological heritage. Although most accept the need for the protection of endangered species or threatened habitats, how many would accept the need to protect the seemingly permanent and unthreatened 'habitats' of cliffs, upland areas, mountains, rock exposures or even quarries?

The purpose of this chapter is to outline the past and present agendas for earth heritage conservation, and to suggest the future: that the battle for the protection of our greatest geological riches lies wholly in the hands of the people to whom geology is nothing more than a boring collection of dry terms and dusty museum specimens. Successful conservation is dependent upon raising public awareness and increasing education; we must take the message to them.

The rationale for conservation

In all aspects of conservation, whether it be human history, as represented by historic sites or buildings, or elements of the natural world, such as the biological and geological resources of the planet, the rationale is effectively the same. The need for conservation relies upon four simple convictions which encompass the worth of the feature to be conserved or protected (Fig. 4.1; Huxley 1947; Nature Conservancy Council 1984, 1990; Wilson *et al.* 1994; Bennett & Doyle 1997).

1. That they be conserved for their own sake.
2. That they provide a basis for economic exploitation.
3. That they form a basis for research, training and education.
4. That they have aesthetic or cultural value.

The first of these tenets is most commonly understood with reference to wildlife and ecological aspects of the natural world. It has its basis in many of the major cultural and spiritual traditions of the world, particularly in the major religions of both the modern and ancient world which have traditionally espoused a principle of empathy with, and respect for, our fellow creatures. This fundamental principle of wildlife conservation is, however, more difficult to reconcile with inanimate or non-sentient features of a landscape, such as landforms and exposures of rock. Unlike wildlife, these features convey an appearance and impression of solidity and permanence which is at odds with their general vulnerability. In general this aspect is the most difficult to convey to the public in general and developers in particular, and therefore the conservation of such features has traditionally lagged behind that of wildlife.

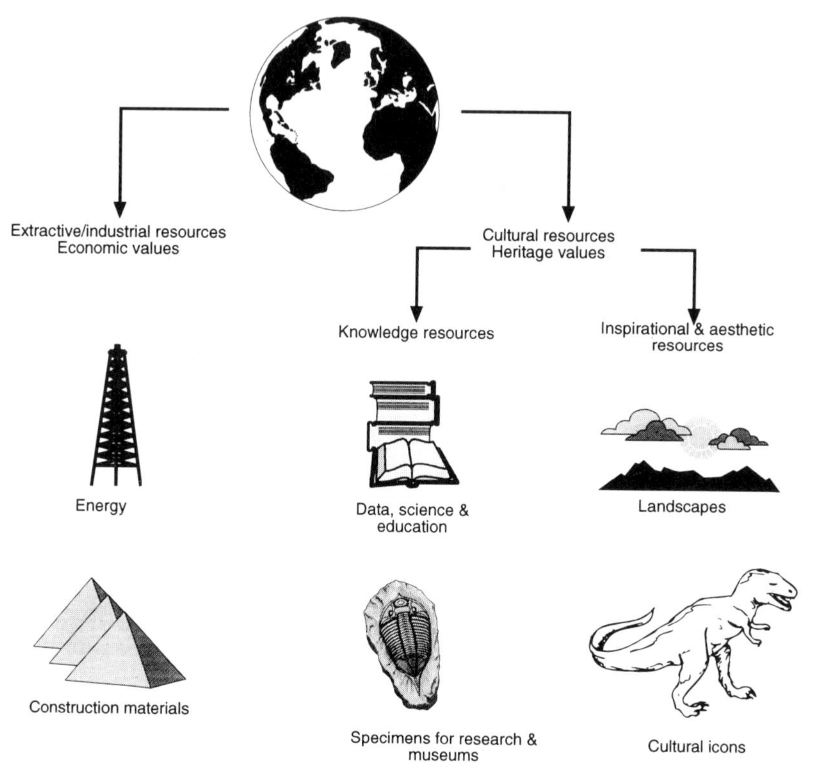

Fig. 4.1. The rationale for earth heritage conservation, based on both economic and heritage issues (modified from Wilson *et al.* 1994).

The other guiding principles of conservation are broadly linked and are all illustrative of the way that aspects of the natural or geological resource may be utilized in some way by humans. Of these, economic exploitation is the most common, and a clear understanding of the nature of a physical, biological or even historical resource is important in order to exploit it. Scientific research, and associated with this the need for the training and education of the next generation of scientists and technologists, is also of great importance to the furtherance of human progress. Such research may be both 'blue-sky' – the ideals of knowledge without direct or obvious economic value – and applied – intended to enhance the life of humanity. Finally, the Earth's natural and physical resources provide inspirational and aesthetic assets, in the form of beautiful landscapes and attractive specimens, both of which contribute to the spiritual and emotional well-being of people.

The past conservation agenda

Development of the boundary concept

Nature conservation in Britain can be traced back to the Middle Ages with the establishment of royal hunting sanctuaries maintaining game for the enjoyment of the aristocracy (Stamp 1969; Nature Conservancy Council 1984; Evans 1997). The rationale behind this type of conservation was mostly exploitative, but was increasingly driven by a concern over declining numbers of rare species – particularly birds – and a denigration of the countryside in the eighteenth and nineteenth centuries. The late nineteenth century saw the birth of the major non-governmental conservation bodies which were effectively to organize the conservation movement in Britain. During the last decades of the nineteenth century and the first of the twentieth, concerns over conservation led to the development of the Royal Society for the Protection of Birds (RSPB) and the National Trust – both now literally with millions of members – and the Society for the Promotion of Nature Reserves (SPNR), now the Royal Society for Nature Conservation (RSNC). These influential bodies, together with other smaller societies, sought to define the conservation agenda in the form of the development of nature reserves – safe havens within which wildlife could exist without interference. This thinking was a major influence on the philosophy of the governmental committee on Nature Conservation, which sat during the Second World War, and which was chaired by Sir Julian Huxley.

The findings of the Committee, published in 1947 (Huxley 1947), identified five salient points which are still at the heart of conservation in this country:

(1) that natural resources should be developed yet sustained;
(2) that research should be carried out to expand knowledge of

natural systems;
(3) that natural phenomena should be valued for its scientific, educational, aesthetic and economic aspects;
(4) that nature conservation encompasses biological, geological and geomorphological features;
(5) that nature reserves should form the mechanism for protection.

It ultimately led to the development of the National Parks and Access to the Countryside Act of 1949 which had at its heart the philosophy of the protection of natural areas through what may be called the 'boundary concept': the recognition of areas of scientific or scenic merit and the representation of each of these as a geographic entity on a map surrounded by a boundary (Fig. 4.2). The boundary effectively defines worth: inside it, the area is worthy of conservation; outside, it is not.

The boundary concept remains the dominant force in defining formal conservation strategies in Britain, at least. It is of great importance in the defence of designated areas at planning public inquiries under the Town and Country Planning Act, as it clearly denotes which areas are of greatest worth in a landscape. This has led to the development of networks of sites which are careful to contain only those areas of the greatest scientific worth – even through their significance may be remote from public perception.

The growth of site networks

Over the last 50 years the conservation of our earth heritage has been the responsibility of the conservation agencies (formerly the Nature Conservancy Council, now English Nature, Scottish Natural Heritage and the Countryside Commission for Wales), and groups of interested amateur geologists and volunteers. During this time these groups have formulated a strategy for the conservation or protection of Britain's earth heritage – its geology and geomorphology – which has at its core the establishment and management of specialist site networks using the boundary concept, designated either as SSSI or Regionally Important Geological/ Geomorphological Sites (RIGS) (Richards 1982; Clements 1984; Black 1988; Allen *et al.* 1989; Nature Conservancy Council 1990;

Fig. 4.2. The boundary concept in conservation. This cartoon demonstrates the principle of relative worth behind the recognition of special wildlife/geological areas. Reproduced with the permission of English Nature.

Wilson *et al.* 1994; Carson 1995; Ellis *et al.* 1996). These site networks are defined on restricted criteria: purely scientific for SSSIs; broader for RIGS, including historical and educational importance.

Statutory earth heritage sites in Britain – ultimately SSSI to be protected under the Wildlife and Countryside Act of 1981 – were selected under the twenty-year process of the Geological Conservation Review or 'GCR' (Ellis *et al.* 1996). This review employed experts to select sites on the strength of their scientific importance alone, and rated them in terms of national/international importance. GCR sites were selected to form a network of mutually supporting sites within which there was intended to be no duplication of interests, in order to strengthen the argument for uniqueness in a public inquiry situation. A total of 2992 GCR sites was selected to represent the geology of Britain and these distil down into around 2200 SSSI. RIGS are selected on the basis of a wide variety of criteria, most of which reflect the agenda of the local RIGS group of academics, museum staff, planners, amateurs and other interested parties (Nature Conservancy Council 1990; Wilson *et al.* 1994; Carson 1995).

With the exception of some RIGS sites which take in cultural heritage, most earth science site networks currently selected give little consideration to the values placed upon them by the wider community – generally considered to be essential for effective conservation (Barker 1996). The protection of Britain's earth heritage effectively follows this pattern, with the recognition by statutory and non-statutory organizations alike of 'networks' of sites, each with a boundary. These exclude much which is of value in Britain's earth heritage, which is then sacrificed to save the precious 'jewels in the crown' – a direct consequence of the application of the boundary concept.

Mechanics of site conservation: selecting and safeguarding networks
Traditionally, the conservation of sites and networks has been site specific, tackling problems directly as they appear (Nature Conservancy Council 1990; Wilson *et al.* 1994). Assessment is on limited, and often scientifically esoteric, criteria. Protection is largely reactive and a function of the planning process which alerts the statutory agencies and other interested parties of impending planning applications. Four basic themes are involved: assessment of value; promotion of site awareness; site protection and management; and site enhancement.

Assessment of value
Assessment of value is primarily subjective and is a function of the comparison of selected areas of the Earth with others. It assumes that some areas of the Earth are more important on the basis of aesthetic, cultural or scientific grounds than others, and in most countries, it leads to the selection of a series of areas of varying size. For the earth sciences, the term 'geotope' has been coined for those spatial areas forming distinct parts of the Earth's crust which are

characterized by their possession of an outstanding geological and/ or geomorphological interest (Stürm 1994). The recognition of such spatial units is central to the concept of conservation. Effectively it leads to the judgement of relative worth, which is often intended to be objective, as demonstrated by those sites selected for the GCR. The rigour by which the criteria are applied is important, as with planning pressure it is essential to be able to demonstrate the worth of one site compared with another with similar attributes (Allen *et al.* 1989; Ellis *et al.* 1996). However, in most cases, selection is highly subjective, and may be heavily observer biased.

Promotion of site awareness
Awareness is, in the majority of cases, the most important aspect of successful conservation. Attempts are made to increase awareness of the scientific value of certain sites through an appropriate use of on-site signage and interpretative boards, guided tours, leaflets and other literature (Badman 1994; McKirdy & Threadgould 1994; Page 1994; Robinson & McCall 1996). The relevance of the scientific aspects of the site may not always be directly apparent, and the site may appear unattractive or uninteresting to the lay person. This, together with a lack of awareness of the ultimate fragility of geological features, has greatly hampered effective conservation in many instances. In such cases, the links between the human experience and the site in question must be emphasized, drawing upon wider themes of scenic, wildlife and historical/ cultural aspects of the site in question.

Site protection and management principles
The effective protection of spatial areas which encompass geological or geomorphological features of importance (geotopes) is dependent on the definition of an appropriate management strategy. In some cases this is parallel to the difference in strategy between 'preservation', protection in an unchanged state or mothballing, and 'conservation', protection with a recognition that the resource is bound by nature to evolve and change with time.

Areas selected on the basis of their scientific importance may be protected with reference to both preservation and conservation principles, but their application depends on the nature of the scientific interest in the site. Here, the distinction is drawn between those sites which are representative of a finite resource – integrity sites – and those which are representatives of a much greater resource which is otherwise obscured elsewhere. These are referred to as exposure sites as they commonly include the exposures of rock outcrops which are not common above surface (Nature Conservancy Council 1990; Stevens *et al.* 1994; Wilson *et al.* 1994). Integrity sites include those with a rare fossil or mineral assemblage, and those which are unreproduceable and unique, including most landform sites. Exposure sites encompass most geological sites selected for conservation on the basis that they represent the typical expression of a particular geological unit. By their nature they include most stratotypes. In effect, the management principle to be applied at integrity sites is the maintenance of the integrity of the site, clearly akin to the concept of preservation. At exposure sites, effective management is needed to keep the exposure open, and if threatened, the site could be duplicated through the selection of an alternative, or through the creation of a new exposure in the same material which is otherwise plentiful underground.

Site enhancement

Effective management and conservation of a site also needs the development of site enhancement techniques. This can vary from the promotion of a greater awareness of a site, through better access, to physical actions such as, for example, the clearance of unwanted vegetation or dumped rubbish. Site enhancement is very often necessary at the small scale, and particularly in exposure sites, where the overriding conservation principle is that of the maintenance of the exposure. Here, particularly where the material is unconsolidated and prone to movement, the periodic cleansing and clearance of the most important faces is an ongoing necessity.

Endangered species and fragile landscape areas

Outside the boundary of nature reserves and protected sites it is commonly recognized that certain species are under threat of extinction. The recognition of the need for the protection of endangered species was a concept central to the development of major organizations such as the RSPB, which originally sought to

Fig. 4.3. Limestone pavement in northern England. This is the nearest geological analogue to an endangered species (photograph: A.F. Bennett).

combat the illegal trade in rare bird feathers for millinery (Evans 1997). The protection of endangered species is therefore outside of the boundary concept. Instead, absolute protection of certain designated wildlife species – plants, animals and birds – is provided for within the Wildlife and Countryside Act of 1981 (Nature Conservancy Council 1984). In such cases, protection is meant to be absolute, and is not boundary constrained.

Geologically speaking, the nearest analogue is that of the protection of limestone pavement. Limestone pavement as developed in the British Isles is unique, and a major contribution to the

Fig. 4.4. Fossil collecting is a valuable aid to the promotion of worth and, if carried out responsibly, rarely has a conservation impact.

world's landscape heritage (Fig. 4.3). Its distinctively weathered limestone is threatened by illegal extraction for use in gardens (Bennett, A. *et al.* 1996), and it has therefore been identified for special protection under Section 34 of the Wildlife and Countryside Act – the only geological resource to be protected in this way (Nature Conservancy Council 1984, 1990; Goldie 1993). However, despite this level of protection, the boundary concept is still in force, as only those areas of limestone pavement which are deemed to be of special merit, as defined by the statutory agencies, are protected (Goldie 1993). The remainder is, therefore, effectively open for exploitation.

The vexed question of specimen collecting

Probably the most often debated aspect of earth heritage conservation is the question of specimen collecting (e.g. Duff 1979; Clements 1984; Crowther & Wimbledon 1988; English Nature 1992; Norman 1992, 1994; Young 1994; Stanley 1995). In

Fig. 4.5. Shops selling geological specimens in Whitby, northern England. This type of retailing brings awareness of geological materials into the high street.

the absence of legislation in Britain which identifies fossil and mineral specimens as valuable artefacts which belong to the state – as in other countries (Norman 1994) – the debate is often sterile. However, most authorities agree that in all but the most exceptional circumstances, and under normal conditions (i.e. without explosives and heavy earth-moving machinery), the collecting of geological specimens has limited adverse environmental impact (Fig. 4.4). In fact, we would argue that the collecting has positive conservation benefits, through the promotion of awareness of value, worth and beauty of geological materials (Fig. 4.5).

Summary

The conservation agenda of the last one hundred years has been boundary dominated: the setting apart of selected areas of the Earth's surface for protection. Much effort has gone into the designation of special sites and site networks, and an infrastructure of protection, which includes detailed assessment, conservation and awareness strategies has grown up around this. In the main this has been successful, but with the desire to be spartan in the selection of sites – to avoid duplication – some fragile areas have escaped. The protection of limestone pavement in this country is an example of a more flexible approach, but even this is ultimately boundary delineated. In our desire to protect sites of scientific worth, a kind of paranoia has grown up with respect to the collection of geological materials, and this still pervades the conservation debate today. Although some sites are exceptionally fragile and unique, in the main responsible collecting can only promote a greater connection between the public and the scientists in the protection of important geological areas.

The present conservation agenda

The present agenda within earth heritage conservation is being set by the statutory conservation agencies which are seeking to divide Britain into a number of Landscape Areas (in England) or Natural Heritage Zones (in Scotland) (English Nature 1993, 1996; Duff 1994; Prosser 1995; Countryside Commission & English Nature 1996; A.P. McKirdy, pers. comm. 1997). These are intended to provide a wider context for site networks, especially since it is unlikely that many new sites will be added since the end of the GCR process (C.D. Prosser, pers comm. 1997).

The concept of Natural Areas

The concept of Natural Areas has grown in response to a perceived demand for a 'holistic' approach to nature conservation, concentrating as it does on the linkages between geology, geomorphology and wildlife. In England, 116 Natural Areas have been recognized, and these correspond to 188 'Character Areas' of the Countryside Commission which identify land-use patterns and features of the landscape which are intimately associated with the nature of the geology, geomorphology and wildlife (Countryside Commission & English Nature 1996).

The rationale behind the development of the Natural Areas or Natural Heritage Zones is primarily that of wildlife conservation in the first instance (English Nature 1993). Natural areas are intended to give context to wildlife areas, particularly vegetational types and their dependent animal species. Typically, within a Natural Area there may be several SSSIs, and its gross boundary corresponds with naturally occurring aspects of the underlying geology, soils, geomorphology and so on. In the southeast of England, where bedrock is closer to the surface of the landscape because of its lack of glacial sedimentary cover, these display direct correspondence with lithological types (Duff 1994). For example, the Weald of Kent and Sussex is defined into Natural Areas on the basis of its chalk outcrop (North Downs, South Downs and Hampshire Downs Natural Areas and corresponding Character Areas), greensand outcrop (Wealden Greensand Natural Area and Character Area), and Weald Clay and Hastings Beds outcrop (Low Weald and High Weald Natural Areas and Character Areas) patterns. However, elsewhere, the relationship is less clear where the variations in the bedrock are obscured by thick glacial deposits.

Earth heritage conservation and Natural Areas

The development of the Natural Area concept has been, as discussed, primarily in response to the need to place wildlife in a local context. In theory this can also work for geology and geomorphology, through the illustration of the importance of bedrock, soil type and landform to the development of specific plant and animal communities. Perhaps the best example is the typical chalk downlands of southern England, which provide an evocative and 'typically English' landscape of geology, geomorphology and wildlife in harmony. The intention, then, is to provide context and to marshal wildlife to come to the defence of geology in demonstrating worth, ultimately increasing public awareness. These are laudable aims, but there are two problems: (1) that Natural Areas are predominantly rural; and (2) that there is a danger in marginalizing geology at the expense of wildlife, in other words, that geology is not worth protecting in its own right.

The problem really lies with the fact the majority of the British public live in urban areas, and therefore, the majority of Natural Areas, with the exception of some classic examples such as chalk downland, are equally remote. The Natural Areas concept suffers because it is once again boundary-concept conservation, and is effectively another site network. Away from the classic examples, geology is also seen as a marginal aspect of the Natural Area – this is clearly demonstrated by English Nature's own consultation document (English Nature 1993). Therefore, although the Natural Area concept is clearly a step forward from the classic site-based approach to conservation, in effect it does little to promote the greater public appreciation of the fragility of Britain,s earth heritage.

The future: promotion of geology in our urban centres

At its heart, effective conservation is dependent upon public awareness and interest – both of which are connected with a commitment to conserve for future generations. However well framed the statutory powers of site protection may be, however much effort is expended on site safeguard and the promotion of Natural Areas, it is clear that these will be ineffective without a public will to conserve (Doyle & Bennett 1997). Erosion of national (SSSI) and regional (RIGS) geological sites is continual. Clearly, the future for earth heritage conservation lies with dramatically increasing public awareness and perception of geology. Unfortunately, although refocused with the Natural Area concept, there has been a tendency for the aims of increasing public awareness to become subsumed into the philosophy of the site network, with its broader aims largely neglected (English Nature 1996). This cannot be effectively achieved through the simple promotion of site networks, since most are the exclusive preserve of the interested amateur geologist and the academic elite, and few cater for the needs or interests of the community as a whole.

Effectively, there should be a twin-strand approach to earth heritage conservation, in which one strand aims to identify, and protect site networks as part of Natural Areas – the 'jewels in the geological crown' – while the other strand involves raising awareness of geology and its conservation, irrespective of its intrinsic or extrinsic value (Fig. 4.6; Doyle & Bennett 1997). Of these two strands one has its focus on communities of interest (i.e. people such as professional and amateur geologists; Robinson 1995) for whom the site network is conserved, while the other encompasses

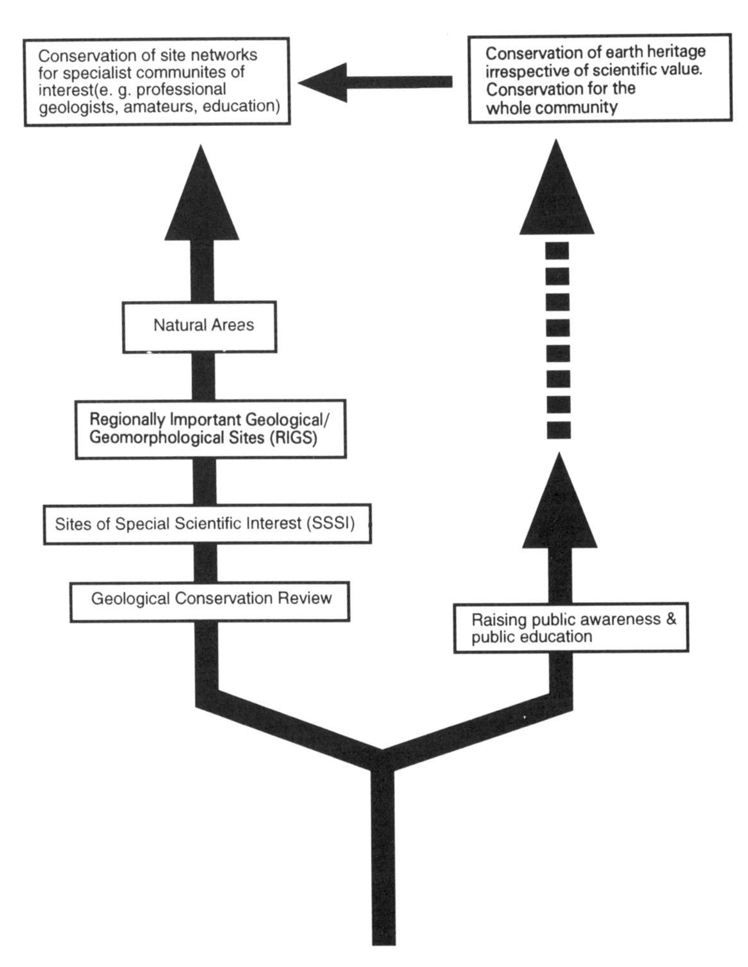

Fig. 4.6. The twin strand approach to earth heritage conservation (modified from Doyle & Bennett 1997).

all types and sectors of the community. Wildlife conservation, particularly in its continuing ascendancy in public interest over geological conservation, illustrates the importance of public awareness for effective conservation, and serves as a model for its effective promotion (Barker 1996). Public awareness of wildlife conservation is high, with a greater ambient concern or interest in wildlife *per se*, leading directly to a recognition of the importance of such sites as ancient woodland, wetlands and peat bogs. Effective site safeguard in wildlife conservation therefore feeds directly from this enhanced public awareness: witness the numerous pressure groups, the activists and the involvement of local people in protecting local trees, birds and sites of ecological interest from recent road development. This is the importance of awareness; earth heritage conservation needs to follow the lead of wildlife conservation.

Raising awareness and the role of urban geology

In theory, the aim is a simple one; to raise general awareness in geology and its conservation and thereby exercise control through popular opinion on the planners, developers and decision-makers who have the potential to threaten or conserve our earth heritage. In practice, however, raising geological awareness is difficult, especially so since many of the most precious and threatened components of Britain's geological fabric are often remote and obscure. To be effective, an awareness strategy must be focused on the local community, that is the community of place, and linked to its social, historical and cultural fabric.

The move towards the development of the Natural Areas concept was in direct answer to this question, and was a means of promoting the relevance of earth science through appealing to the wildlife that lives upon them. This is only partially successful, as it draws upon sometimes tenuous and tortuous links which are not easily examined by the general public, away from the chalk downlands and other more clear-cut examples. Since over 80% of the population in Britain live in urban areas away from Natural Areas it is to these people that we must take the message of geology. The most appropriate message is that of urban geology: the nature and fabric of the built environment and its intimate relationship with the original and vestigial landscape. Urban geology – building stones, graveyards parks and gardens – provides an important resource through which we can attempt to foster interest in geology and demonstrate its links with the community (Bennett, M. *et al.* 1996; Robinson & McCall 1996; Doyle & Bennett 1997). This, in

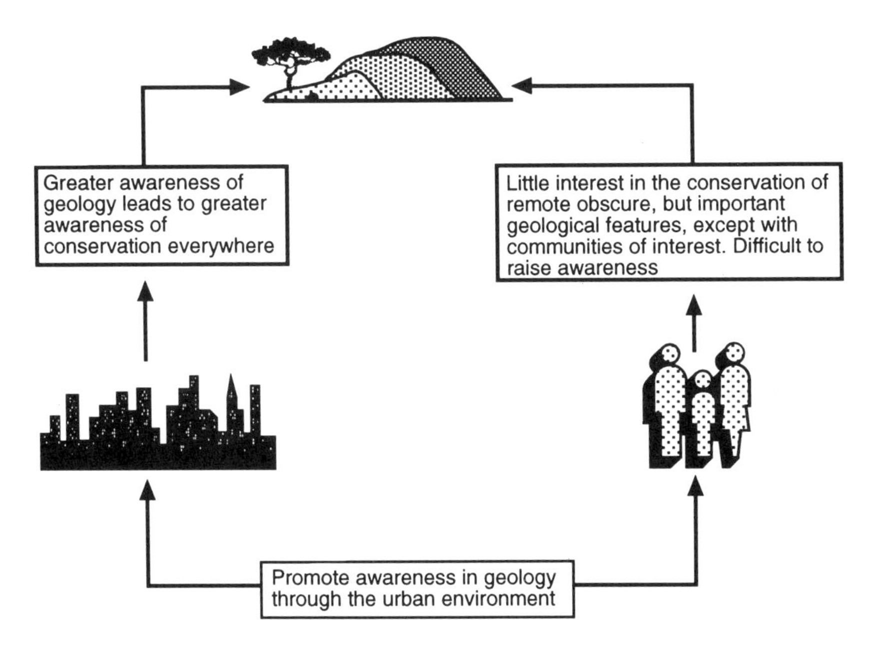

Fig. 4.7. Conceptual model of the role of urban geology in increasing public awareness of earth heritage conservation (modified from Doyle & Bennett 1997).

turn, provides the means to illustrate the value and importance of the earth sciences, and effectively promote greater awareness of the need for its conservation everywhere (Fig. 4.7).

The nature of the urban geological resource

The built environment of our towns and cities is a reflection of the earth science resource. It has two basic components: (1) remnants of the primary geomorphology of the region prior to construction; and (2) the buildings, roads and other constructions which are composed primarily of materials derived from the geological resource, in the form of building stones and other construction materials (Bennett, M. *et al.* 1996; Robinson & McCall 1996; Doyle & Bennett 1997).

In many cases the original siting of a town or city was controlled by the nature of the geomorphology. For example, Roman, Norman or Medieval defensive positions on high ground or within river meanders are typical in Britain. The growth of urban centres which engulfed the countryside following the Industrial Revolution has left a legacy of remnant geomorphology in the centre or suburbs of many towns and cities. This is often preserved in a semi-natural state in city parks or in the grounds of large private houses which have since been incorporated into the fabric of the city. These remnants were retained primarily as the 'lungs' of the great industrial cities in the nineteenth century and after, and have an aesthetic and cultural significance which is, in many cases, equivalent to that of the naturally occurring landscape, despite the fact that often, very little in the way of unaltered landscape is left. The scientific importance of such areas is muted, but there is no doubt that there is some educational benefit in having such open spaces located within the inner city, close to schools and other educational centres.

Building stones similarly have an important aesthetic and cultural association. Often, the early, pre-Industrial Revolution and transport-age building stones were locally derived and are in harmony with the local landscape and geology. The built environment therefore mirrors the local geology, and the vernacular architecture of villages, towns and cities, using local stone, has a strong cultural connection and aesthetic appeal. Educationally, the built environment offers a great challenge, in displaying a wide range of geological materials which are subjected to stringent weathering effects of the often hostile urban atmosphere.

If we are to raise the public profile of geology in urban areas then we must exploit the fundamental links between geology, urban development and social history. This geological influence may not always be clear, yet in almost all cases urban growth will have been guided or shaped by geology at some stage in its development.

The exploitation of urban geology will differ from location to location and will be a function of the nature of the resource. In particular, in raising the profile of geology the approach adopted will naturally vary, and will usually be dependent upon whether an urban area has any visible, naturally occurring geology. Typically, it is possible to group urban areas on the basis of visible geological resources, and three groups are apparent (Bennett & Doyle 1996; Doyle & Bennett 1997).

1. Those areas with a striking or visible geological resource. A classic example is Edinburgh: Arthur's Seat and Castle Rock are pre-eminent. Lyme Regis and Whitby are smaller towns whose economy has been influenced by geology both in the past and present (Fig. 4.5). Public awareness, amongst both the locals and visitors, of geology in all three examples is strong.
2. Those areas with some naturally outcropping geology. These towns are typically small and located within a countryside setting, particularly in upland areas. In these towns public awareness of geology may be low, present only as a subliminal recognition of geological features which the local population may take for granted.
3. Those areas with little or no exposed or visible geology. Most inland towns in the southeast of England fall into this category, along with major cities such as Cardiff, Birmingham and London. Awareness of geology in these large urban conurbations is usually low and the clarity of the link between a town and its geology may be completely lost.

The recognition of the existence of the three types of urban area is important in guiding strategies for the promotion of earth science awareness. For example, in urban areas of Types 1 and 2, although the nature of the built environment will also be important, the focus should be the promotion of the existing natural geology. The aim of this approach is to demonstrate the importance of easily observed and often well known geological features in the development of a town and city. For Type 3 urban areas, with little or no visible geology, the importance of the built environment is paramount and the focus must be: (1) the creation and introduction of appropriate geological features into the urban environment; and (2) the exploitation of existing artificial resources.

The classification presented above provides the initial foundation upon which to base a strategy for increased awareness of geology in the urban environment. In most cases, previous attempts at demonstrating geology in the urban environment have been presented piecemeal. Many valuable attempts have been made by

Table 4.1. Urban Geological Promotional Plan for Type 1 and 2 towns

Aim: to promote the natural geological resource within the town and to demonstrate the links with its historical, social and cultural development.

Objectives

- The geology/geomorphology of a town should be incorporated into every leaflet, display or exhibition depicting the history of urban development. This may be subtle, or may be stressed and developed in greater detail, especially where the location or commercial development of a town was particularly governed by the nature of the local geology.
- Wherever possible, use of the traditional and local building stone should be used in street furniture. This will be in keeping with the vernacular architecture, and will usually enhance harmony with its local environment. These links can be made stronger by the inclusion of occasional stone inscriptions giving the stone name and local source.
- Landscaped ground should be in harmony with the local geology. In particular rockeries should use local stone wherever possible, and should be organized so that the dip and strike of the emplaced stone mimics or reflects that of the naturally outcropping geology which surrounds the town.
- The education potential of local quarries, natural exposures and other site resources should be exploited. Educational packages for local schools can be developed through consultation with teachers and developed with the needs of appropriate key stages of the National Curriculum.

local groups. For example, there are numerous building stone and graveyard stone trail guides (Robinson & McCall 1996). However, the valuable opportunity presented by their message is lost through a lack of co-ordinated planning. One of the most useful ways of co-ordinating this process is by the use of Urban Geological Promotional Plans by RIGS groups, local geological societies and other interested parties (Doyle & Bennett 1997). Urban Geological Promotional Plans are intended as manifestos for raising the profile

Table 4.2. Urban Geological Promotional Plan for Type 3 towns

Aim: to create a geological resource were no obvious one exists, and to encourage the exploitation of the urban fabric as a resource.

Objectives
- To create geology where no obvious evidence of it exists, through the use of artificial reconstructions, sculptural features which are made from geological materials, or which are in harmony with geological structure, and in problem walls intended to demonstrate the richness of lithology or geological structure.
- Encourage the link between geology and architecture through promotion of the good practice of stone labelling, or in the development of interpretative leaflets distributed to local schools and museums.
- Development of high street building stone trails and cemetery visitor centres in which the geology and weathering of gravestones can be interpreted and understood.
- The construction of problem walls to demonstrate geology in parks and play grounds.

of urban geology within a town or city. Each one sets out a list of generic objectives focused towards the single aim of promoting public awareness of the geology within a town. Each plan will be different depending on the nature of the geological resource present and opportunities which are available. The aim of these plans is simply to get geology promoted within a town at the same level, and in the same way, that cultural heritage or tourist attractions are packaged (Tables 4.1 and 4.2). Critical in the equation is the acceptance of the concept of the Urban Geological Promotional Plan by developers and, more importantly, by local planning authorities.

In summary: the future

The development of site networks and their recent refocus into Natural Areas spotlights those areas which should be protected within the four basic conservation convictions. The long and tortuous road in gaining recognition for these areas has been the lot – often ungratefully received – of the statutory conservation agencies. There have been significant achievements, but the debate has, in many ways, become sterile while erosion of the site base is continual. In reality there is only one route to effective conservation: the promotion of the importance of geology to the urban majority.

The future agenda for earth heritage conservation should, therefore, focus around the following three objectives:

1. maintenance, management and enhancement of the existing site resource;
2. education, investing in the future generation of decision-makers and planners, and raising the profile of geology and of conservation;
3. raising public awareness of geology through the promotion of the urban geological resource.

References

Allen, P., Benton, M. J., Black, G. P., Cleal, C. J., Evans, K. M. *et al.* 1989. The future of earth sciences site conservation in Great Britain. *The Geological Curator*, **5**, 101–109.

Badman, T. 1994. Interpretting earth science sites for the public. *In:* O'Halloran, D., Green, C., Stanley, M. & Knill, J. (eds) *Geological and Landscape Conservation*. Geological Society, London, 429–432.

Barker, G. M. A. 1996. Earth science sites in urban areas: the lessons from wildlife conservation. *In:* Bennett, M. R., Doyle, P., Larwood, J. G. & Prosser, C. D. (eds). *Geology on your Doorstep: the role of urban geology in earth heritage conservation*. Geological Society, London, 181–193.

Bennett, A. F., Bennett, M. R. & Doyle, P. 1996. Paving the way for conservation? *Geology Today*, **11**, 98–100.

Bennett, M. R. & Doyle, P. 1996. The introduction of geology into the urban environment: principles and methods. *In:* Bennett, M. R., Doyle, P., Larwood, J. G. & Prosser, C. D. (eds) *Geology on your Doorstep: the role of urban geology in earth heritage*

conservation. Geological Society, London, 239–261.
——— & ——— 1997. *Environmental Geology: geology and the human environment*. John Wiley, Chichester.
———, ———, Larwood, J. G. & Prosser, C. D. (eds) 1996. *Geology on your Doorstep: the role of urban geology in earth heritage conservation*. Geological Society, London.
Black, G. P. 1988. Geological conservation: a review of past problems and future promise. *In:* Crowther, P. R. & Wimbledon, W. A. (eds) *The use and conservation of palaeontological sites*. Special Papers in Palaeontology, **40**, Palaeontological Association, London, 105–111.
Carson, G. 1995. Education or legislation? The role of RIGS in geological conservation. *Geoscientist*, **5**, 17–20.
Clements, R. G. (ed.) 1984. *Geological Site Conservation in Great Britain*. Geological Society, London, Miscellaneous Papers, **16**.
Countryside Commission & English Nature 1996. *The Character of England: landscape, wildlife and natural features*. Countryside Commission and English Nature, Cheltenham and Peterborough.
Crowther, P. R. & Wimbledon, W. A. (eds) 1988. *The use and conservation of palaeontological sites*. Special Papers in Palaeontology, **40**, Palaeontological Association, London.
Doyle, P. & Bennett, M. R. 1997. Earth-heritage conservation in the new millennium: the importance of urban geology. *Geology Today*, **13**, 29–35.
Duff, K. L. 1979. The problems of reconciling geological collecting and conservation. *In:* Bassett, M. G. (ed.) *Curation of palaeontological collections*. Special Papers in Palaeontology, **22**, Palaeontological Association, London, 127–136.
——— 1994. Natural areas: an holistic approach to conservation based on geology. *In:* O'Halloran, D., Green, C., Stanley, M. & Knill, J. (eds) *Geological and Landscape Conservation*. Geological Society, London, 121–126.
Ellis, N. V. (ed.), Bowen, D. Q., Campbell, S., Knill, J. L., McKirdy, A. P., Prosser, C. D., Vincent, M. A. & Wilson, R. C. L. 1996. *An Introduction to the Geological Conservation Review*. GCR Series **1**, Joint Nature Conservancy Committee, Peterborough.

English Nature 1992. *Fossil Collecting and Conservation*. English Nature, Peterborough.

——— 1993. *Natural Areas. Setting nature conservation objectives: a consultation paper*. English Nature, Peterborough.

——— 1996. *Conserving England's Earth Heritage. Taking earth heritage conservation forward into the next millennium*. English Nature, Peterborough.

Evans, D. 1997. *A History of Nature Conservation in Britain* (2nd edn). Routledge, London.

Goldie, H. S. 1993. The legal protection of limestone pavements in Great Britain. *Environmental Geology*, **21**, 160–166.

Huxley, J. E. 1947. *Report of the Committee on Nature Conservation in England and Wales*. HMSO, London.

McKirdy, A. & Threadgould, R. 1994. Reading the landscape. *In:* O'Halloran, D., Green, C., Stanley, M. & Knill, J. (eds) *Geological and Landscape Conservation*. Geological Society, London, 459–462.

Nature Conservancy Council 1984. *Nature Conservation in Great Britain*. Nature Conservancy Council, Peterborough.

——— 1990. *Earth Science Conservation in Great Britain. A strategy*. Nature Conservancy Council, Peterborough.

Norman, D. B. 1992. Fossil collecting and site conservation in Britain: are they reconcilable? *Palaeontology*, **35**, 247–256.

——— 1994. Fossil collecting: international issues, perspectives and a prospectus. *In:* O'Halloran, D., Green, C., Stanley, M. & Knill, J. (eds) *Geological and Landscape Conservation*. Geological Society, London, 63–68.

Page, K. N. 1994. Information signs for geological and geomorphological sites: basic principles. *In:* O'Halloran, D., Green, C., Stanley, M. & Knill, J. (eds) *Geological and Landscape Conservation*. Geological Society, London, 433–439.

Prosser, C. D. 1995. Conserving our earth heritage through natural areas. *Geoscientist*, **5**, 19–20.

Richards, L. 1982. Earth science conservation in Britain. *British Geologist*, **8**, 4–9.

Robinson, E. 1995. The role of the voluntary sector in earth science conservation. *Geoscientist*, **5**, 13–14.

——— & McCall, G. J. H. 1996. Geoscience education in the urban setting. *In:* McCall, G. J. H. & Marker, B. R. (eds) *Urban Geoscience.* A.A. Balkema, Rotterdam, 235–252.

Stamp, D. 1969. *Nature Conservation in Britain.* Collins New Naturalist, London.

Stanley. M. F. 1995. Earth science conservation. *Geoscientist*, **5**, 11–12.

Stevens, C., Erikstadt, L. & Daly, D. 1994. Fundamentals in earth science conservation. *Memoire de la Société Géologique de France*, **165**, 209–212.

Stürm, B. 1994. The geotope concept: geological nature conservation by town and country planning. *In:* O'Halloran, D., Green, C., Stanley, M. & Knill, J. (eds) *Geological and Landscape Conservation.* Geological Society, London, 27–32.

Wilson, C. (ed.), Doyle, P., Easterbrook, G., Reid, E., & Sklipsey, E. 1994. *Earth Heritage Conservation.* Geological Society and the Open University, London and Milton Keynes.

Young, B. 1994. Mineral collectors as conservationists. *In:* O'Halloran, D., Green, C., Stanley, M. & Knill, J. (eds) *Geological and Landscape Conservation.* Geological Society, London, 439–442.

5 Our 'green and pleasant land': loving it to death?

Andrew F. Bennett, MP

Summary

- In 1995, the House of Commons Environment Select Committee, began to investigate how far using the countryside for leisure pursuits was in conflict with conservationist principles.
- Only a small minority of the very large amount of evidence received suggested that long-term irreparable damage was happening as a result of leisure and tourism.
- The evidence also showed that the car presented a major issue of concern within these recreational areas.
- This chapter focuses on sustainable tourism, the changing role of the Countryside Commission, new legislation affecting National Parks and the government's response to the Select Committee's Report.

My somewhat folksy title is intended to draw attention to the question of sustainable development – that fashionable, but vague, concept – as it relates to tourism and countryside issues. When the House of Commons Environment Select Committee set out to enquire into the environmental impact of leisure activities, sustainable development was a key buzz word for a variety of organizations who gave evidence. In this chapter, I want to ask the question 'can the historic English countryside (the 'green and pleasant land' of my title) cope with an increase of visitors without being irreparably damaged?' and answer it from the material we collected and published in our report of 1995 (House of Commons Environment Select Committee 1995). First, however, I want to begin by asking, not how far the 'historic English countryside' is at risk, but how real is it? To what extent are the various lobbying groups attempting to defend a myth?

Cultural historians such as Raymond Williams, Alun Howkins and Martin Wiener have argued that historic, rural, peaceful England was a cultural invention born of centuries of the English pastoral tradition in the arts and the special conditions of the last quarter of the nineteenth century (Williams 1974; Wiener 1985; Howkins 1986). As the towns became industrialized and overcrowded, and as workers were herded together in insanitary slums, the countryside came to represent the antithesis of all that was seen as ugly and unwholesome in the urban experience. It was invested with aesthetic and moral values, and despite all evidence to the contrary, came to be seen, not only as the representative of a purer, healthier way of life, but also as the 'real' England. Today we have absorbed those attitudes so completely that they are never challenged. So it has come about that the country is seen both as a spiritual resource and as a national heritage. This notion finds its expression at all levels of society, including government documents. For example, the Department of the Environment's 1990 statement of Britain's environmental strategy states that: 'Our countryside and coasts are a central part of our heritage ... Our landscapes have been an inspiration for centuries for poets, painters and nature lovers. They help us form our sense of national identity' (Department of Environment 1990).

Unsurprisingly, these concepts covertly informed much of the evidence presented to the Environment Select Committee. They underlay the assumption of most conservation groups and leisure-users that the countryside was, and obviously should be, 'peaceful' -— a place to find a space for reflection that could not be found in towns. They encouraged fears that the countryside would be spoilt by a mass of visitors; and were expressed in judgements that noisy sports and machines were an 'intrusion', in arguments about the 'right to roam', and in the language of 'threat', 'pollution' and 'danger' it was so easy to fall into when examining the issues.

It only takes a moment's reflection, however, to remember that virtually none of the English rural landscape is natural, and that there is nothing inherently peaceful, let alone spiritual, about it. It is, above all, moot how far the language of 'preservation' is relevant to a discussion of landscape. It is no real trouble to preserve the mythical landscape, though even here, witnesses to our enquiry were sometimes at loggerheads with each other – some urging the strictest conservation practices, others complaining about the 'museumization' of the countryside (what one group called 'the brown signpost syndrome'). The real landscape however, as earth scientists know better than most people, is the product of aeons of constant, sometimes violent, change. Even in the relatively cosy manufactured natural world, some of the most treasured features of the rural landscape are the result of catastrophes which have wiped out the old, or of gradual changes in land use, or of often strenuously opposed building projects. I doubt Stonehenge would have got past the public consultation, planning inquiry and inspection by the Secretary of State for the Environment that is the norm now!

The famous patchwork landscape was, of course, shaped, not only by centuries of taming the land for food production, but also by aesthetic choices and by leisure and sporting activities. Large country estates were created both to express the owners' wealth, power and taste, and to cater for their recreation. What is a modern phenomenon, however, is the increasing leisure use of rural space by the urban working- and middle-classes, who are demanding a share in what they have been encouraged to believe is, not only a natural resource, but a part of their rightful heritage.

There is, in fact, a certain element of class in most discussions of sustainability. Attitudes encapsulating covert social, political and class claims surfaced remarkably often in our inquiry. Some of those we most frequently encountered in our informal visits were the possessive, 'We don't want all these people swarming over our land' and the judgmental, 'If they cannot walk, they shouldn't be able to get to it'. In formal evidence, too, similar attitudes were encountered: 'The sun brings them [the visitors] out like summer flies'; 'If all they want is a pub lunch or "nick-knacks" then there are less environmentally sensitive areas where these can be obtained'; 'My family have followed this way of life since time immemorial, thereby conserving this national asset for you and the whole nation'.

As a consequence, perhaps, of attitudes like this, many groups encountered specific difficulties in pursuing their chosen activity; the problems encountered ranged from ramblers' complaints about obstructed paths, to the British Orienteering Federation's complaint that permission to stage large events was being turned down because of concerns for 'visual impact'. In campaigning for their right to use rural space, however, many organizations hijacked the concept of 'sustainable development' to argue for priority for their own pastime at the expense of others'. So we heard anglers complain that power boats were ruining lake- and river-banks, walkers complain that riders were destroying footpaths, and birdwatchers complain that walkers were causing cliff erosion. Our conclusion was that cultural problems are just as real as physical problems – indeed, they are often the root cause of the various tensions and dissatisfactions that are expressed in terms of threats to the environment.

As far as physical problems were concerned, we actually found substantial evidence that, contrary to popular perceptions, leisure and tourism pose no inevitable and intrinsic threat to the environment. English Nature concluded that, overall, 'leisure activities do not in themselves represent the greatest threat to nature conservation' (English Nature 1995). In particular, leisure use of rural land is less of a threat to Sites of Special Scientific Interest (SSSI) and Areas of Outstanding Natural Beauty (AONB) than industry or other developments. According to English Nature's records, no SSSI have been lost through tourist activity and most of any damage done is short-lived rather than long term. Moreover, on the whole the trend between 1989 and 1994 was downwards; five cases of long term damage to SSSI by leisure use and fifty cases of short-term damage were reported in 1989/90 against four cases of long-term damage and twenty-seven cases of short-term damage in 1993/4 (English Nature 1995). Likewise the Countryside Commission noted that damage to the environment by industrialization, farming and urbanization heavily outweighed any damage caused by recreational activity (English Nature 1995).

A final point is that, though the National Parks have been keen to foster the impression that visitor numbers have increased to danger-level in the past years, there is actually little evidence to support this claim as yet. Our own impression was that the number of visitors was not growing exponentially year by year, but was subject to fluctuations which depend on factors such as the state of the economy, the weather and changing fashions in leisure pursuits. Rather than making any judgements on this point, therefore, we preferred to await publication of the Countryside Commission's 1994 survey, *Visitors to National Parks*. When it came out in 1996, the figures showed rather more academic ingenuity than useful data. However, even after the massage, the figures did not appear to back apocalyptic claims. For example, the Peak Park estimate of around 22 million visitors per year proved to be a gross exaggeration, the actual figure quoted by the Countryside Commission being about 12.5 million (Countryside Commission 1996).

Specific issues

Three representative topics will now be considered: footpath erosion, forests and traffic.

Footpath erosion

Footpath erosion was one of the largest single issues, particularly in upland regions. On our site visits the Select Committee found that erosion had made some paths extremely unsightly or almost impassable; and in a few cases the erosion was spreading outwards over a very wide area. According to the evidence we were given, the causes are complex. Contributing factors are: high stock levels on the land crossed; frequency of leisure use, particularly in bad weather; bad drainage of the surrounding land and the path itself; and the nature of path surfaces. The universal opinion seemed to be that erosion was a 'problem' and it had to be dealt with. We saw various solutions in operation. In reality, however, some of the solutions were worse than the 'problem'. Some were inappropriate or ineffective; others only led to worse erosion a few years later or a few yards further away; and in others, the very substantial physical work that had been undertaken had changed the nature of the path from hillside track or riverside path to a paved way not unlike a roadside pavement.

The situation was most graphically illustrated on national long-distance footpaths. Neither Tom Stevenson in campaigning for the Pennine Way, nor A. R. Wainwright in promoting the coast-to-coast path, can have envisaged how popular such routes would become or what the effect of that popularity would be. Owing to the fragile nature of many path surfaces, some sections of these routes have not been able to cope with the high levels of use to which they have been subjected. Considering the natural tendency of peat to erode and turn to bog, and the fact that walkers are quite capable of finding alternative routes – considering, too, that the repairs are often more unsightly than the 'damage' – the work done on these paths cannot be 'sustainable development' under any definition.

Incidentally, the cost is hardly sustainable either: restoration of the Pennine Way, for example, has been estimated to cost £7 million and requires £250 000 each year to maintain. Personally, I believe this is a waste of money and a ridiculous interpretation of the principle of sustainability. An alternative would be to insist that any trails designated in the future should follow naturally sustainable paths or that a wide variety of alternative routes should be provided to disperse visitors. As far as erosion of the existing paths goes, I believe that the main reason for the concern is once again principally aesthetic, and that, here as elsewhere, a 'sustainable' landscape is interpreted as a 'neat and tidy' one. Many of the witnesses to the inquiry pointed out that eroded paths were not necessarily lifeless paths and that the destruction of one sort of habitat made room for the creation of another.

Forests

In response to government requirements, the Forestry Commission has opened up large tracts of its land for woodland walks, nature trails, picnic sites and so on. The Select Committee was concerned that public access to forests should not be reduced when land is transferred from public to private ownership. Guarantees were given concerning land held by privatized water companies, and so far those promises have been upheld at least in the National Parks, but in evidence to us, the Royal Town Planning Institute pointed out that the proposed arrangements for maintaining rights of access are inadequate to guarantee that this will continue. On a visit to the Lake District, however, we were told that North West Water had adhered to the pre-privatization policy of opening up more and more of its water catchment areas to public access. This initiative is useful and it would be good to see the practice extended.

There is, though, perhaps a more serious issue. The government aims to double English forest cover from 7% to 15% to answer arguments put forward by countries such as Brazil and Malaysia who, criticized for reducing the world's carbon store by clearing tropical forests, have pointed out that the UK has itself in the past destroyed quantities of hardwood trees. The government's reforestation plan actually means planting an area equivalent to the county of Cambridgeshire. Mile after mile of conifers is not only an unattractive sight, but it has potentially serious environmental repercussions. During a more recent inquiry into water conservation, the Select Committee was told by the Centre for Hydrology and Ecology that: 'the proposed doubling of the area under forestry by the year 2045 ... may well have important and detrimental consequences on water resources which will be magnified if and when climate change results in a hotter drier climate in lowland England'. The problem is complex, but it seems that coniferous tree cover in upland areas would reduce the run-off to reservoirs by 20%: if global warming results in the sorts of climate changes that are presently envisaged, this problem will extend to lowland areas and broad-leaved tree cover. Evaporative differences between trees and short crops will become more marked; whereas short crops are 'water soil limited', trees are not, and will increase their transpiration rates to accommodate the new situation (House of Commons Environment Select Committee 1996).

Traffic

The private car has given people access to large areas of the countryside, and brought with it a good deal of relaxation and pleasure. Ironically, however, the resulting cars, roads and car parks present a challenge to the quality of the landscape that people have set out to enjoy. Many witnesses to the Select Committee inquiry picked out fumes and congestion as central issues. It is not only moving traffic that causes problems; parked cars are just as problematic. In the 1960s and 1970s parking increased to meet demand, with extra car parks being built. At that time, these were often designed to hide cars from sight behind trees and bushes.

Thanks to fears about crime, the trend is no longer to keep cars hidden from view, so the familiar criticisms about untidiness and intrusiveness surface here too. The response has been to reduce parking provision. This has suited many organizations because it has rationed access to popular beauty spots.

The Committee concluded that National Parks, County Councils, authorities covering AONB, and all other relevant authorities should develop a rural transport strategy, the aims of which should include.

1. the provision of new sport and leisure facilities as close to good public transport as possible and near to urban centres;
2. the development of public transport to reduce the need for cars for leisure;
3. the encouragement of cycling;
4. the recognition that leisure traffic might have to be restricted in some places and under some circumstances;
5. measures to ensure that lorries and other heavy vehicles are restricted to major routes except for access.

Conversely, the strategy should recognize that some people do enjoy looking at the scenery from a car window (or, if they are frail or disabled, can only see the countryside this way); also that some sports and leisure activities necessitate large amounts of equipment and that public transport is unsuitable for these groups.

The Committee looked specifically at the feasibility of traffic restriction schemes. Personally, I think that removing all but essential local traffic from some country lanes might work, but there are very few schemes actually in operation, and too many of them seem doomed to failure. Though there are successful schemes in the Goyt Valley and the upper Derwent Valley and at the Roaches in the Peak District National Park, their experience cannot be easily translated to other areas because of their unique topographical suitability. What is needed is some sound experimental evidence, as suggested in the Edwards Report, to assess the problems and benefits of this sort of scheme in typical areas and where the interests of local tourist businesses (Bed & Breakfasts, tea shops etc.) have to be accommodated. Sadly, no one will grasp this nettle.

It is useless, though, to try a blanket ban on cars or attempt to squeeze them out by sky-high parking charges. Already in many places, car park charges are higher than needed to meet running costs. Effectively this is a policy of rationing by price, which is not only unacceptable, but often simply moves the problem on. The roads get clogged with prowling cars or the cars stack up on roadside verges. In the end, the press of visitors has to be acknowledged and catered for.

Conclusion

The evidence given to the Environment Select Committee showed that the central question is not whether tourism and leisure activities are damaging the environment, but how conflicts about land use can be resolved. These conflicts include practical disputes between incompatible user-groups about what are, or are not, ‘suitable’ country activities; aesthetic arguments; cultural conflicts; economic disputes between different groups of country dwellers; and conflicts caused by the exacerbation of pre-existing problems such as access, traffic and pollution.

It is important to emphasize this point because of calls for restrictions on recreational activities. The rationale of restrictions is often, I believe, based on a desire for a tidy, picturesque landscape in keeping with largely mythical and romantic notions of ‘Englishness’ and ‘heritage’, though it is expressed in terms of threat to the environment and the buzz word of ‘sustainability’. Many things said and done in the name of ‘sustainability’ – a concept I support – or concern for ‘the environment’ are, in fact, the result of aesthetic considerations, a precise vision of the way the countryside ‘should be’.

It must be remembered that much of Britain’s treasured ‘heritage’ was not sustainable in its originally designed form, anyway. Traditional stone walls in upland farming areas were never sustainable as field boundaries; stone castles, after a very short military life, crumbled to picturesque ruins. These, and countless other features, have lived a much longer life as landscape features and tourist attractions than as functional artefacts. Tracks over peat, field paths and riverside walks of their nature come and go, shift and change. The landscape is not unchangeable and the ideal of sustainability should not lead to trying to preserve it in an artificial time-capsule.

Large-scale footpath reconstruction, calls to ban noisy recreations in national parks, the homogenization and 'prettifying' of rural villages may all come into this category. Perceptions about damage and threat should obviously be taken seriously and explored (the precautionary principle is an excellent one), but action restricting any particular activity should be taken on good scientific and planning advice, not out of a desire to preserve a vision of an England that never was.

It is necessary, then, to separate the real from the imaginary problems and to step back from seeing change as necessarily a 'threat'. Within this outlook, the sustainability principle could mean, not the fossilization of the countryside, but a way of meeting leisure needs which uses recreational activities as a way of increasing people's appreciation of the world about them.

References

Countryside Commission. 1996. *Visitors to National Parks: Summary of the 1994 Survey Findings*. Countryside Commission.

Department of Environment. 1990. *This Common Inheritance: Britain's Environmental Strategy*. HMSO, London.

English Nature. 1995. *The Environmental Impact of Leisure Activities*. English Nature, Peterborough.

House of Commons Environment Select Committee, 1995. *The Environmental Impact of Leisure Activities*. HMSO, London.

——— 1996. *Water Conservation and Supply, Minutes of Evidence and Appendices*. HMSO, London.

Howkins, A. 1986. The Discovery of Rural England. *In:* Colls, R. & Dodd, P. (eds) *Englishness: Politics and Culture 1880–1920*. Croom Helm, London, 62–88.

Wiener, M. J. 1985 *English Culture and the Decline of the Industrial Spirit 1850–1980*. Pelican, Harmondsworth.

Williams, R. 1974. 'Pastoral Voices.' *New Statesman*, **27 September**, 13–39.

Part Two

The built environment

Some of the most important issues in modern environmental geology stem from the built environment. The built environment acts as a machine: consuming inputs of energy and natural materials from the urban hinterlands; producing outputs of wastes (gases, solids and liquids) and manufactured goods; demanding maintenance of its buildings and infrastructure. Part Two collects individual contributions on aspects of this urban machine: resource implications; aspects of the engineering geology of its infrastructure; issues of domestic, industrial, commercial waste disposal. A major issue is the characterization of wastes and dealing with the difficult legal and technical issues of contaminated land, nuclear wastes and airborne particulates.

6 Environmental geology of Gibraltar: living with limited resources

Edward P. F. Rose

Summary

- Gibraltar is a British dependent territory which forms a narrow peninsula 6 km^2 jutting south from Spain at the western entrance to the Mediterranean Sea.
- Geology, rugged topography, seasonal rainfall, dense population and frequent political isolation are major factors which influence the local environment.
- Urban development has extended from low ground on to land reclaimed from the sea and a 50 km system of tunnels and chambers complements the narrow roads.
- Potable water is provided mainly by desalination, demand is reduced by a separate seawater supply for sanitary purposes; Jurassic limestone has been used for building, scree breccias for land reclamation, and Quaternary aeolianites for fine aggregate. Incineration is currently the main means of waste disposal.
- Since the former fortress is being increasingly developed as a centre for tourism safety measures are being enhanced, with particular regard to the stability of slopes, cliffs and tunnels.

Gibraltar is a narrow, rocky peninsula – often referred to as 'the Rock' – which juts south from Spain at the western entrance to the Mediterranean Sea (Fig. 6.1). Seized from the Visigoths by an Islamic (Moorish) army in AD 711, it was finally captured by Spaniards in 1462, and then by an Anglo-Dutch force in 1704. Ceded to Britain by the Treaty of Utrecht in 1713, it was granted a large measure of internal self-government in 1969, and in 1973 it formally joined the European Community under Article 227(4) of the Treaty of Rome as a European territory for whose external affairs a member state is responsible.

Historically, Gibraltar's fame rests upon its role as a fortress, developed under British rule as a naval base. The British armed forces have therefore provided the mainstay of the Gibraltar economy for about 280 years. However, Ministry of Defence contributions to Gibraltar's Gross Domestic Product will have fallen in recent years from a peak of 65% to around 5% by the year 2000 – promoting rapid adaptation to a new economic environment.

Environment

Additional to stratigraphy and geological structure, factors which influence the natural environment of Gibraltar are: geographical position, tectonic setting, size, topography, climate, population density and periodic political isolation.

Geographical position

The narrow (24 km) Strait of Gibraltar provides the shortest crossing of the Mediterranean between Europe and Africa. If, as discussed by Warren & Varley (1996), proposals to link the two continents by a tunnel are implemented, Gibraltar will lie close to, if not directly on, the main land route through Spain to Morocco. Additionally, the Strait is one of the busiest maritime traffic thoroughfares in the world, and Gibraltar's port has been developed into one of the most important bunkering facilities in the region.

Tectonic setting

Gibraltar is situated close to the European–African plate boundary. In the past 90 years earthquakes of Modified Mercalli Scale Intensity 4 to 7 have been experienced. In 1905 an intensity 7 earthquake was recorded with computed acceleration of 0.1 g. This has been taken as a 100 year return value for recent construction design (Gibraltar Public Works Department, pers. comm. 1987).

Size

The Gibraltar peninsula is 5.1 km long, and 1.6 km in natural maximum width, so total natural land area is only about 6 km^2.

Topography

The Isthmus (Figs 6.2 & 6.3) which links the peninsula northwards to mainland Spain is a low-lying neck of land less than 3 m above present sea-level which is now largely covered by the Gibraltar airfield and associated buildings. The Main Ridge of the Rock extends south for 2.5 km from the precipitous North Face which overlooks the Isthmus – as an asymmetric, sharply ridged crest with peaks over 400 m high, whose western dip-slope contrasts with the near-vertical scarp slopes to the east. The Southern Plateaux truncate the end of the Rock: Windmill Hill Flats sloping south from 130 to 90 m elevation, succeeded by Europa Flats at 40 to 30 m elevation, bordered by steep cliffs to the sea.

Climate

The climate is mild Mediterranean, with warm dry summers alternating with cooler, wetter winters. Mean maximum daily temperatures are 24°C for the warmest month (August) and 13°C for the coldest month (January).

Population density

The population exceeds 30 000: some 20 000 Gibraltarians, 6000 British expatriates, and 4000 Moroccans. It fluctuates daily with migration of workers from and to Spain, and seasonally with tourism.

Fig. 6.1. Aerial view of the Gibraltar peninsula, from the southeast. The Southern Plateaux form the foreground, the wave-cut platform of Europa Flats (with sports field) at 30–40 m above sea-level, succeeded northwards by Windmill Hill Flats at 90–130 m. The Main Ridge extends further northwards, as a sharp crest with peaks over 400 m in height, and with relatively gentle dip-slope westwards to the town and harbour contrasting with the steep cliffs of the east coast. On the east coast, a breccia/sand slope whose upper surface has been covered to form a water catchment area is just visible to the south of a major partly-quarried scree breccia slope. On the west coast, the harbour area is enclosed successively northwards by the South Mole, Detached Mole, and North Mole breakwaters. The Isthmus can be seen to the top right of the figure, extending from the North Face of the Rock into mainland Spain as a flat sandy plain <3 m above sea-level. The airfield runway extends on made ground into Algeciras Bay just beyond the North Mole, top centre of the figure. (Reproduced courtesy of the Institution of Royal Engineers from Rose & Rosenbaum 1990.)

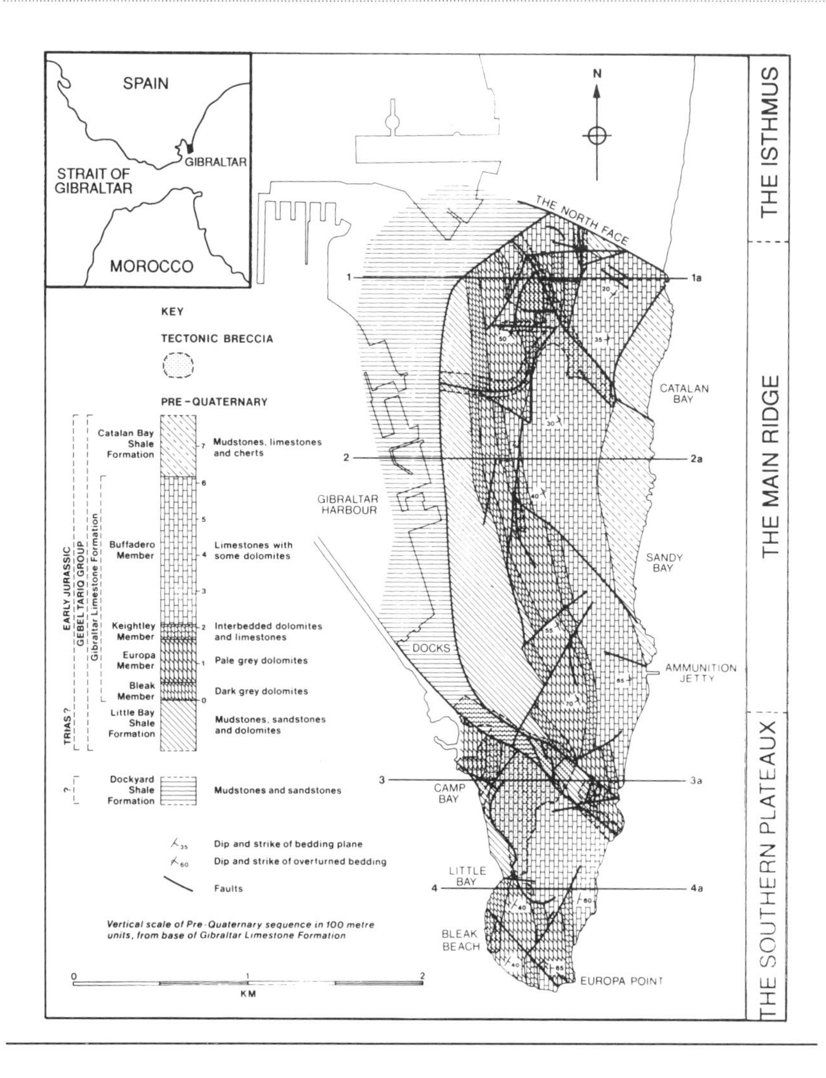

Fig. 6.2. Map of the pre-Quaternary geology of Gibraltar, with major topographic features sidelined to the right, and locality diagram inset top left. For cross-sections drawn along the lines of section shown on the map, see Rose & Rosenbaum (1990, 1991*a,b*). (Reproduced courtesy of the Institution of Royal Engineers from Rose & Rosenbaum 1990.)

Fig. 6.3. Map of the Quaternary sediments on Gibraltar. (Reproduced courtesy of the Institution of Royal Engineers from Rose & Rosenbaum 1990.)

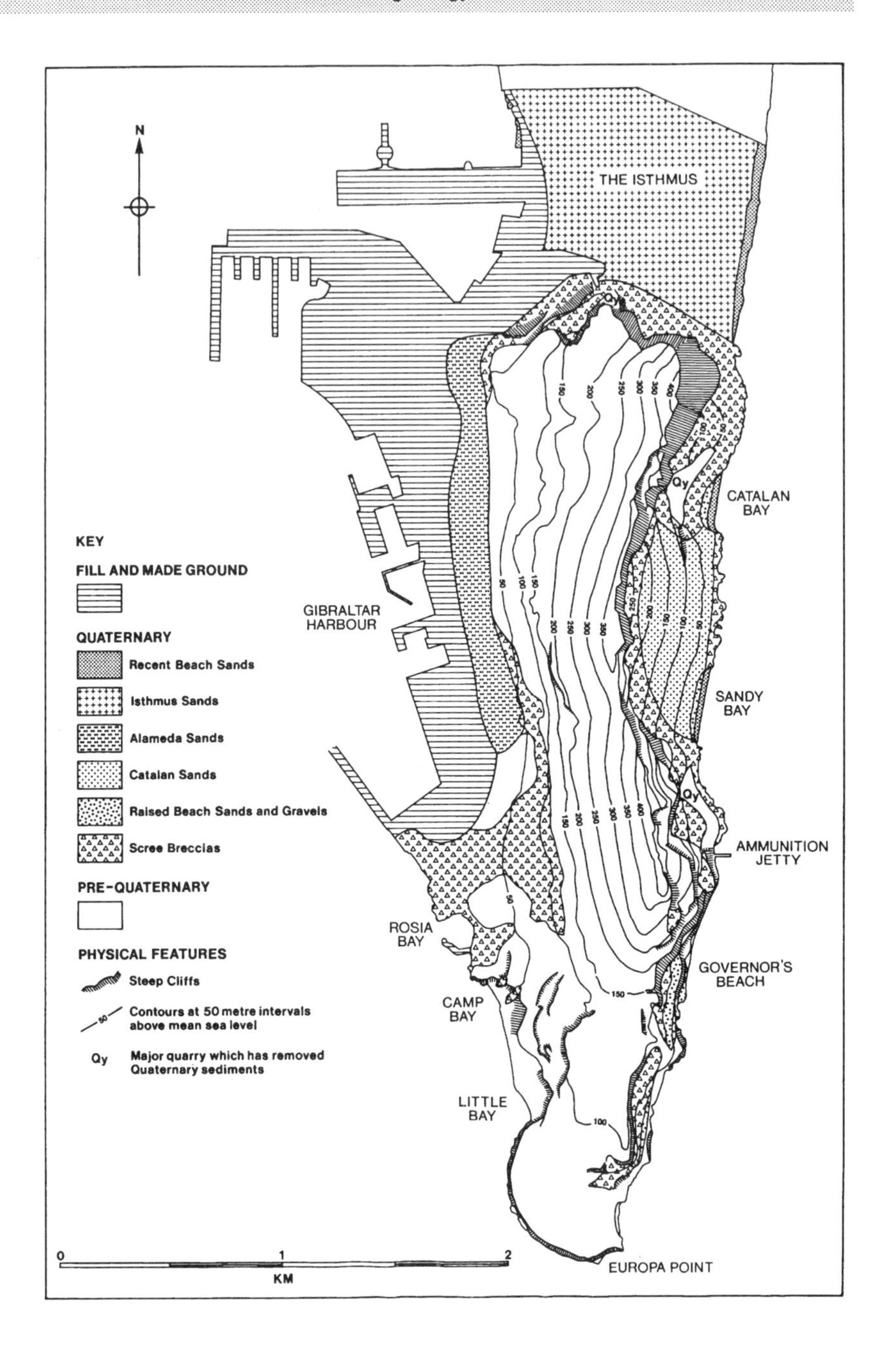
N
THE ISTHMUS
CATALAN BAY
SANDY BAY
AMMUNITION JETTY
GOVERNOR'S BEACH
GIBRALTAR HARBOUR
ROSIA BAY
CAMP BAY
LITTLE BAY
EUROPA POINT
KEY
FILL AND MADE GROUND
QUATERNARY
Recent Beach Sands
Isthmus Sands
Alameda Sands
Catalan Sands
Raised Beach Sands and Gravels
Scree Breccias
PRE-QUATERNARY
PHYSICAL FEATURES
Steep Cliffs
Contours at 50 metre intervals above mean sea level
Qy Major quarry which has removed Quaternary sediments
0 1 2
KM

Political isolation

Gibraltar is 'probably the most fought over ... place in Europe' (Chapple *in:* Hughes & Migos 1995, p. vii). Since 1309 it has been besieged fifteen times, the fifteenth siege a long, if non-violent, isolation from Spain, from 1969 to 1985. Subsequently, political disputes have intermittently caused disruption to traffic across the land frontier.

Stratigraphy and geological structure

Aspects of the geology of the Rock have been of continuing interest to scholars since at least the mid-eighteenth century (Rose & Rosenbaum 1992), but the first geological survey as such took place only in 1876, generating a 1:2500 scale map (Ramsay & Geikie 1876) and complementary report (Ramsay & Geikie 1878). Later work by Royal Engineer geologists generated unpublished maps in 1943 and 1947 (Rose & Rosenbaum 1990; Rose & Cooper 1997), and a new interpretation of the geology by Bailey (1952). However, the first detailed geological map has been published only recently (Rosenbaum & Rose 1991*a*), complemented by a detailed field guide (Rose & Rosenbaum 1991*b*) and associated papers (e.g. Rose & Rosenbaum 1991*a*; 1994; Owen & Rose 1997). Features of the bedrock geology have been most comprehensively reviewed by Rose (1998), and of the superficial deposits by Rose & Hardman (1998).

Gibraltar is currently interpreted as a klippe – the eroded remnant of a nappe thrust into its present position during the Betic-Rif orogeny in the late Cenozoic. The main mass of the Rock is composed of a thick (>600 m) sequence of well-cemented, cyclic, sparsely fossiliferous peritidal dolomitic limestones of Early Jurassic age – the Gibraltar Limestone Formation. This is a medium to massively bedded, strong crystalline rock much resembling the Carboniferous Limestone of England and Wales in general appearance. Four members have been distinguished and mapped within the formation on the basis of their gross lithology and weathering characteristics (Fig. 6.2). The sequence, now overturned in the Main Ridge area, was originally underlain by a thinly bedded dolostone/mudstone/sandstone sequence of unproven but presumed Late Triassic age – the Little Bay Shale Formation – now poorly exposed on the western flank of the Rock. The Limestone was overlain by more thinly bedded limestones and mudstones of Late Pliensbachian (Mid Lias) age – the Caleta Member of the Catalan Bay Shale Formation – and these in turn by argillaceous radiolarian cherts of probable Late Jurassic age: the North Face Chert Member of the Catalan Bay Shale Formation. Catalan Bay Shales currently crop out at the base of the North Face, at four localities along the east coast of the Rock and within the Lower Waterworks Tunnel.

A major NW–SE fault occurs as a zone of fractured rock separating the inverted, westward-dipping sequence in the Main Ridge from an uninverted, eastward-dipping sequence in the Southern Plateaux. Within these regions major faults trend NW–SE, and less extensive faults NNE–SSW to NE–SW, and N–S. An eastward-dipping thrust plane is inferred to lie beneath the Rock as a whole.

Following erosion and karstic solution, massive slopes of Quaternary scree breccia were deposited along much of the flanks of the Rock (Fig. 6.3), together with extensive windblown sands – most obviously the Catalan Sands to the east of the Main Ridge, and the Alameda Sands to the west – and the Quaternary and contemporary sands and clays which now form the Isthmus. The two Southern Plateaux are wave-eroded platforms, and a dozen displaced shorelines intermittently notch the Rock to a height of at least 210 m – with raised beach sediments lying upon narrow wave-cut platforms, backed by former marine cliff lines. The Rock is honeycombed with caves, supposedly at least 143 of them above sea-level and many more below it. Some of the caves preserve a thick sedimentary infill, and record of human occupancy intermittently dated back to at least 40 000, potentially up to about 120 000 years BP.

Fortification

For much of its history Gibraltar has been regarded as 'the type and ideal of a stronghold' (Hughes & Migos 1995, p. 164). Its strength is derived partly from the conveniently-defendable size of the Rock,

Fig. 6.4. Parson's Lodge Battery, north of Camp Bay on the southwest coast of Gibraltar – first fortified in the eighteenth century, abandoned in the 1950s, and one of the sites recently restored under the auspices of the Gibraltar Heritage Trust.

being only 11 km in circumference, and partly from its topography – its precipitous northern, eastern and southern cliffs forming major obstacles to attack. Military landscaping and massive fortifications enhanced the natural features bordering the town area, and massive artillery firepower provided the ultimate deterrent.

Fortification was begun by the Moors and extended by the Spanish. The British made use of existing fortifications with relatively minor alterations until the Great Siege by French and Spanish troops from 1779 to 1783. Thereafter there were major construction works. Initially, defence was concentrated against an attack from the land: flooding of part of the Isthmus to create a major obstacle; scarping and scaling of potential access routes across the base of the North Face to make them too difficult to climb; fortification of lowland areas by tower, wall and bastion to deny ready access; and tunnelling to provide vantage points for cannon fire. Later the perceived threat was greater from the sea, as screw-driven, armoured steamships began to ply the Mediterranean in place of wooden sailing vessels. Under the protection of its own developing, formidable armament (Fig. 6.4), the harbour facilities adjacent to the town on the lowland area to the west of the Rock were considerably extended: docks were excavated through the underlying weak 'shale' formation (Scott 1914), coastal land was reclaimed from the sea with fill quarried from the Quaternary scree breccias, and an enlarged anchorage was created by greatly extending both North and South Moles, and by constructing an additional Detached Mole (Fig. 6.1). In the Second World War airpower increasingly replaced seapower – an airstrip was constructed across and beyond the Isthmus, guns were additionally directed skywards, and the garrison was largely accommodated underground to ensure safety from air attack. However, following the close of hostilities in 1945 the Royal Navy has been progressively reduced, and the traditional role of Gibraltar and comparable bases, that of repairing and refuelling a large British fleet, has been lost. With the end of the Cold War in 1989, army and Royal Air Force as well as naval elements of the garrison have been largely withdrawn, and their facilities have been increasingly transferred to civilian use. There has been no recent attempt to enhance the fortifications of the Rock and only selective attempts to conserve them.

Urbanization

The town was founded by the Moors in AD 1160 and has grown principally on the western side of the Rock, around the harbour, taking advantage of the more extensive flat and gently sloping ground there, formed by erosion of the western 'shales' (Rosenbaum & Rose 1994), which are less resistant rocks than the strong Gibraltar Limestone, and probably eroded as Quaternary marine platforms prior to tectonic elevation. Small settlements have developed elsewhere but the early town was largely confined to the fortified area south from the castle and extending uphill on to the lower slopes of the Main Ridge. Buildings are variously sited on Alameda Sands or on scree breccia, or directly on the underlying 'shale' or Gibraltar Limestone bedrock – many on the rubble of the

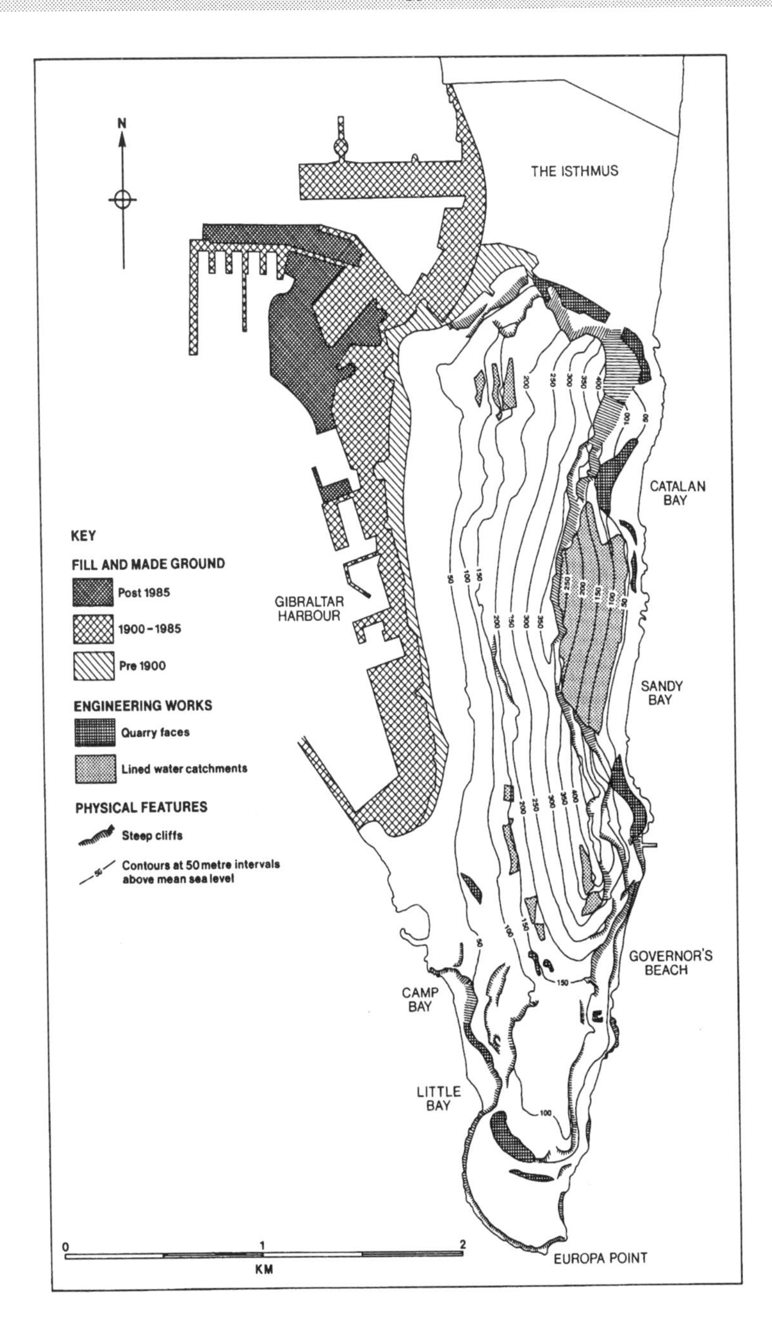
N
THE ISTHMUS
CATALAN BAY
SANDY BAY
GIBRALTAR HARBOUR
GOVERNOR'S BEACH
CAMP BAY
LITTLE BAY
EUROPA POINT
KEY
FILL AND MADE GROUND
Post 1985
1900–1985
Pre 1900
ENGINEERING WORKS
Quarry faces
Lined water catchments
PHYSICAL FEATURES
Steep cliffs
Contours at 50 metre intervals above mean sea level
0
1
2
KM

old town largely destroyed in the Great Siege of 1779–1783. The Limestone provides an excellent foundation, except where weakened by solution cavities. 'Shales' pose more of a problem, since the top 3–6 m are commonly weakened by weathering, and this zone needs removal or piling through to create foundations for large buildings. Similar procedures are required for the relatively uncompacted Alameda Sands which underlie most of the town.

Quaternary wave-cut platforms have provided natural, relatively flat ground used for building construction, such as the former South Barracks (Ramsay & Geikie 1878). Building has also extended on to artificially levelled ground provided by the floors of disused quarries.

However, post-1900, the most significant development has been on to land reclaimed from the sea (Fig. 6.5). Some reclamation took place prior to 1900, to provide foundations for the coastal fortifications and later through infill of the Inundation, an artificial brackish lake created in the eighteenth century from a morasse in the southwest corner of the Isthmus to restrict access to the town. Major reclamation took place to facilitate expansion of the dockyards in 1895–1905. Further reclamation took place from 1942 to facilitate extension of the airfield. All these works used fill quarried from the scree breccias. In contrast, sea-dredged sand was used in 1989 to infill part of the northern harbour area, creating significant new space for high-rise housing and office development. Further, more limited reclamation has been effected from 1995 to create a marina and shore facilities to the south of Eastern Beach, at the northeast corner of the Rock.

Access routes

Land

Roads on Gibraltar are almost invariably narrow, a consequence of the restricted land area. Many are steeply inclined, a consequence of

---

Fig. 6.5. Map showing the engineering use of major geological features on Gibraltar. (Reproduced courtesy of the Institution of Royal Engineers from Rose & Rosenbaum 1990.)

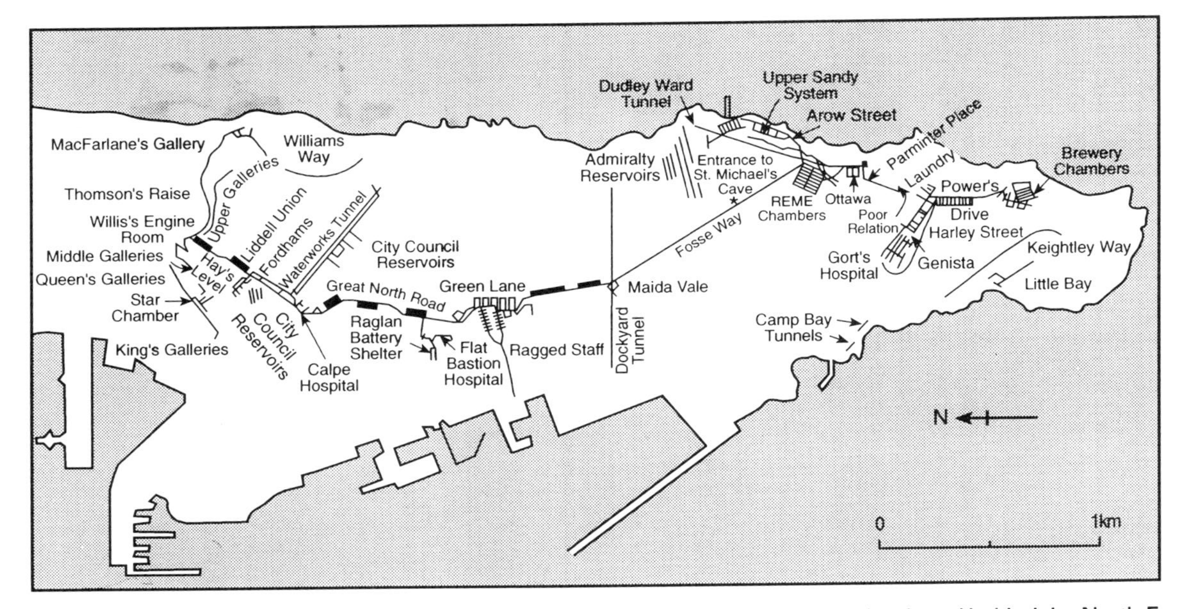

Fig. 6.6. Plan of the main tunnels on Gibraltar. A northern system was built first, largely developed behind the North Face of the Rock, which overlooks the Isthmus extending northwards into Spain. A southern system was developed later, in the then main military base area on the southeast side of the Rock, facing the Mediterranean Sea. These two systems were later joined by a north–south access tunnel, the Great North Road, and its continuation as the Fosse Way, completed in 1944. (Other tunnels cannot conveniently be illustrated on this scale.) (Modified from Rose & Rosenbaum 1991*b*)

the Rock's rugged topography. Traffic can be congested, and parking in the town is often a problem. However, widening of the roads on the Rock itself necessitates excavation through the strong Gibraltar Limestone, or at least through patches of well-cemented breccia. The Europa New Road was widened by Royal Engineers in 1973 using controlled blasting to fragment adjacent bedrock. Formerly, the technique was widely practised in military areas of the Rock, but is increasingly constrained by the spread of civilian-inhabited buildings – building proximity increasing the risk of damage to property, and civilians being less amenable to inconvenience than the military. The single road which provides land access from Spain crosses the airfield runway, and is closed during aeroplane take-off and landing. A tunnelled alternative has been considered.

Tunnels and underground chambers constructed by the military are a major feature of Gibraltar and now total over 50 km in length (Fig. 6.6) (Wilson 1945; Haycraft 1946*b*; Cotton 1948; Lauder 1963; Ramsey 1978). Most date from five periods (Rosenbaum & Rose 1991*b*, 1992, 1994), outlined below.

1782–1799

The early tunnelling period, beginning when Gibraltar was besieged by Spanish and French forces entrenched across the Isthmus. The British garrison tunnelled behind the North Face to achieve vantage points from which cannon fire could be brought to bear on enemy batteries.

1880–1915

When, after a long gap in which there had been no tunnelling, the Admiralty, the Army and the City Council all began underground excavation, to provide ammunition magazines, communications tunnels or water reservoirs.

1933–1938

When construction of reservoirs for water storage, and more particularly the construction of air raid shelters and underground hospitals, caused much tunnelling activity.

1939–1945
When a greatly increased garrison during the Second World War required new accommodation space, and space for reserves of food, equipment and ammunition.

1956–1968
When the existing tunnels were linked together in a more efficient manner, and additional storage chambers and reservoirs were hewn out of the Rock.

Stable tunnels have all been cut through the Gibraltar Limestone. Early construction methods were very slow but they caused minimal damage to the unexcavated rock, and this has been repaid by the long-term stability of the smooth walls so created. By the Second World War tunnelling techniques had become much quicker. However, blasting with high explosives produced large quantities of gas and consequent large scale fragmentation of the surrounding rock. Stress relief now develops fractures and consequently loose rock around the tunnels, requiring periodic scaling for safety (see Fig. 6.14) (Rosenbaum *et al.* 1994). Parts of the tunnel system are now being increasingly converted to civilian use. For example, a new road tunnel to divert traffic from the town has been considered.

Sea
Since Gibraltar is a peninsula and politically isolated from neighbouring Spain, the development of the harbour and docks has been of fundamental importance to the prosperity of the town (Rosenbaum & Rose 1994). Considerable expansion of these facilities took place from the end of the nineteenth century (Scott 1914). The former South Mole was extended northward, the North Mole westward and then southward, and a Detached Mole was sited in the intervening gap. These breakwaters were constructed using masonry blocks of Gibraltar Limestone around a core fill of limestone rubble. Three new docks were excavated at this time at the south end of the harbour and a graving dock further to the north, through the Dockyard Shale Formation, a sequence of weak

rocks. Excavation was therefore relatively easy but the anisotropic strength of the 'shales' and the steep dip of the bedding necessitated substantial timber supports to maintain the sides of the excavation until the final dock lining could be installed. The lining was of trimmed blocks of granite imported from Britain by sea.

Air

Construction on the Isthmus of an 'emergency' landing ground for service aircraft was begun on 3 September 1934, and the strip was fully available by 10 March 1936 (Ramsey 1978). On the outbreak of war in 1939 the runway extended nearly 900 m east–west from coast to coast, through a race track. Under favourable conditions it gave a take-off barely long enough for stripped Wellington bombers staging between England and Malta (Haycraft 1946*a*). Soon waves lapped over wrecked aircraft at both ends of the runway: a sight, together with the adjacent city cemetery, giving little cheer to pilots coming in to land. Preliminary work on an extension began at the end of 1941: a project facilitated by the availability of adequate raw materials. Spoil from tunnel excavation, and considerably larger amounts of scree breccia quarried from the base of the North Face, were emplaced westwards into Algeciras Bay. Large rock fragments were deposited so as to form the sides of the extension, accumulating at their natural angle of rest. Smaller material was used to infill the extension core, and tunnel spoil with a water-bound finish was used to surface the strip. By 3 April 1942 the runway had been extended to a length of 1050 m 27 days ahead of schedule, 'one of the great constructional achievements of the war' (Ramsey 1978, p. 14) which allowed aircraft to take off with enough fuel to fly directly to Egypt, bypassing beleaguered Malta. An extension to almost 1417 m was completed by November, allowing the necessary air cover to support Operation Torch, the Allied landings in North Africa commanded by General Dwight D. Eisenhower. By July 1943 the full planned extension, to 1645 m, was complete, with a width of 137 m, raised to a height of about 3.05 m above high water level. By that time 1.15×10^6 m^3 of fill had been placed in position. After the war a further 183 m were added to the airfield length.

Raw materials

Building stone

Stone has been the traditional building material used through much of Gibraltar's history. Its use by the Moors prior to 1462 can be observed in the remaining walls of their castle, and by the Spaniards in the Charles V and Phillip II walls built in the sixteenth century to protect the town from southerly attack. These appear to be constructed of local stone but quarries of appropriate early date cannot now be verified on the ground because of later quarrying or construction works. Considerable quantities of stone were used by the British for their fortifications. James (1771, p. 327), on the basis of observations made during his residence on the Rock between 1749 and 1755, recorded a 'fine quarry of stone, nearly resembling the Portland [of Dorset, England]' in the Southern Plateaux, between Nun's Well and Europa Advance, and other 'Large quarries of freestone' – evidently the Gibraltar Limestone. Parts of the Rock have been scarped 'for the purpose of making it more inaccessible' (Smith 1846, p. 43) so old cut faces are not necessarily evidence of quarrying primarily for building materials. However, the Gibraltar Limestone has been quarried north as well as south of the town. Both Smith (1846, p. 49) and Ramsay & Geikie (1878, p. 508) refer to a 'quarry at the North Front'. Additionally, a 'red sandstone' being 'extremely hard and tough, [was] used in preference to the ... limestone for lining the [gun] embrasures' (Smith 1846, p. 43). Ramsay & Geikie (1878, p. 522) report the 'quarry opened in calcareous sandstone' to be above No. 3 Europa Advance Battery on Monkey's Cave Road, where a disused quarry is currently used for temporary storage of household waste. Cut blocks of the Quaternary Monkey's Cave Sandstone (Rose & Hardman 1998) are still visible in gun embrasures fringing the cliffs of the southwest Europa coast. No building stone is currently being quarried on Gibraltar: cheaper or more versatile materials for building construction are all imported.

Gibraltar Limestone was formerly burnt for lime, and employed in a variety of uses: white-washing buildings; painting of water cisterns; pouring over corpses in mass graves after epidemics to prevent the spread of disease. The last remaining kiln, thought to

Fig. 6.7. Quaternary scree breccia exposed in a disused quarry north of Catalan Bay (cf. Fig. 6.1). The large size and sharp angularity of the Gibraltar Limestone clasts have been used to infer formation by freeze–thaw action under a more extreme climate than presently exists on Gibraltar. (Modified from Rose & Rosenbaum 1991*a,b*.)

date from the late nineteenth or early twentieth centuries, is sited near the Great Siege Tunnels at the northern end of the Upper Rock Nature Reserve. There were other kilns in the past: Lime Kiln Street still remains, but no working kilns.

Rock fill

Extension of the harbour area at the end of the nineteenth century required considerable quantities of fill. During the first phase of extension, in about 1880, substantial quantities of rock were required for the construction of the South Mole. Material was first obtained from a quarry specially opened in the hitherto inaccessible Camp Bay. The rock fill needed was transported the relatively short distance northward along the coast by a railway constructed for the purpose. This needed two tunnels (under Parson's Lodge Battery and in Camp Bay), later used to provide road access.

A second phase of expansion, between 1894 and 1903, needed more rock fill than the Camp Bay quarries alone could provide and so a new railway tunnel was driven eastwards through the Rock to the south end of Sandy Bay where a large new quarry was opened. Additional smaller quarries were opened up immediately west of Catalan Bay village and along the North Face of the Rock. All these quarries began by exploiting the Quaternary scree breccias (Fig. 6.7) before tackling the intact Gibraltar Limestone beneath. This was in part due to the greatest volume of stone being required only for general fill, and in part due to the necessity of removing the scree before the higher quality bedrock could be accessed.

From early 1942 until mid-1943 construction of the airfield brought a new demand for fill. The old quarry west of Catalan Bay was reopened and worked both southwards and northwards, and a new quarry, adjacent to the airfield, was developed in the Quaternary scree slope at the base of the North Face – calculated to have a potential volume of 1 338 000 m^3 (Haycraft 1946*a*). Initially, this was worked by diamond drilling to insert gelignite for blasting. However, due to the lithologically variable and fissured nature of the cemented scree, the drills tended to jam and the boreholes tended to collapse when the bit was withdrawn. Consequently, progress was slow. Moreover, it proved difficult to judge the correct charges for use in the anisotropic rock. Either the force of the explosion was dissipated through fissures in the breccia, or the spoil was thrown all over the Isthmus (Haycraft 1946*a*). A Canadian officer solved these problems by adapting the pumps supplied to operate static flame throwers to propel sea water at high speed to extract the material. Two pumps were installed, one at either end of the scree. A jet of water was directed so as to develop a vertical rock face from below, and the two ends of the scree worked on alternate days so as to alternate hydraulicing with spoil removal. In time it was shown that volume of water, rather than head of pressure, controlled the effectiveness of the process.

Recent pressures for land development have renewed interest in land reclamation, but remaining sources of breccia are inadequate for large-scale developments. The major reclamation of the northern harbour area in 1989 utilized sand dredged from just off the southeast shore, transported for hydraulic placement behind rock fill bunds. Progress was, however, slowed by the need to deal with unexploded ordnance dumped earlier in the source area.

Sand

Some sand has been extracted from a quarry developed to the south of the Botanic Gardens in the Alameda Sands, which is a Quaternary windblown deposit of medium-grained quartz sands whose red coloration is evidence of significant pedogenesis under humid conditions (Rose & Hardman 1998). The quarry is now disused, and the former quarry floor occupied by tennis courts. The remaining outcrop of the sands is largely overbuilt, and their known maximum thickness is < 16 m, so any future development is unlikely.

Fine-grained aggregate suitable for concrete cannot be produced by crushing the Gibraltar Limestone because of the angularity and high dust content that normally result from the crushing process, but other sand sources are very restricted. During the 'siege' of 1969–85, when importation of building materials had to be by sea rather than by land, considerably increasing their cost, some aggregate was produced from limestone quarried at Europa Advance Road and crushed at an older quarry site at Catalan Bay, but this short-term expedient was abandoned when the 'siege' ended.

The Catalan Sand slope on the east side of the Rock (Fig. 6.8) was also used as a resource during the 1969–85 'siege', initially as a by-product of safety measures. The natural surface layers of the slope are of Quaternary windblown medium-grained quartz sands, to significant thickness. About 30 truncated dune cross-bedded units are currently visible in a roadside quarried face, the rock varying from weakly to uncemented sands to a strongly cemented sandstone, their deposition punctuated by a period of stability and formation of a palaeosol (Rose & Hardman 1998). Occasional rockfalls from the cliffs above the water catchment cladding on the upper slope always impact within 25 m of the base of the cliffs, before bouncing or sliding down the catchments to the road, hotel and beach below. To prevent continuing damage to the catchments and to protect the development of tourist facilities, it was proposed to excavate a 50 m wide rock trap at the top of the slope. Excavation was expected to generate large quantities of sand, with grading (zone 4) suitable for use in concrete by careful mix design, and eminent suitability for plastering, at a time when importation

Fig. 6.8. Disused quarry in the Quaternary Catalan Sands, at the toe of the main east coast water catchment slope, viewed from the north. The large-scale cross-bedding, together with the roundness and surface features of the individual sand grains, have been used to infer deposition as windblown dunes. (Modified from Rose & Rosenbaum 1991*a,b*.)

of such materials was expensive. The idea of a rockfall trap quickly became a secondary objective, and a company was created to develop the sand. However, problems were encountered in its excavation (large boulders) and transport (by conveyor, since gravity chute proved unsuccessful). Excavation at the top of the slope was therefore abandoned in favour of quarrying the more accessible material at the toe. Quarrying here was stopped by 1982, when its potential effect on the stability of the slope above had given rise to concern.

The Isthmus Sands are largely overbuilt and as such are 'sterilized' and inaccessible. Contemporary beach sands are unsuitable because of their amenity value, and the presence of salt. Sea-dredged sand would also be contaminated with salt, and so be unsuitable for concrete where strength was required unless washed, increasing its production cost, since freshwater is also a costly resource. High quality sand for concrete is therefore imported rather than produced locally.

Water supply

The supply of adequate potable water at Gibraltar has been a long-standing and increasing problem, due to population growth and improvement to amenities. Current demand is over $1 \times 10^6\,m^3$ per year, and is expected to rise to about $1.5 \times 10^6\,m^3$ by the year 2000 (Wright *et al.* 1994). The city's potable water is stored within 13 reservoirs excavated in the Rock (Rose & Rosenbaum 1991*b*; Rosenbaum & Rose 1991*b*). Construction of the principal twelve (Fig. 6.9) began in 1898, and continued intermittently to 1961, to provide a total capacity of $73\,000\,m^3$.

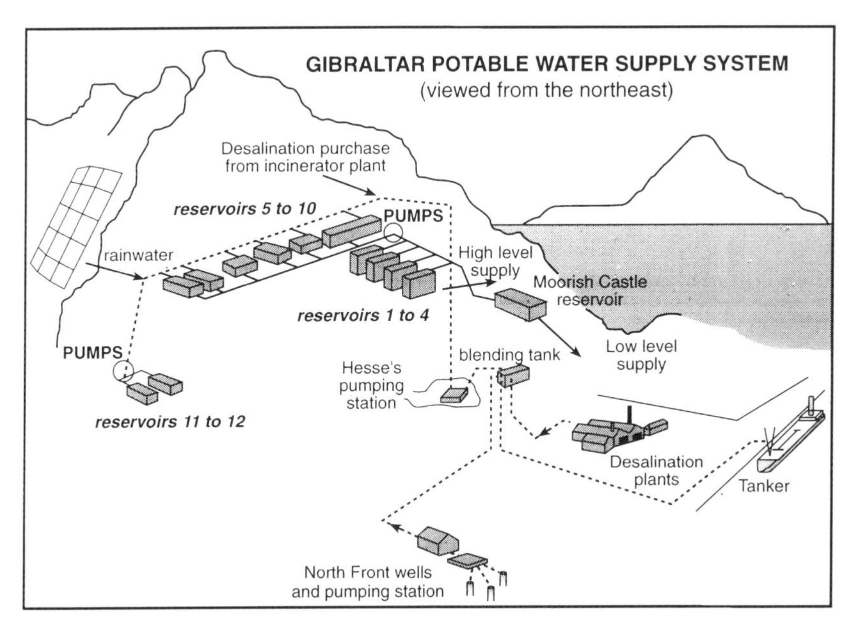

Fig. 6.9. Cross-section diagram of collection, storage and supply of domestic potable water, showing main underground reservoirs and associated tunnel systems. Major reservoirs are about 60 m long, 15 m high, and 6 m wide, and each contains about $4500\,m^3$. (Modified from data supplied by Lyonnaise des Eaux (Gibraltar) Ltd.)

The distribution system is divided into districts, and each district is metered, the district meters being read automatically via telemetry so as to quickly detect leakages and initiate repairs. Initially, the public system was complemented by separate supplies for the army and navy. Largely between 1949 and 1986, when other sources proved inadequate due to a prolonged dry season or low rainfall during the wet season, water was imported – taking advantage of Gibraltar's position on a major sea route. Supplies were shipped from the United Kingdom, via Portsmouth (Anon. 1949), Southampton or Newcastle via 20 000 to 30 000 tonne tankers, and occasionally from Holland (Rotterdam) or France (Dunkirk). For a while, water was imported daily from Tangier in Morocco (Doody 1981), by two 1200 tonne tankers (*Gungadin 1+2*). Additionally, oil tankers on their maiden voyage to the Middle East and so uncontaminated by bulk petroleum were used to convey 9000 to 36 000 m^3 of water at a time – but this was very expensive. In the long term, significant quantities of domestic water have been derived from three sources: rainwater, groundwater and sea water (Gonzales 1966; Doody 1981; Wright *et al.* 1994).

Rainwater

Daily rainfall records have been kept for one or more low-level stations on the Rock since 1790. The frequency distribution of annual values is normal, with a mean of 838 mm, a standard deviation of 262 mm, and a range from 381 to 1956 mm. Secular plots of raw and smoothed data of departures from the mean show no long-term trends to wetter or drier conditions but do show some obvious periodicities (Wright *et al.* 1994, fig. 4).

The Moors derived water supplies for their developing town from roof catchment of rainwater, wells in the Quaternary sands which fringe the Rock (James 1771), and storage by cistern of local surface runoff. These supplies were enhanced during subsequent Spanish control by construction (or refurbishment) in 1571 of a 1200 m aqueduct to convey seepage percolating through the Alameda Sands to the south of the town to two underground reservoirs. Two further reservoirs were excavated about 1694. Roof catchments and the aqueduct continued to provide the main means of public supply after the town and fortress were ceded to Britain in

Fig. 6.10. The main east coast water catchments above Sandy Bay, viewed from the south, prior to decommissioning in 1993. Sheets of corrugated iron fastened to wooden battens cover the upper surface of a 250 m high Quaternary scree breccia/ Catalan Sand slope. Hotel developments at the toe of the slope indicate scale. (cf. Figs. 6.1, 6.8, 6.11). (Modified from Rose & Rosenbaum 1991*a,b*.)

1713 until in 1865 the aqueduct was identified as the source of infection causing a major epidemic of cholera and consequently closed. In 1869 the first Public Health Ordinance on Gibraltar made it mandatory for all new dwellings to have underground tanks. Until at least 1863, there were no water pipes in the town apart from the aqueduct (Finlayson 1994).

From 1903 onwards increased efforts were made to collect rainfall from surfaced catchment areas, the main government catchments eventually occupying an area of 243 000 m^2. The most spectacular were constructed on the east side of Gibraltar (Fig. 6.10), on the Catalan Sand slope – an idea conceived by the then City Engineer. The first 40 000 m^2 were constructed in 1903 (Gonzales 1966). Rockfall boulders embedded in the slope were blasted away, the surface was trimmed as evenly as possible, and a channel and footpath were constructed at the lower perimeter of the collecting area. Timber piles 150 mm×38 mm×1.52 m long were driven through the sand for their full length and a timber framing of 76×64 mm purlins and 76×50 mm rafters was nailed on to them. This framing was then covered over with 2.44×0.91 m corrugated iron sheets secured by means of drive screws round their edges. Rain falling on this catchment area flowed down into the channel at the foot of the catchments, and thence via a tunnel into reservoirs within the Rock (Rosenbaum & Rose 1991*b*). The catchment area increased by a further 56 000 m^2 between 1911 and 1914, and by another 40 000 m^2 below the existing catchments between 1958 and 1961. Collection and distribution were therefore designed to work solely by gravitation. However, these catchments contributed a declining percentage of total potable water in recent years, from 12% in 1976 to 1% in 1992, with increasing maintenance costs. On a cost per m^3 basis, maintenance costs made corrugated iron catchments the most expensive source of potable water on Gibraltar – and they yielded no more than 60 000 m^3 in a year of good rainfall – so their use was discontinued in October 1993. With increasing age, they constitute a potential hazard, unless maintained or dismantled.

Other catchments still in use on Gibraltar cover a much smaller surface area. They were constructed from 1863 onwards on natural slopes of the Gibraltar Limestone exposed on the upper Rock, by clearing the surface vegetation and sealing fissures with a cement or mortar grout.

Another spectacular, but more transient, local feature is the Levanter cloud. The prevailing wind directions are either westerly or easterly, normal to the strike of the Main Ridge, with directional frequencies about the same but with the east winds more prevalent during the summer and proportionally of higher force than the west winds. The Levanter is a cloud which results when the moisture-laden east winds strike the precipitous eastern cliffs and are deflected upwards, to condense over the town and harbour to the west. A pilot scheme to collect water from these winds by condensation on metallic meshes was activated for 8 months from mid-1957 (Hurst 1959). Experience at Table Mountain in South Africa (Nagel 1956) and in Tasmania (Twomey 1957) had indicated that Gibraltar might be an ideal location in which to obtain water from cloud condensation. Galvanized mosquito netting of 1.59 mm mesh spacing and 33 gauge was made up into two frames nominally 1.83×0.91 m but effectively 1.49 m^2 in area. One had a single thickness of mesh, the other a double thickness with 50 mm spacing. These frames were mounted on the crest of the Main Ridge above Catalan Bay, in the Middle Hill gap. The single-meshed frame proved more efficient at generating condensation from winds of low speed (≤ 6 m s^{-1}), the double-meshed at higher speeds (≥ 10 m s^{-1}), with equal efficiency at about 8.7 m s^{-1}. It was concluded that at least 0.342 $m^3 m^{-2}$ could be collected from cloud in a typical summer, rising to 0.587 m^3 for a full year, plus a further 0.391 m^3 from rainfall rather than cloud condensation. A 929 m^2 catch to produce an estimated 90.92 m^3 per year was recommended for Middle Hill, but not constructed.

Groundwater

During most of the late Spanish and early British occupation of the Rock, shallow (largely privately-owned) wells which yielded brackish water, and water from the aqueduct, provided the only significant sources of supply additional to rainwater. There were no major springs. These sources were broadly sufficient during the eighteenth century when the population did not exceed 3000, but became increasingly inadequate during the nineteenth century as the population increased. Sanitary commissioners were appointed in Gibraltar in 1866, tasked with ensuring an adequate supply of potable water to the town and garrison. Their efforts included support for a geological survey, planned for 1865 but successively deferred until 1876, directed by A. C. Ramsay. Attempts to locate potable groundwater by shallow borings through the Quaternary Isthmus Sands failed (but led to development of brackish waters for 'sanitary' purposes to supplement the available potable water). Attempts to penetrate an aquifer beneath the sands by deeper boreholes in 1878, 1892 and 1943 also failed.

However, Wright *et al.* (1994) identified an aquifer within the bedrock of Gibraltar, as a thin lens of fresh water overlying sea water. Recharge rates were calculated by soil moisture and chloride balance to be of the order of 400 000 m^3 per year, which are significant in relation to the present demand for potable water of some $1 \times 10^6 m^3$ per year. Exploration drilling focused on fracture zones, but tested vertical boreholes generally demonstrated a rapid decline in water quality, even where drawdowns were very small. Exploration also revealed oil pollution of the groundwater within the Rock, most significantly near the Naval Dockyards to the southwest and less so beneath the North Face. The greatest potential for sustained discharge was found to occur in shallowly inclined boreholes. Discharge quality was clearly sensitive to a range of factors including immediate rainfall events, tidal fluctuations, permeability of the aquifer, borehole design and anthropogenic effects. Taking further exploratory and capital costs into account together with other strategic factors it has so far been deemed not to be cost-effective to pursue further exploration and development of the main aquifer of the Rock, but the possibility is subject to periodic review.

Further, as yet unpublished, studies have investigated the possibility of storage and/or artificial recharge by surface runoff from the airport tarmac within the Isthmus Sands. Unpublished studies in 1933 had demonstrated the presence of two aquifers in the Quaternary sands which overlie shale/brecciated limestone bedrock: an upper unconfined aquifer of fresh water, and a lower confined aquifer of brackish water. Currently, only some 9% of Gibraltar's water is supplied from 19 shallow wells developed in the upper aquifer, whose water is high in chloride content and has a high degree of hardness but is useful for blending with water produced by desalination.

Sea water

Nearly all the water currently used on Gibraltar comes from the sea. Within recent years, from 1953 but particularly from 1966, desalination of sea water has been developed as the main source of potable water. This method contributed some 49% of total production in 1976, which had increased to about 90% by 1992. Current output is achieved by means of two multi-stage flash distillation and two reverse-osmosis plants. However, on running costs alone, water derived from desalination is around three times as expensive as water derived from wells. Within the last few years, heat generated by a new refuse incinerator has been used to desalinate additional quantities of sea water. The Gibraltar Government entered into an agreement in 1990 for construction of an incinerator that would produce steam from waste heat and in turn 650 000 m^3 a year of potable water via three distillation plants.

To reduce demand for potable water there is a separate sanitary water supply. Sea water is pumped, stored and distributed to all households for flushing toilets, firefighting, street cleaning, and any purpose where the use of potable water is not essential. Water is pumped from the sea at two sources, because the system of reservoir storage and supply is divided into two districts, one for the north and town, the other for the south. Nearly 4×10^6 m^3 are pumped annually – about four times the consumption of potable water.

Waste disposal

Traditionally, waste disposal was by direct discharge to the sea. In 1845, because of the growing problem of disposing of refuse from the town, the governor had a rubbish dump marked out with four posts on the Isthmus, regarded at the time as politically neutral ground. His action was justified in a letter from the UK Secretary of State for Foreign Affairs to the Spanish Ambassador in London on health grounds (Ramsey 1978). However, since 1942 the Isthmus has been largely overbuilt by the airfield and associated structures. Until very recent years unserviceable motor cars and other large items of non-combustible waste were jettisoned from the Europa Flats, at a car shute on the east coast, where steep cliffs rise near vertically from the sea. Tunnel spoil and some building waste have been used for land reclamation from the sea, but there are currently no landfill sites for waste disposal as such. The incidental rather than planned dumping of building rubble into the sea 'has resulted in the loss of some of the remaining natural rocky shoreline both at Camp Bay and on the east side over the past five or six years. The negative effect has been reduced somewhat as a result of pressure from the Gibraltar Ornithological and Natural History Society (GONHS) which managed to achieve that some areas were not used as dumps' (J. Cortes, pers. comm. 1996).

Incineration has long been used for disposal of combustible

waste, and a former incinerator sited at the foot of the North Face has recently been superseded by a new incinerator sited near the southeast coast. However, GONHS has commented (J. Cortes, pers. comm. 1996) that it 'does not work well (it is often broken down), emits fumes and questionable fluids and toxic fly ash that no one knows what to do with. Some has been locked away into the tunnels. There are ideas of mixing with concrete and building sea defences, but ... GONHS is not in favour of this due to the likelihood of leaching.' The adjacent, disused Europa Advance Quarry is currently used for temporary storage of waste pending incineration. Gibraltar government engineers are more sanguine in their perception than GONHS, whose views are disputed.

Tourism

According to the Gibraltar Tourist Board, over four million tourists visit the Rock every year. Of the twenty key tourist attractions, seven are based on the natural environment: the Barbary apes (probably introduced by the British in the early eighteenth century from the mountains of Morocco); views from the crest of the Main Ridge (Fig. 6.11) (which can be reached by cable car); views across the Strait to the south from Europa Point, and from Jew's Gate higher on the rock; St Michael's Cave; the Alameda Botanic Gardens; the Catalan Bay area. The other thirteen sites are largely parts of the fortification system and of historical interest. An 'Official Rock Tour' provides the opportunity to view or visit nearly all of these attractions in less than four hours – indeed the highlights can be covered in approximately 90 minutes. Most tourists are therefore day visitors from Spain: there are but seven hotels on Gibraltar for those who wish to stay longer.

As military land has been decommissioned, so from 1991 an Upper Rock Nature Reserve has been created to protect much of the flora and fauna as well as some of these key natural and historical sites (Cortes *et al.* 1993). Condensation from the Levanter cloud keeps vegetation on the Upper Rock fresh and green even during the hot dry summer months. Over 600 species of flowering plants are known from Gibraltar, and over 270 species of birds, as well as other wildlife. Migrating birds of prey are a feature well

Fig. 6.11. View north along the crest of the Main Ridge from O'Hara's Battery, with south-facing 9.2 inch gun at Lord Airey's Battery in the foreground at a height of 418 m. Gibraltar Limestone here dips at a high angle to the west. The main east coast water catchments are clearly visible to the right of the photograph (cf. Figs. 6.1, 6.10); a hotel at the toe of the slope gives some indication of scale; and the initiation of land reclamation by dumping of fill along the foreshore can be seen further to the north.

known to naturalists. In spring and autumn they can be seen from most vantage points and as many as 12 000 birds can be seen in a day, the main species including Black Kite, Honey Buzzard, Short-toed Eagle and Booted Eagle. The Gibraltar Ornithological and Natural History Society has developed a field centre at Jew's Gate, and has many research and conservation projects in hand.

The Gibraltar Heritage Trust was set up at a similar time to the Nature Reserve to co-ordinate conservation and development of sites of historical interest. O'Hara's Battery, the 100 Ton Gun at Napier of Magdala Battery, the Heritage Centre at Princess Caroline's Battery, and the extensively restored Parson's Lodge Battery (Guerrero 1994) are all spectacular locations opened to tourism only within recent years.

In contrast, St. Michael's Cave has been known to visitors since classical times. Its seemingly bottomless extent may have given rise

Fig. 6.12. View within part of New St Michael's Cave system, showing the lake (modified from Rose & Rosenbaum 1991*b*).

to the belief that the entrance to Hades, Pluto's under-world of the dead, lay in the western face of the Rock (James 1771; Jackson 1987). Shaw (1955*a*,*b*), has provided an extensive account of the history of exploration of the cave and the numerous bibliographic references to it since its earliest citation by the Roman geographer Pomponius Mela in AD 45. Rose & Rosenbaum (1991*b*, figs 4.2 & 9.3) have provided cross-section and plan-view diagrams. The site is now managed as a major tourist attraction, and its main chamber as a concert hall. A lower sequence of caves (New St Michael's) was discovered only in 1942 (Shaw 1953). Adults may now gain access by arranging evening guided tours to view extensive speleothem deposits and an underground lake (Fig. 6.12). Gorham's and Vanguard caves, opening from Governor's Beach on the southeast coast of the Rock, are sites of major archaeological interest currently being excavated under the direction of Professor C.B. Stringer (The Natural History Museum, London) and Dr J.C. Finlayson (Director, Gibraltar Museum). Artefacts from Gorham's (Waechter 1964; Cook *in:* Rose & Rosenbaum 1991*b*, p.71) indicate that the cave was successively occupied by Neanderthals and anatomically-modern humans, and Neanderthal skulls have been recovered from Forbes' Quarry (Keith 1911) and Devil's Tower rock shelter (Garrod *et al.* 1928) at the North Front. Recent dating suggests that Gibraltar and adjacent regions of southern Spain may have been the Neanderthals 'last refuge before extinction' (Bryson 1996) – an inference being used to attract scientific as well as tourist interest in Gibraltar's prehistory (Rose & Stringer 1997).

Sandy beaches are not a major feature of the peninsula, but do occur along the east coast: on the Isthmus south of the airfield, in Sandy Bay, and in Catalan Bay. Bathing facilities have also been developed along the southwest coast, in Camp Bay and Little Bay, and to a lesser extent along the northwest shore of the Isthmus. Along the east coast, movement of beach sand is predominantly northwards (Rose & Rosenbaum 1991*b*).

Slope and tunnel stability

The massive cliffs and slopes which flank the Rock by their very size may give concern for the safety of building structures and roads upon or beneath them, but most appear to be relatively stable. The stability of rock slopes is governed principally by the surface topography, rock strength, the orientation of fractures within the rock and the water pressure within the rock (Rose & Rosenbaum 1991*b*). The three basic solutions employed generally to prevent mass-movement on slopes have all been applied on Gibraltar: (1) restraint, (2) redistribution of forces and (3) improvement of strength. Restraint is commonly achieved by a retaining wall at the

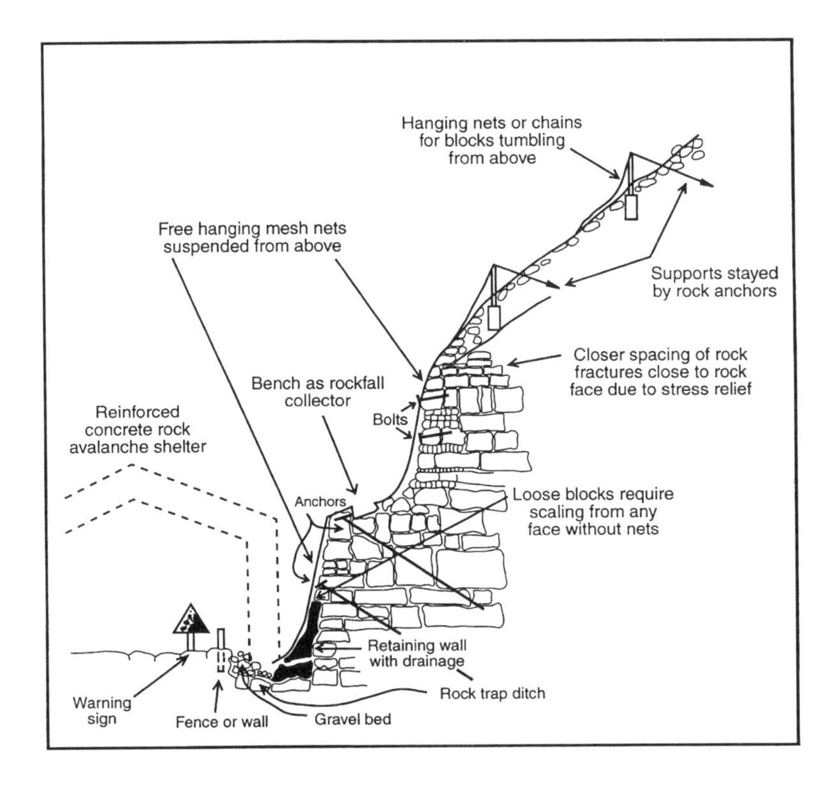

Fig. 6.13. Rock slope stabilization techniques to be seen on Gibraltar (modified from: Rose & Rosenbaum 1991*b*, after Fookes & Sweeney 1976).

toe of a slope, but ground anchors and piles can also be used to pin the potentially unstable mass to firm rock, and rockfall areas are frequently restrained by the use either of individual rock bolts to hold large blocks, or of a steel net bolted at regular intervals over the face (Fig. 6.13). Redistribution of forces commonly involves reducing the angle of the slope, for example by scraping material from the top and/or tipping it at the bottom. Improvement in strength can most readily be achieved by reducing pore water pressures by drainage, and in Gibraltar the extensive network of tunnels has achieved this fortuitously by lowering the pressure head of groundwater.

The equable climate with a general absence of frosts makes serious rockfalls very rare events on Gibraltar, but one fall did occur in 1989, bringing to light instability problems in 500 m long cliffs above Camp Bay and Little Bay (Anon. 1992), potentially serious because these bays are popular bathing areas. Mountaineering techniques were used for work on a major rock stabilization project thought to be on a scale larger than anything before in this field. Stabilization was effected by grouting of the limestone, colluvium and breccia exposed in natural cliff and old quarry faces. It was also achieved by inserting rock anchors using specially designed cliff face drilling rigs, four down-the-hole-hammer (DTH) rigs and a specially made rotary percussive drifter rig. Drilling conditions were very difficult, the voids which riddled the Gibraltar Limestone causing three DTH drill bits to break, an 'unheard of' occurrence, which meant that both hammers and holes had to be abandoned. The original contract was for 120 permanent anchorages, of 4 to 15 m length. Additionally, boulders up to four tonnes were removed by crowbar or by winches and hydraulic jacks, and larger boulders were retained in place by a combination of steel cable strapping and rock bolts, plus steel mesh where required. Other areas thought to be unstable, such as breccia faces and gullies, were consolidated using rock bolts and shotcrete, reinforced with mesh. In the worst zone, where the initial failure occurred, a grid pattern of rock bolts was used, large jammed boulders were identified and stabilized using anchorages, and the whole area encapsulated in shotcrete. However, as reported in *The Sunday Times* (24 February 1997), further rockfalls occurred along the cliff line above Camp Bay following heavy winter rainfall, and a survey to assess the scope for further stabilization measures was rapidly put in hand.

Rosenbaum *et al.* (1994) have drawn attention to the factors influencing the integrity of unlined tunnel walls in Gibraltar. Extensive tunnel inspection work has been carried out by consultants, notably Mott MacDonald, initially acting for the Property Services Agency of the UK Department of the Environment. Less extensive studies have been carried out by Royal Engineer geologists of the Engineer Specialist Pool, and its re-titled successor, the Royal Engineers Specialist Advisory Team (V) (Rose

Fig. 6.14. Scaling of tunnels, to remove the loose rock which in time develops by stress relief. Communication tunnels were cut to about 2.5 m diameter, main tunnels to about 4 m. Early tunnels had flat soffits, but these proved prone to rock falls, so later tunnels and chambers exceeding 4 m span were driven with arched soffits as above, the rise normally being about 20% of the span. (Modified from Rosebaum & Rose 1991*b*. © British Crown copyright 1991/MOD. Reproduced with the permission of the Controller of Her Britannic Majesty's Stationery Office.)

& Hughes 1993). Areas deemed potentially unsafe have either been rendered safe by scaling of loose blocks (Fig. 6.14) or by emplacement of roof support structures, or the tunnels have been closed to access.

Quarrying at the toe of the Catalan Sand slope, both by construction of the Sir Herbert Miles Road and subsequently for fine aggregate extraction, has diminished the stability of the natural slope – which is subject to occasional slippage. Major slope failures closed the road for some time in the 1940s, and in 1995. There is concern that once the deteriorating corrugated iron catchment covering is removed, allowing water into the sands beneath, increased pore water pressure will lead to increased slope failure and increased closure of the only east coast road. Current plans are to stabilize the slope by means of plastic matting and sowing of locally collected seed, from parts of the slope currently vegetated. An idea to plant exotic sand stabilizing plants, such as the Hottentot Fig *Carpobrotus edules,* a plant extensively used for slope stabilization in the past, has been discarded on the grounds that such plants tend to obliterate native vegetation (J. Cortes, pers. comm. 1996). A current road-widening scheme includes construction of a major retaining wall.

Conclusion

Gibraltar's limited natural resources have necessitated some policies very different from those in the UK, notably incineration rather than landfill as a primary means of waste disposal; use of desalination rather than surface water or groundwater to provide drinking water; dual supplies for potable and sanitary water; and land reclamation for building construction. However, there are some similarities in broad trends of environmental policy. Gibraltar is investing millions in pursuit of quality tourism and developing its role as a financial centre (Searle 1996). For example, the sum of £5.2 million out of a £72 million government budget is currently being invested to promote the Rock and to refurbish access points and hotels.

As part of the plan to go 'upmarket' in tourist attraction, aspects of the natural environment, ultimately controlled by the geology, are being more effectively managed. The year 1991 marked the initiation on the upper Rock of a nature reserve to contain some 1 220 000 m^2 (about 60% of the territory still in a fairly natural state), and many of the best-known tourist attractions (including Apes' Den, St, Michael's Cave, the Tower of Homage). On the lower Rock restoration of the 60 700 m^2 Alameda Gardens as the Gibraltar Botanic Gardens began at the same time, with determination by a new director to enhance not only their attractiveness but also their significance as a centre for conservation and research.

In 1991 the Gibraltar Museum also came under new directorship, since when it has expanded in terms of building, area publications and research initiatives. In 1996 the Museum joined with the John Mackintosh Hall (an educational facility) and the Gibraltar Government Archives to set up the Gibraltar Centre for Mediterranean Studies, to promote studies in the biology, geology, geography and history of the Straits region. Also in 1996, the Gibraltar Trust for Earth Sciences and Archaeology, based in the Museum, was founded 'for the enhancement of earth sciences, archaeology, geology, palaeontology and geography' (preamble to Declaration of Trust). Gibraltar geology as well as flora and fauna has featured amongst aspects of environmental conservation given attention both by the UK Ministry of Defence (Rose & Hardman 1994), and the many organizations grouped together to form the Instituto de Estudios Campogibraltareños (Rose & Hardman 1996). The quest to manage the natural resources of the Rock so as to generate wealth is complemented by renewed academic and conservation interest.

Acknowledgements

Information used in this paper has been derived from work begun on Gibraltar in 1973, and still continuing, during which period I have received help from too many people to list satisfactorily here. However, I am especially grateful for help through recent correspondence to J. Cortes (Director, Gibraltar Botanic Gardens), J.C. Finlayson (Director, Gibraltar Museum), M. Perez (Managing Director, Lyonnaise des Eaux (Gibraltar) Ltd.), M.S. Rosenbaum (Imperial College London), and the Gibraltar Information Office, London. Figures were drawn by C. Flood, S. Muir and L. Blything, photographs printed by K. Denyer, and the typescript prepared by J. Pickard, all of the Department of Geology, Royal Holloway, University of London. J.D. Mather kindly commented on a draft of the manuscript, as did C.P. Nathanail.

References

Anon. 1949. Portsmouth water for Gibraltar. *Journal of the Institution of Water Engineers,* **3**, 514.

——— 1992. Abseil anchoring saves Gibraltar cliff face. *Ground Engineering,* **March issue**, 10–11.

Bailey, E. B. 1952. Notes on Gibraltar and the Northern Rif. *Quarterly Journal of the Geological Society of London,* **108**, 157–175.

Bryson, B. 1996. Gibraltar. *National Geographic*, **190**, 54–71.

Cortes, J., Shaw, E., Perez, C., Linares, L., Harper, A. & Santana, T. 1993. Wildlife in Gibraltar. *Sanctuary*, **22**, 38–41.

Cotton, J. C. 1948. The tunnels of Gibraltar. *In:* Anon. *The Civil Engineer in War,* **3**, Institution of Civil Engineers, London, 229–248.

Doody, M. C. 1981. Gibraltar's water supply. *Journal of the Institution of Water Engineers and Scientists,* **35**, 151–154.

Finlayson, T. J. 1994. The history of Gibraltar's water supply. *Gibraltar Heritage Journal*, **2**, 60–72.

Fookes, P. G. & Sweeney, M. 1976. Stabilisation and control of local rock falls and degrading rock slopes. *Quarterly Journal of Engineering Geology*, **9**, 27–35.

Garrod, D. A. E., Buxton, L. H. D., Smith, G. Elliot & Bate, D. M. A. 1928. Excavation of a Mousterian rock-shelter at Devil's Tower, Gibraltar. *Journal of the Royal Anthropological Institute,* **58**, 33–113.

Gonzales, F. J. 1966. The water supply in Gibraltar. *Aqua: The Quarterly Bulletin of the International Water Supply Association,* **2**, 58–67.

Guerrero, E. 1994. Gibraltar. *Sanctuary*, **23**, 61–62.

Haycraft, T. W. R. 1946*a*. The Gibraltar runway. *Royal Engineers Journal*, **60**, 225–230.

——— 1946*b*. The Gibraltar tunnels. *Royal Engineers Journal*, **60**, 310–320.

Hughes, Q. & Migos, A. 1995. *Strong as the Rock of Gibraltar.* Exchange Publications, Gibraltar.

Hurst, G. W. 1959. Collection of water from cloud at Gibraltar. *Journal of the Institution of Water Engineers,* **13**, 341–352.

Jackson, W. G. F. 1987. *The Rock of the Gibraltarians.* Associated University Presses, Cranbury NJ.

James, T. 1771. *History of the Herculean Straits, commonly called the Straits of Gibraltar.* Rivington, London.

Keith, A. 1911. The early history of the Gibraltar cranium. *Nature*, **87**, 313–314.

Lauder, J. G. 1963. Tunnelling in Gibraltar. *Royal Engineers Journal*, **77**, 339–369.
Nagel, J. F. 1956. Fog precipitation on Table Mountain. *Quarterly Journal of the Royal Meteorological Society*, **82**, 452–460.
Owen, E. F. & Rose, E. P. F. 1997. Early Jurassic brachiopods from the Rock of Gibraltar and their Tethyan significance. *Palaeontology*, **40**, 497–513.
Ramsay, A. C. & Geikie, J. 1876. *Geological Map of Gibraltar*. Scale 1:2,500. British Geological Survey. Unpublished.
——— & ——— 1878. On the geology of Gibraltar. *Quarterly Journal of the Geological Society of London*, **34**, 505–541.
Ramsey, W.G. (ed.) 1978. *After the Battle No. 21: Gibraltar*. Battle of Britain Prints International, London.
Rose, E. P. F. 1998. The pre-Quaternary geological evolution of Gibraltar. *In:* Finlayson, J. C. (ed.) *Gibraltar during the Quaternary: the southernmost part of Europe in the last two million years*. Gibraltar Government Heritage Publications, Gibraltar.
——— & Cooper, J. A. 1997. G. B. Alexander's studies on the Jurassic of Gibraltar and the Carboniferous of England: the end of a mystery? *Geological Curator*, **6**, 247–254.
——— & Hardman, E. C. 1994. The caves, tunnels and rocks of Gibraltar. *Sanctuary*, **23**, 16–17.
——— & ——— 1996. Conservation and the geology of Gibraltar. *Almoraima*, **15**, 35–51.
——— & ——— 1998. The Quaternary geology of Gibraltar. *In*: Finlayson, J. C. (ed.) *Gibraltar during the Quaternary: the southernmost part of Europe in the last two million years*. Gibraltar Government Heritage Publications, Gibraltar.
——— & Hughes, N. F. 1993. Sapper Geologists. Part 3. Engineer Specialist Pool Geologists. *Royal Engineers Journal*, **107**, 306–316.
——— & Rosenbaum, M. S. 1990. *Royal Engineer Geologists and the Geology of Gibraltar*. Gibraltar Museum, Gibraltar. (Reprinted from the *Royal Engineers Journal*, **103** (for 1989), 142–151, 248–259; 104 (for 1990), 61–76, 128–148.)
——— & ——— 1991*a*. The Rock of Gibraltar. *Geology Today*, **7**, 95–101.

——— & ——— 1991*b*. *A Field Guide to the Geology of Gibraltar*. The Gibraltar Museum, Gibraltar.

——— & ——— 1992. Geology of Gibraltar: School of Military Survey Miscellaneous Map 45 (published 1991) and its historical background. *Royal Engineers Journal*, **106**, 168–173.

——— & ——— 1994. The Rock of Gibraltar and its Neogene tectonics. *Paleontologia i Evolució*, **24–25** (for 1992), 411–421.

Rosenbaum, M. S. & Rose, E. P. F. 1991*a*. *Geology of Gibraltar. Single sheet 870 x 615 mm: Side 1, cross-sections and solid (bedrock) geology map 1 : 10,000, Quaternary geology, geomorphology, and engineering use of geological features maps 1 : 20,000; Side 2, illustrated geology (combined bedrock/Quaternary geology) map 1:10,000, plus 17 coloured photographs/figures and explanatory text.* School of Military Survey Miscellaneous Map 45.

——— & ——— 1991*b*. *The Tunnels of Gibraltar*. The Gibraltar Museum, Gibraltar.

——— & ——— 1992. Geology and Military Tunnels. *Geology Today*, **8**, 92–98.

——— & ——— 1994. The influence of geology on urban renewal. *In*: Oliviera, R., Rodrigues, L. F., Coelho, A. G. & Cunha, A. P. (eds) *Proceedings of the Seventh International Congress of the International Association of Engineering Geology, Lisbon, 5–9 September 1994*, **3**. A.A. Balkema, Rotterdam. 2283–2291.

———, ——— & Wilkinson-Buchanan, F. W. 1994. The influence of excavation technique on the integrity of unlined tunnel walls in Gibraltar. *In*: Oliviera, R., Rodrigues, L. F., Coelho, A. G. & Cunha, A. P. (eds) *Proceedings of the Seventh International Congress of the International Association of Engineering Geology, Lisbon, 5–9 September 1994*, **6**. A.A. Balkema, Rotterdam. 4137–4144.

——— & Stringer, C. B. 1997. Gibraltar Woman and Neanderthal Man. *Geology Today*, **13**, 179–184.

Scott, A. 1914. The New Harbour Works and Dockyard at Gibraltar. *Minutes of the Proceedings of the Institution of Civil Engineers*, **198**, 1–78.

Searle, D. 1996. New challenge as the colony goes upmarket. *The Times*, November 28, 36.

Shaw, T. R. 1953. New St. Michael's Cave, Gibraltar. *Cave Science*, **3**, 249–266.

——— 1955*a*. Old St. Michael's Cave, Gibraltar. *Cave Science*, **3**, 298–313.

——— 1955*b*. Old St. Michael's Cave, Gibraltar. *Cave Science*, **3**, 352–364.

Smith, J. 1846. On the geology of Gibraltar. *Quarterly Journal of the Geological Society of London,* **2**, 41–51.

Twomey, S. 1957. Precipitation by direct interception of cloud-water. *Weather*, **12**, 120–122.

Waechter, J. D'A. 1964. The excavation of Gorham's Cave, Gibraltar. *Bulletin of the Institute of Archaeology*, **4**, 189–212.

Warren, C. & Varley, P. 1996. What a bore! Tunnels under the Channel and Gibraltar Straits. *Down to Earth*, **15**, 2–6.

Wilson, W. H. 1945. Tunnelling in Gibraltar during the 1939–945 War. *Transactions of the Institute of Mining and Metallurgy,* **55**, 193-269.

Wright, E. P., Rose, E. P. F. & Perez, M. 1994. Hydrogeological studies on the Rock of Gibraltar. *Quarterly Journal of Engineering Geology*, **27**, S15–S29.

7 A 'whole rock approach' to the engineering geology of the Chalk of Sussex

John Lamont-Black

Summary

- The construction of the Brighton Bypass through the chalk downs of Sussex facilitated a major case study to investigate the effects of a wide range of geological parameters on the engineering process in chalk.
- The investigations concentrated on chalk stratigraphy, sedimentology, structural geology and geomorphology in conjunction with a detailed study of earthworks, foundations and slope stabilities.
- The distribution of syn-sedimentary domes and troughs was confirmed and various surface features were related to the underlying structural geology, correlated to structural domains or linked to sedimentation and tectonics.

Environmental geology was considered by Woodcock (1994) to concern itself with relating the science of the Earth to the activities of humans. He expanded this description by highlighting the subject's concern with the ways in which geological processes influence the urban environment and also its focus on both the inputs to, and outputs from society. Likewise, most other definitions of environmental geology cast a very wide net with the principal concern being that of an integrated geological approach to study or solving the problems associated with human interaction with the landscape.

With the above definition in mind, the subject of engineering geology may be viewed as a subdivision of environmental geology. This paper aims to show how an integrated approach was used to gain a fuller understanding of the engineering geology of the Chalk with respect to a major road scheme in the South Downs of Sussex and is drawn from a more extensive study made by the author (Lamont-Black 1995).

What is a 'whole rock approach'?

Mortimore (1990) introduced the concept of scientific frameworks to advance understanding of the engineering behaviour of the Chalk. These frameworks include stratigraphy, structural geology and tectonic setting, sedimentology and geomorphology. On their own, data collected within each of these frameworks are of only limited value in understanding and predicting the likely engineering behaviour of chalk. For example, Spink & Norbury (1990) noted that a knowledge of the stratigraphic horizon does not allow the assessment of design parameters. However, if rock mass and material data, collected as part of a site investigation programme, are interpreted with reference to all the scientific frameworks and the natural processes that operate therein, a deeper understanding of the meaning of these data is achieved. This provides greater confidence to predict the rock's likely behaviour between and beyond data points and consequently reduces the risks associated with gaps in knowledge.

The 'whole rock approach' therefore, requires the cross linkage of data collected within the scientific frameworks as shown in Fig. 7.1. This approach allows 'ground models' to be developed in which ground conditions can be predicted according to typical assemblages of geological characteristics described within the scientific frameworks. It is a concept which is being introduced by the CIRIA (Construction Industries Research and Information Association) project on construction in the Chalk.

It is important to emphasize that different scientific frameworks are needed in different engineering settings. For example, the design and construction of both a shallow foundation and a tunnel in chalk require a knowledge of a wide variety of rock mass and material characteristics. Three of the most important of these characteristics are hardness, fracture spacings and fracture apertures. For the shallow foundation these characteristics will generally be dominated by near-surface weathering effects such as meteoric dissolution and periglacial processes. On the other hand, the setting of a deep tunnel isolates it from near-surface weathering effects and the hardness and fracture characteristics of chalk are more likely to be influenced by lithostratigraphy, sedimentology

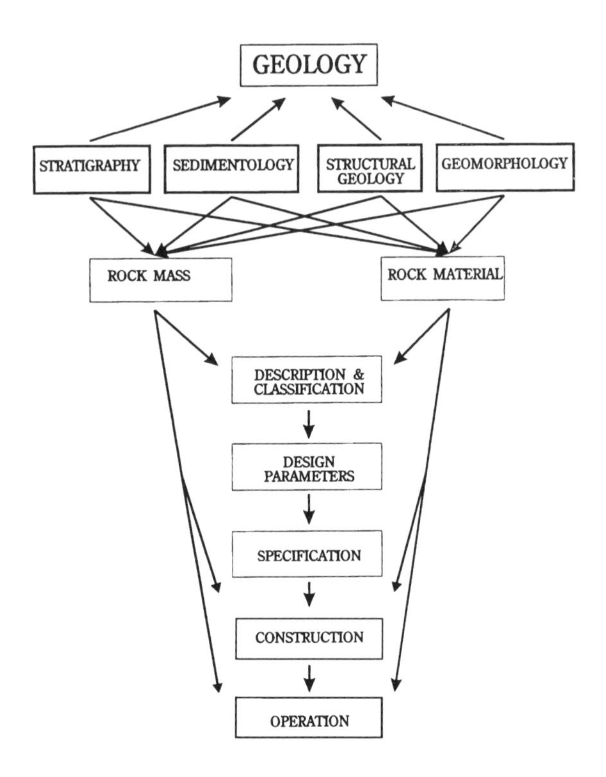

Fig. 7.1. Conceptual position of scientific frameworks within the environment of engineering geology.

and tectonic setting. This is not to say, however, that weathering processes are unimportant in tunnelling as deep phreatic karstification is well known in the Chalk and surface effects become more dominant as the tunnel portals are approached.

The Chalk is an ideal medium for the use of a 'whole rock approach' due to the complex range of variables which affect its engineering characteristics, these include physical, chemical and biological processes such as sedimentation, brittle fracture, freeze–thaw cycles, elastic and inelastic strain, dissolution, cementation, precipitation, fossilization, benthic and pelagic activity all of which have combined at different times since the beginning of the Late Cretaceous to produce the complex material that is the Chalk.

The aim of this chapter is to illustrate the value of the 'whole rock aproach' and show how a detailed knowledge of the stratigraphy on the Brighton Bypass helped identify: large geological structures (structural geology and tectonic setting framework); enabled the interpretation of lithostratigraphy, intact dry density, and the intensity and style of fractures (sedimentology framework); and the distribution and effects of dissolutional weathering between different geomorphological domains (geomorphological framework).

Overview of the geology

The A27 Brighton Bypass was completed in 1995 and comprises 14 km of dual and triple carriageway highway with central reservation (Fig. 7.2). Constructed under four contracts, it comprises 14 major structures and involved approximately $4 \times 10^6\,m^3$ of earthmoving through gently dipping chalks of the Seaford, Newhaven and Tarrant members of the Upper Chalk Formation (Fig. 7.3). Identification of fossil zones and key marker

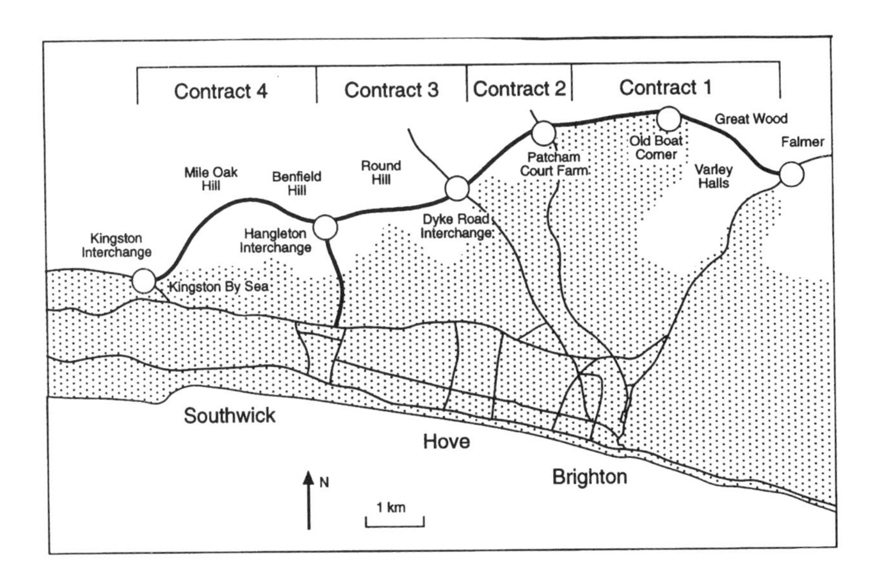

Fig. 7.2. Route of the A27 Brighton Bypass.

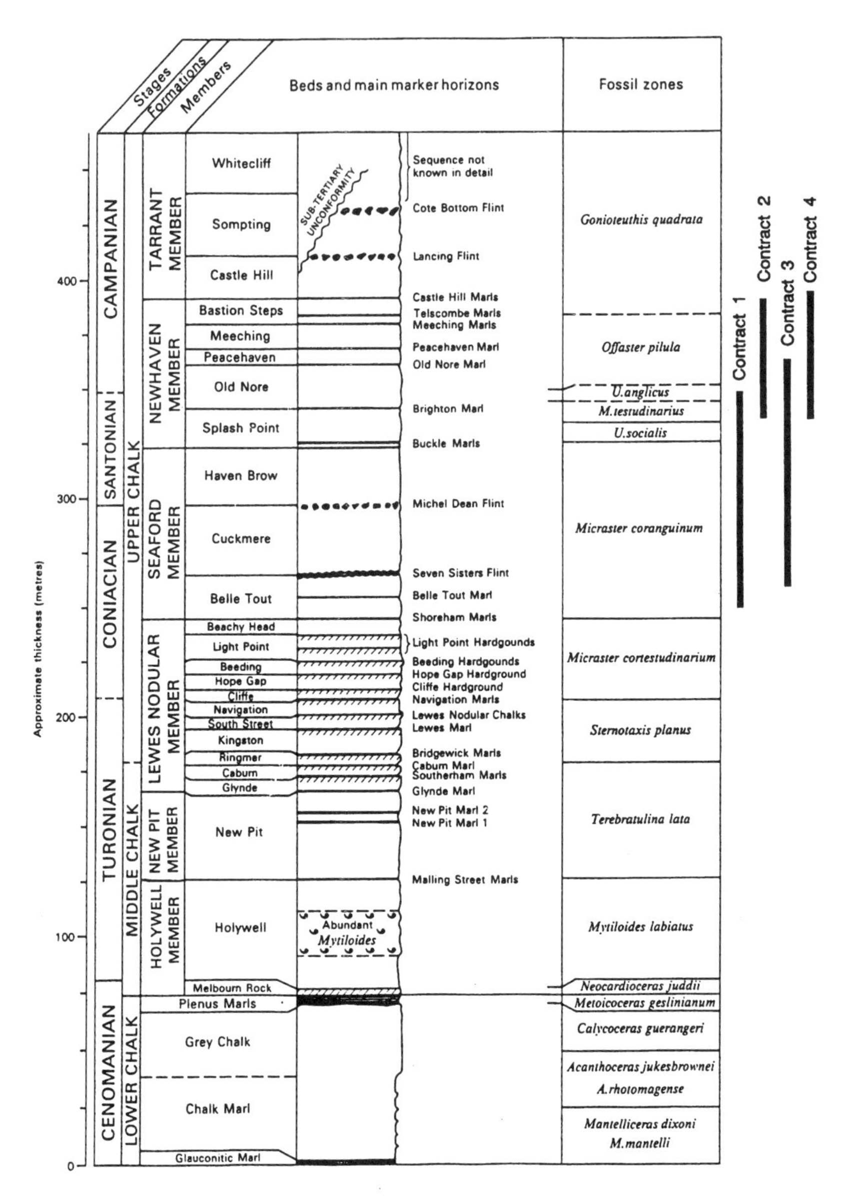

Fig. 7.3. Stratigraphy of the Chalk Group based on Mortimore (1986) and Bristow *et al.* (1997) showing the stratigraphical ranges of chalks encountered in the four contracts of the Brighton Bypass.

Table 7.1. Visual description of chalk after Lord *et al.* (1994)

A. Discontinuity aperture classification

Visual Grade classification		Brief description
A		Discontinuities closed
B		Typical discontinuity aperture < 3 mm
C		Typical discontinuity aperture > 3 mm
D	D_c	clast supported mélange chalk
	D_m	matrix supported mélange chalk

B. Discontinuity spacing

Suffix	Typical discontinuity spacing
1	> 600 mm
2	200–600 mm
3	60–200 mm
4	20–60 mm
5	< 20 mm

horizons such as marl seams, flint bands and hardgrounds allowed correlation of eight major cuttings on Contracts 1, 2 and 3 (Fig. 7.4). The geology of these cuttings is illustrated in Figs 7.5 to 7.12 and they will be referred to throughout the text. In addition to the geology, Figs 7.5 to 7.12 include chalk engineering grades (described in accordance with the forthcoming CIRIA scheme) which are explained in Table 7.1. Although optimum rock mass characteristics will vary according to engineering design, CIRIA rock mass, Grade A1, corresponds to massive tight chalk, while Grade C4 corresponds to highly fractured and loose chalks (i.e. the best and poorest quality intact chalks). Chalks which have lost recognizable intact structure and comprise a mixture of intact lumps and degraded silty chalk matrix are referred to in the CIRIA

Fig. 7.4. Correlation of cuttings on the Brighton Bypass. Note the condensation of the Peacehaven Beds at Old Boat Corner and Great Wood where the Old Nore Marl is replaced by the Old Boat Corner Hardground.

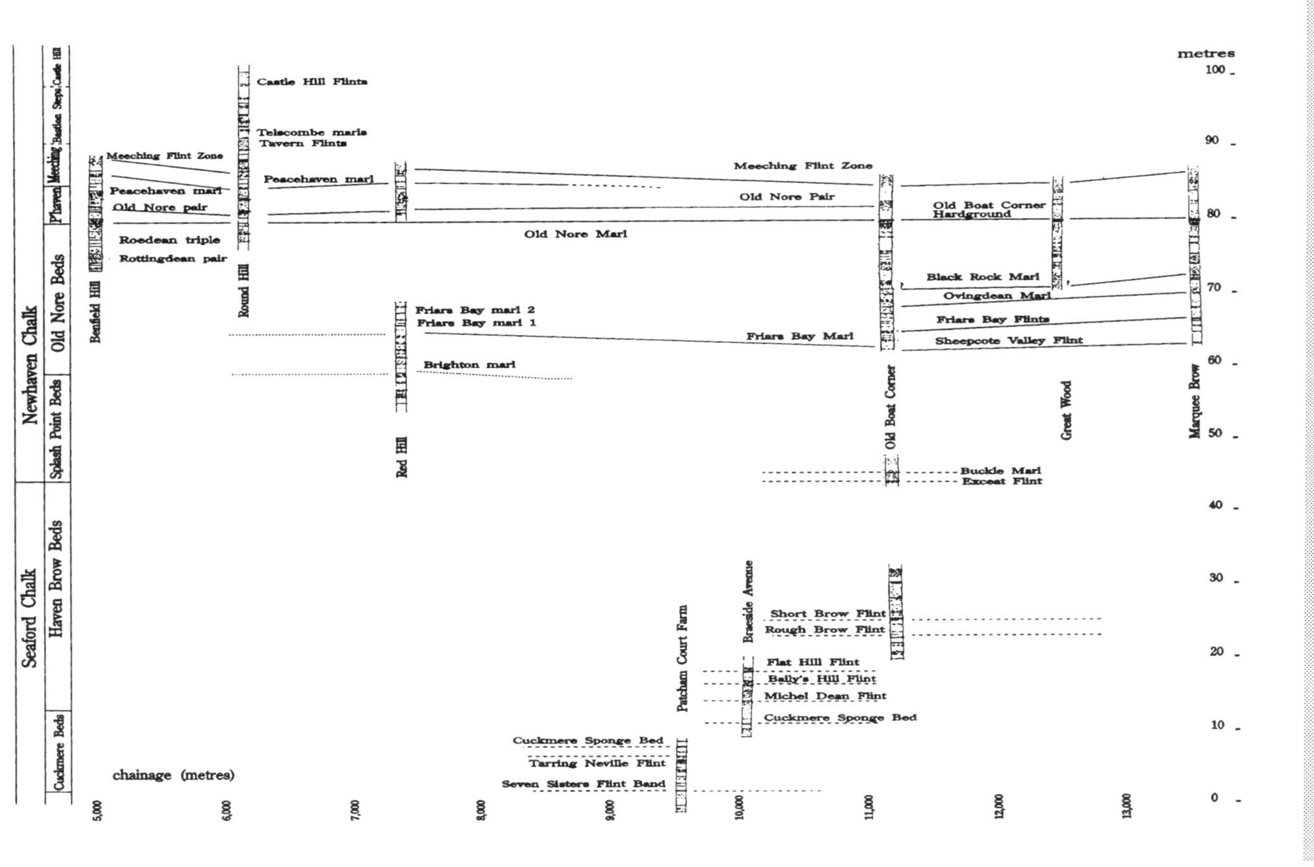
metres
100
90
80
70
60
50
40
30
20
10
0
Newhaven Chalk
Seaford Chalk
Castle Hill
Steps
Bastion
Meeching
P'haven
Old Nore Beds
Splash Point Beds
Haven Brow Beds
Cuckmere Beds
Benfield Hill
Round Hill
Red Hill
Patcham Court Farm
Braeside Avenue
Old Boat Corner
Great Wood
Marquee Brow
Meeching Flint Zone
Peacehaven marl
Old Nore pair
Roedean triple
Rottingdean pair
Castle Hill Flints
Telscombe marls
Tavern Flints
Peacehaven marl
Meeching Flint Zone
Old Nore Pair
Old Nore Marl
Old Boat Corner Hardground
Black Rock Marl
Ovingdean Marl
Friars Bay Flints
Sheepcote Valley Flint
Friars Bay marl 2
Friars Bay marl 1
Friars Bay Marl
Brighton marl
Buckle Marl
Exceat Flint
Short Brow Flint
Rough Brow Flint
Flat Hill Flint
Baily's Hill Flint
Michel Dean Flint
Cuckmere Sponge Bed
Cuckmere Sponge Bed
Tarring Neville Flint
Seven Sisters Flint Band
chainage (metres)
5,000
6,000
7,000
8,000
9,000
10,000
11,000
12,000
13,000

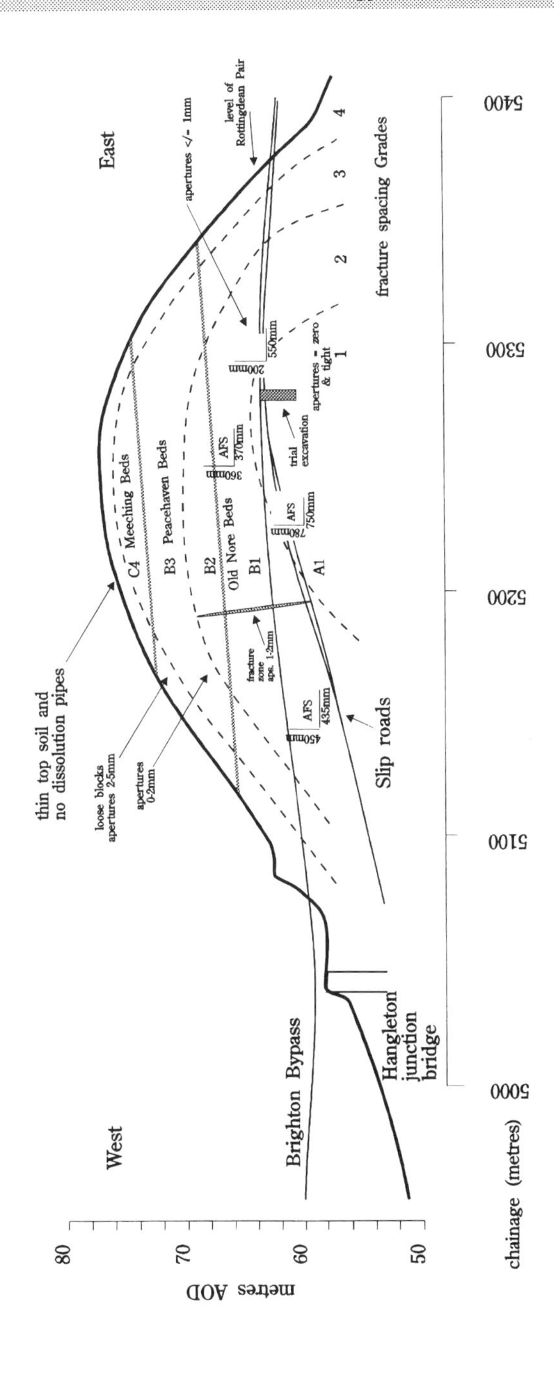

Fig. 7.5. Centreline section through Benfield Hill cutting.

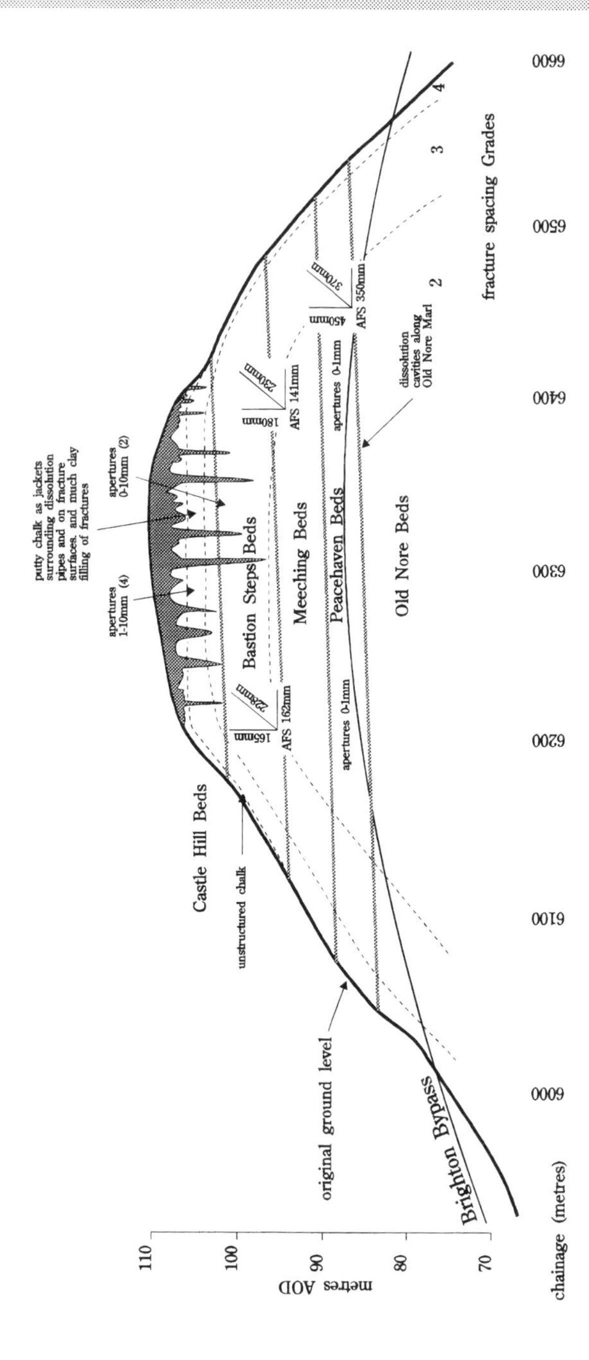

Fig. 7.6. Centreline section through Round Hill cutting.

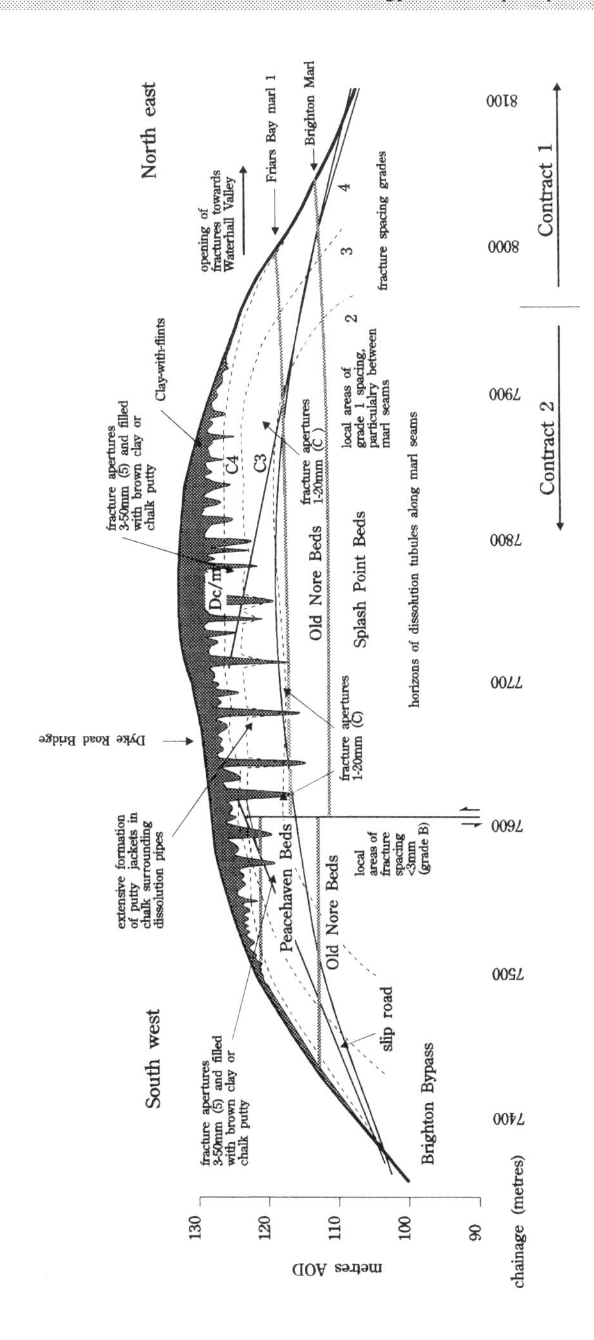

Fig. 7.7. Centreline section through Red Hill cutting (Dyke Road interchange).

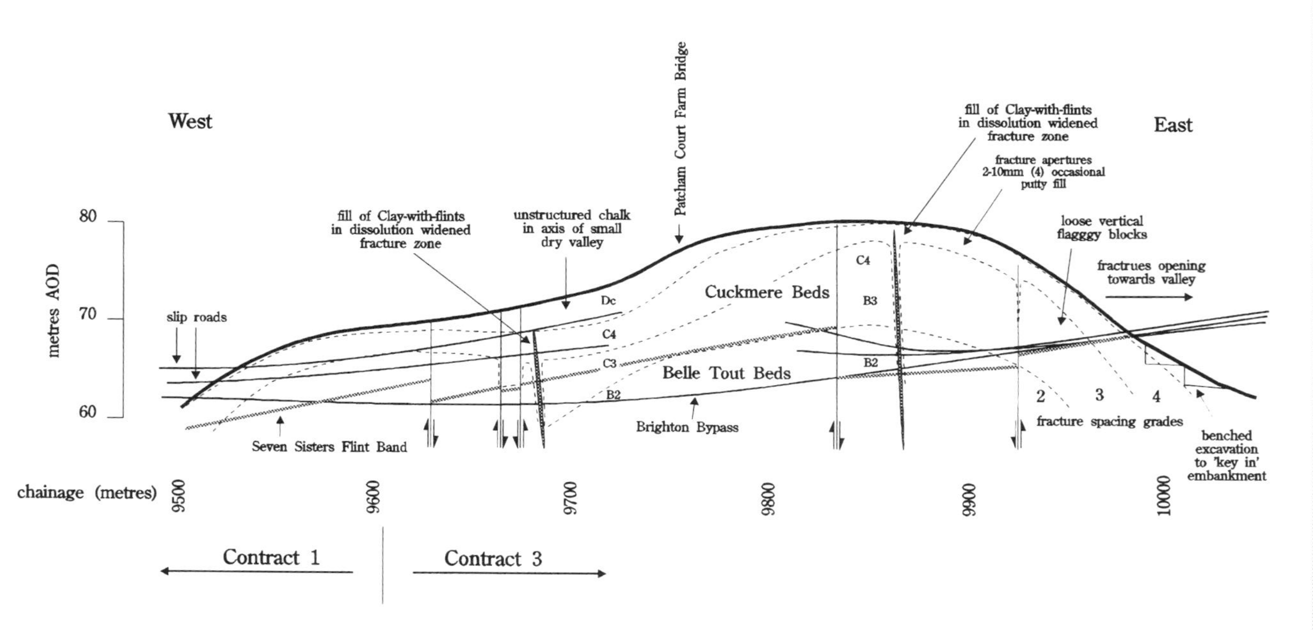

Fig. 7.8. Centreline section through Patcham Court Farm cutting.

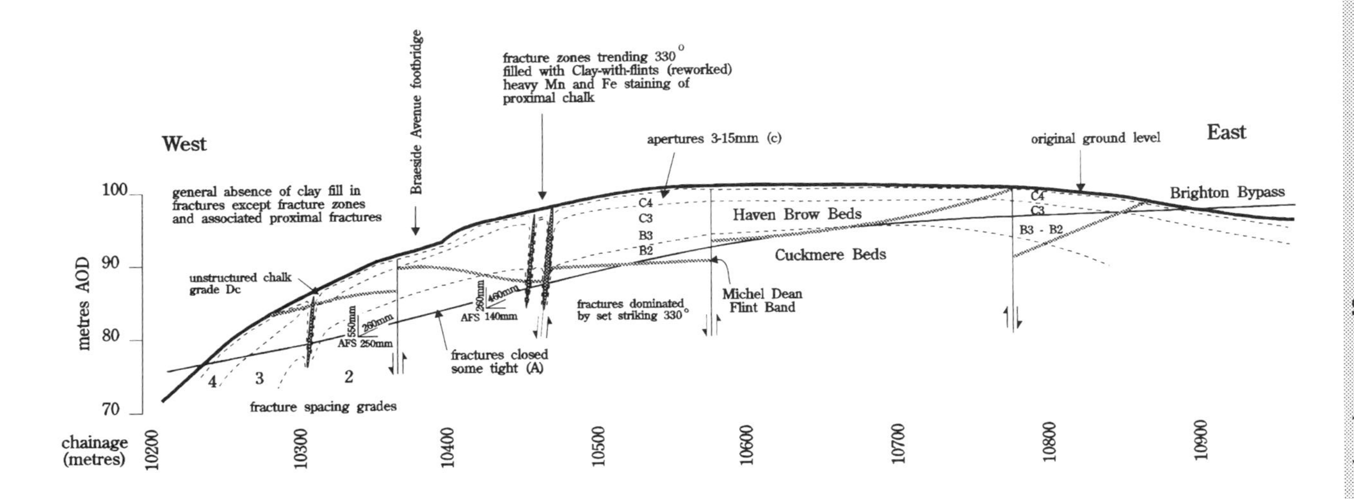

Fig. 7.9. Centreline section through Braeside Avenue cutting.

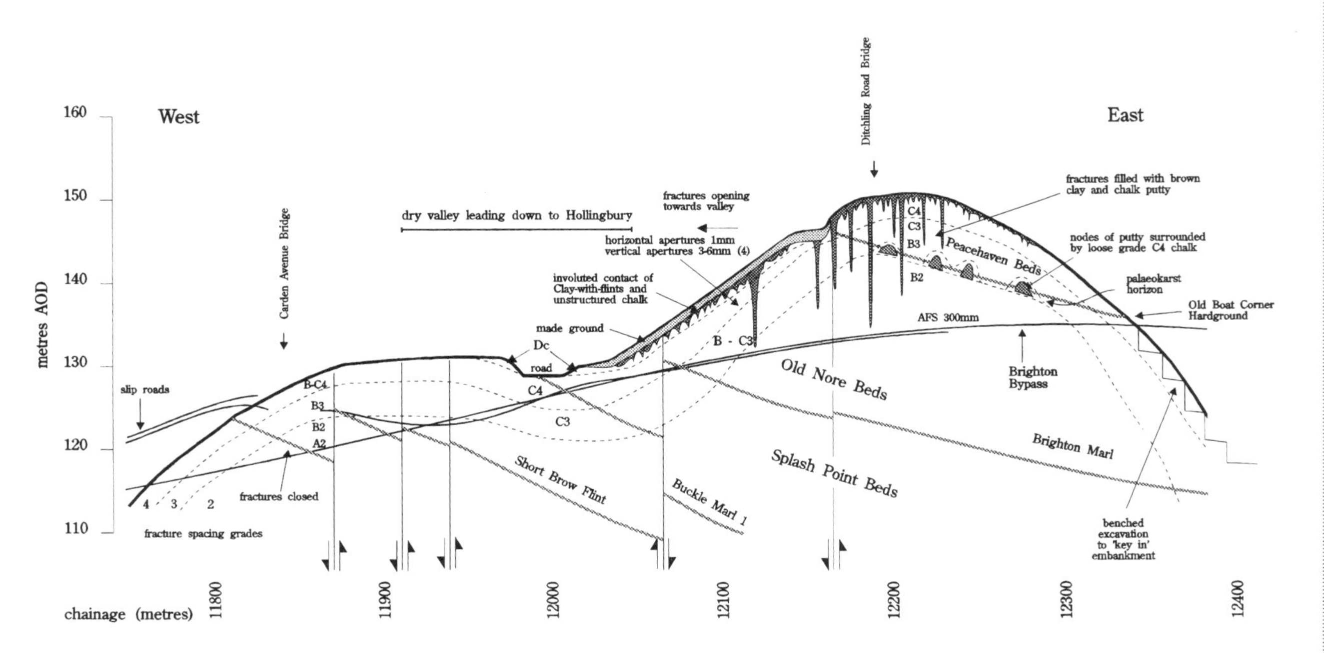

Fig. 7.10. Centreline section through Old Boat Corner cutting.

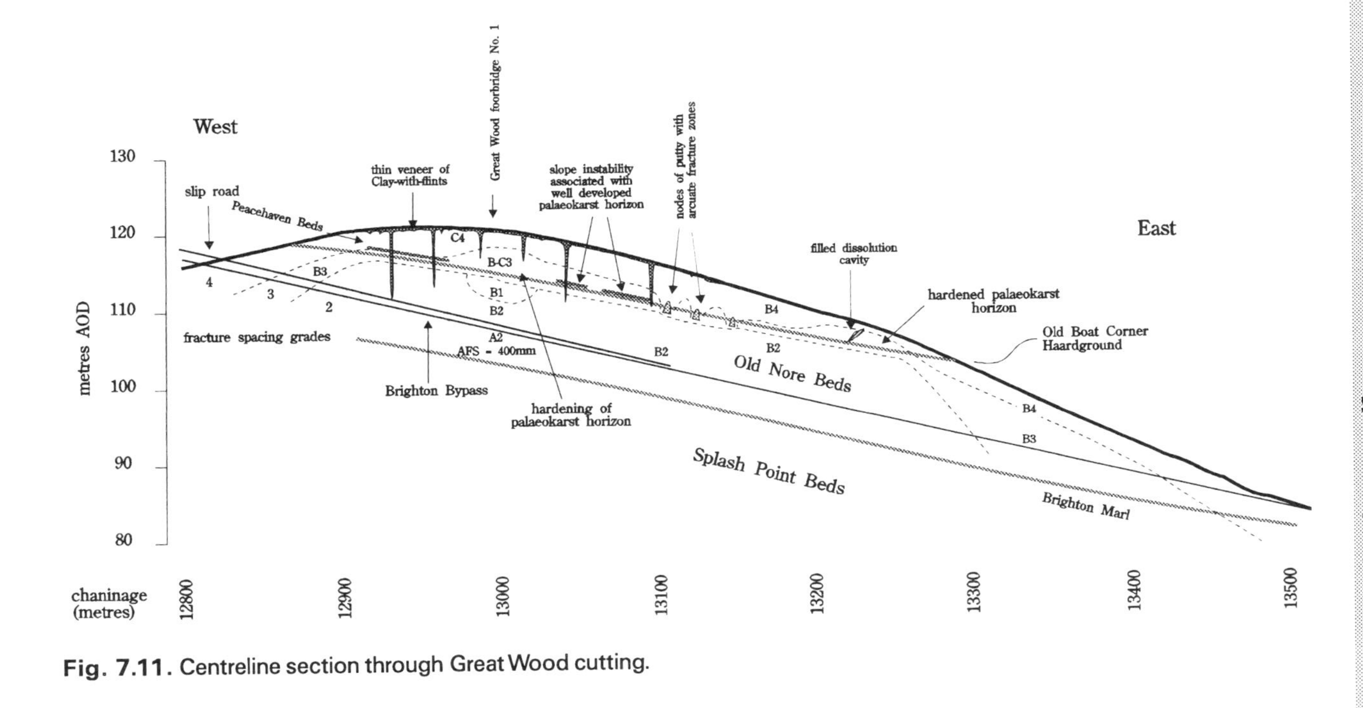

Fig. 7.11. Centreline section through Great Wood cutting.

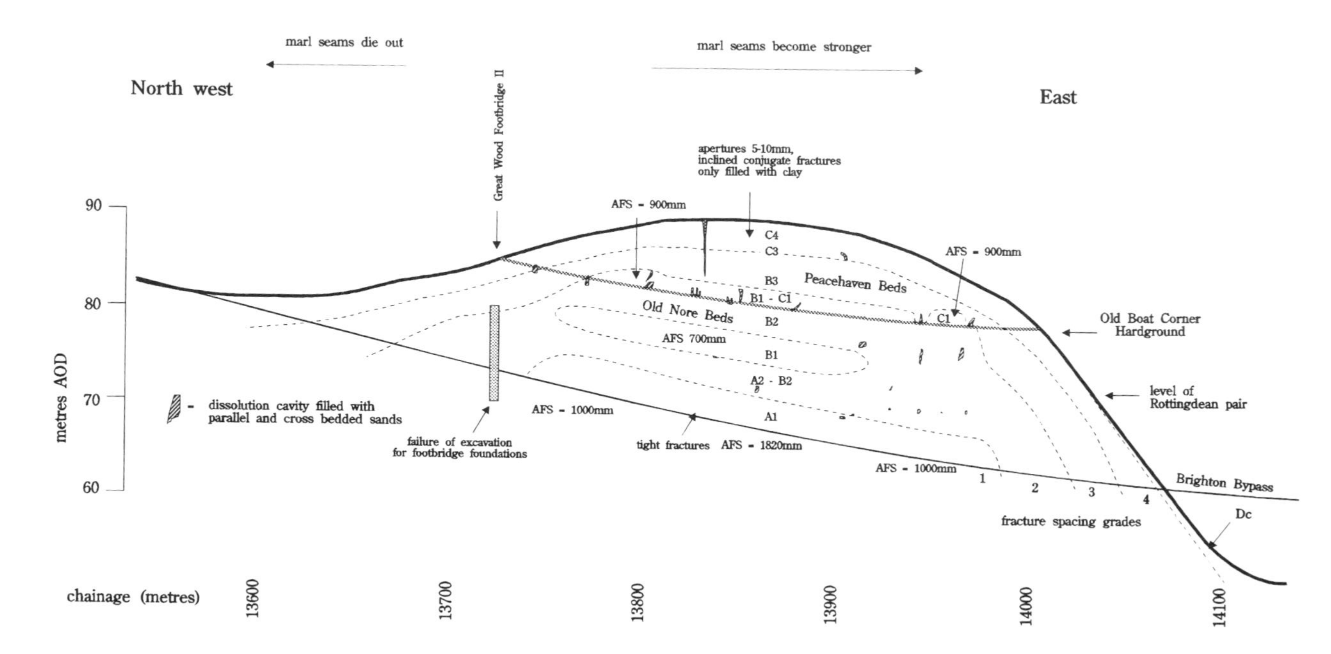

Fig. 7.12. Centreline section through Marquee Brow cutting.

heme as Grade D. Grade D chalk is differentiated according to whether it is clast supported, Dc, or matrix supported, Dm.

Although bedding dips on the Brighton Bypass rarely exceeded 6°, the area around Brighton preserves a complex structural geology. This is manifest as a combination of troughs and domes with occasional large-scale faults (Gaster 1951). As is shown below, the structural geology affects stratigraphy, sedimentology and landscape evolution. Structural features relating to post-lithification tectonic deformation were, however, rarely recorded.

Cuttings through some of the higher altitude hills such as Round Hill, Red Hill, Old Boat Corner and Great Wood encountered clay-with-flints and remnants of Palaeogene sands and clays filling dissolution pipes. During the design stage the landscape of the proposed route was simply classified according to a number of geomorphological domains (Mortimore 1979*a*). These included: 'unweathered' hill tops; hill tops with dissolution pipes; valley slopes; and dry valleys. These domains were found to be useful in correlating moduli from plate loading tests.

The Brighton Bypass was, for the most part, constructed along the strike of the South Downs dip-slope and was consequently cut by numerous dry valleys which preserve solifluction deposits and deep weathering profiles in the Chalk. Cuttings provided the material for earthworks as well as large exposures and therefore their geology is given priority here.

For the purpose of design, engineers need to know the characteristics and properties of both the intact rock (i.e. the rock material) and the ground including discontinuities (i.e. the rock mass). The way in which scientific frameworks helped understanding of variations in some rock material and rock mass characteristics on the Brighton Bypass is explained after a consideration of data interpreted within the scientific frameworks.

Stratigraphy

The Brighton Bypass was constructed through the Lewes Nodular, Seaford, Newhaven and Tarrant members. The Lewes Nodular Chalk, however, was present only in borehole investigations for the piled foundations of the Patcham Viaduct in Patcham Valley (Fig.

7.2) and the Tarrant member was rarely present capping high ground on Contract 4 at the western end on the Bypass. Therefore, chalks of the Seaford and Newhaven members were the dominant strata and these units are focused upon here.

Stratigraphy can control both mass and material characteristics of the Chalk. Ward *et al.* (1968) noted a gradual increase in chalk hardness, fracture spacing and a decrease in fracture apertures with depth in the Middle Chalk at Mundford, Norfolk. On a larger scale, Mortimore (1993) reported a correlation of dominant, weathering independent, fracture styles with broad stratigraphic units.

Whilst a very crude reduction in intact dry density has been noted ascending the whole chalk stratigraphy (Meigh & Early 1957; Carter & Mallard 1974), there is no systematic increase in intact dry density with depth of cover materials (Higginbottom 1966), while Clayton (1983) found no unique relationship between stratigraphical level and intact dry density.

Effect on intact dry density

Mean intact dry densities of units of the Seaford and Newhaven members are shown in Table 7.2 for cuttings in Contracts 1, 2 and 3 of the Brighton Bypass. These data show that the Belle Tout, Cuckmere and Haven Brow Beds of the Seaford Chalk have a consistent mean intact dry density (reflecting its general homogeneity), whilst the beds of the Newhaven Chalk vary widely showing the marked rhythmicity of this unit (Mortimore 1986).

These data also reveal that intact dry densities of the Newhaven Chalk are not constant for a particular stratigraphic unit and that a perceptible trend of increasing mean intact dry density with decreasing stratigraphic level exists in each cutting. However, there was no significant relationship with depth below present ground surface.

Effect on rock mass structure

Mortimore (1979*b*, 1983) noted that chalks containing numerous marl seams such as the Lewes and Newhaven members contain abundant large-scale inclined conjugate fractures with different

Table 7.2. Intact dry densities for stratigraphic units in the main cuttings of the A27 Brighton Bypass (Contracts 2 & 3)

Cuttings	Benfield Hill	Round Hill	Red Hill	Patcham Court Farm	Braeside Avenue	Old Boat Corner	Great Wood	Marquee Brow
Geomorphological domain[1]	A	B	B	A	A	B	B/C	C
Castle Hill		1.50						
Bastion Steps		1.61						
Meeching	1.57	1.58	1.52					
Peacehaven	1.65	1.62	1.60			1.53	1.60	1.62
Old Nore	1.70	1.67	1.59			1.55	1.61	1.66
Splash Point			1.59			1.55	1.65	
Haven Brow					1.63	1.61		
Cuckmere				1.63	1.63			
Belle Tout				1.63				

[1] Geomorphological domains were those recognized by Mortimore (1979*a*, 1990)
A = unweathered hill top
B = deeply weathered hill top with dissolution pipes
C = valley slope

aquifer properties between these units. In general, fracture characteristics of the Seaford and Newhaven Chalks on the Brighton Bypass supported these observations.

In order to compare fracturing characteristics between these lithological units it is necessary to concentrate on exposures in the deepest part (core) of cuttings where fracturing is largely unaffected by dissolution and frost-shattering effects.

Seaford Chalk was exposed at a maximum depth of approximately 10 m in the core of Braeside Avenue cutting (Fig. 7.9) and approximately 12 m in Patcham Court Farm and the western part of Old Boat Corner (Figs 7.8 & 7.10). The fracture characteristics in the cores of these cuttings were remarkably similar. Apart from a few discrete fracture zones, the fracture pattern was entirely orthogonal with two sub-vertical sets at approximately right angles to one another and mutually at right angles to bedding parallel, sub-horizontal fractures. Fracture spacing never exceeded Grade 2 and the dominant aperture grade was B on the new CIRIA scheme (Table 7.1). Some Grade A2 chalk was present in Old Boat Corner cutting and the absence of Grade A chalk in Braeside Avenue and Patcham Court Farm is probably owing to a lack of sufficient depth of cut in the former and the effect of a small dry valley in the latter.

The largely uninterrupted and gradual improvement of rock mass grade with depth in these cuttings reflects the lithostratigraphy of the Seaford Chalk which lacks marl seams or hardgrounds and is characterized by a series of strong nodular flint bands in otherwise bland chalk. The strongest lithological marker of the Seaford Chalk in Sussex is the Seven Sisters Flint Band and where this was parallel to the ground surface as in Patcham Court Farm cutting (Fig. 7.8) it formed a very prominent rock mass grade boundary. This effect has been noted with other flint bands on numerous occasions (Mortimore, pers. comm. 1995) and is probably related to the interruption of previously existing otherwise smooth hydrogeological and cryogenic gradients.

Newhaven Chalk was exposed in cuttings through Benfield Hill (Fig. 7.5), Round Hill (Fig. 7.6), Red Hill (Fig. 7.7), Old Boat Corner (Fig. 7.10), Great Wood (Fig. 7.11) and Marquee Brow (Fig. 7.12). Fracture characteristics, as influenced by stratigraphy, are best understood form Benfield Hill and Marquee Brow as these lack extensive development of dissolution pipes. In both cuttings the style was dominated by marl-smeared slickensided curved and planar inclined fractures which generally took the form of conjugate pairs of normal faults (shear and hybrid shear of Hancock 1985). Throws ranged form about 50 to 200 mm and slip directions were highly variable. Sub-horizontal bedding parallel fractures were very rare in the Newhaven Chalk even along marl seams or sub-horizontal sheet flints and only became apparent at depths less than about 10 m which would have been influenced by freeze–thaw effects. The orthogonal pattern so characteristic of the Seaford Chalk was almost entirely absent.

Rock mass quality in the cores of Marquee Brow and Benfield Hill cuttings was Grade A1. At 27 m, Marquee Brow (Fig. 7.12) was the deepest cutting on the Brighton Bypass and showed a markedly non-linear profile in fracture gradings. Bands of massive, anomalously high fracture spacing Grade 1 chalk were found at levels coincident with the Old Boat Corner hardground and the Rottingdean Pair (pair of marl seams). Both these levels presented problems for earthworks in terms of excavatability and large block size arriving for compaction. The band above the Rottingdean pair could only be excavated in Benfield Hill by prior loosening with a bulldozer fitted with a single ripping hook (tyne).

These variations in the fracture gradings of the Newhaven Chalk reflect its rhythmicity (marl seams, soft and hard horizons and discrete hardgrounds) and can be seen as corollary to the more gradual fracture gradients of the more homogeneous Seaford Chalk.

Sedimentology and tectonic setting

The effects of tectonic structures on chalk lithostratigraphy are quite well known (Hancock 1975; Hancock & Scholle 1975; Fletcher 1978; Gale 1980; Mortimore & Pomerol 1987, 1991; Mortimore *et al.* 1996). The Brighton Bypass offered a good opportunity to study the lithological characteristics associated with zones of uplift and depression and to trace their effects through to the engineering geology. Structure contours for the Bridgewick

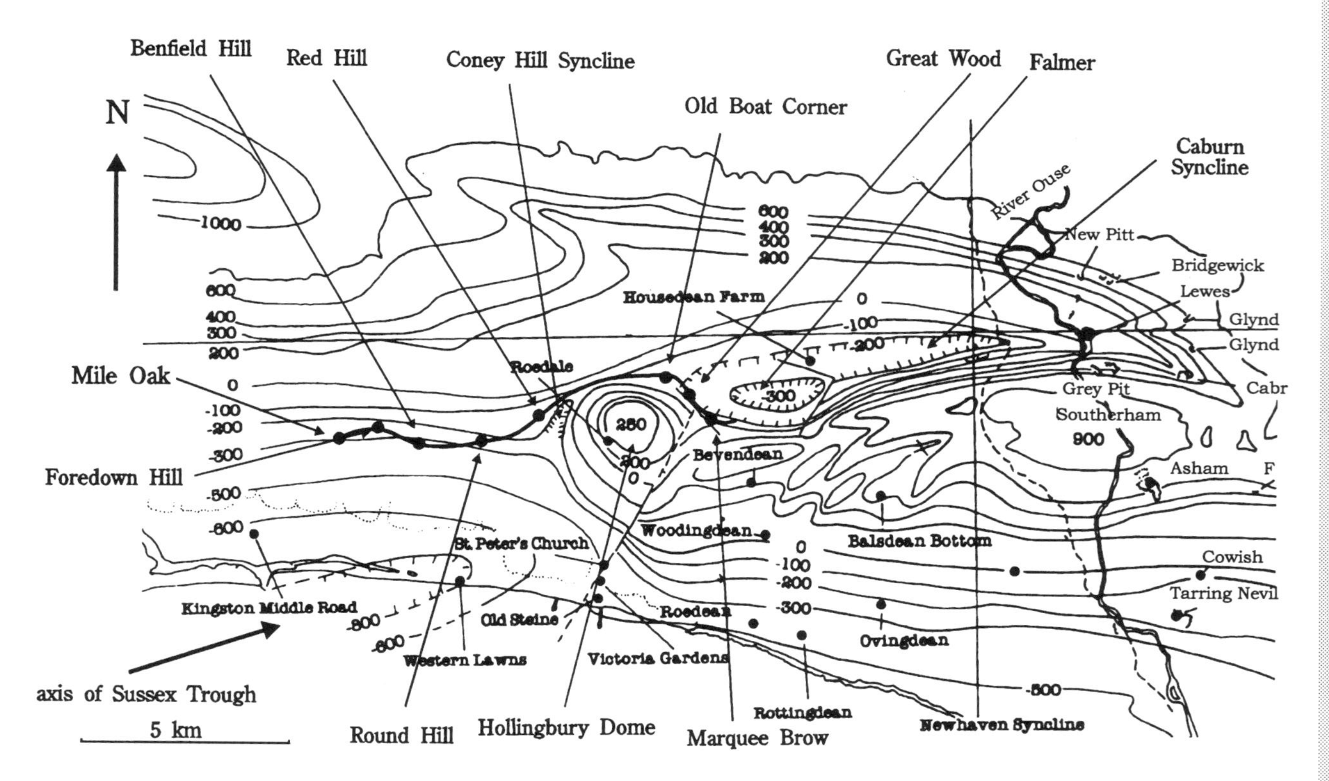

Fig. 7.13. Structure contour map of the Bridgewick Marl 1 (Lewes Nodular Chalk) showing the positions of major syn-sedimentary structures in the Brighton area.

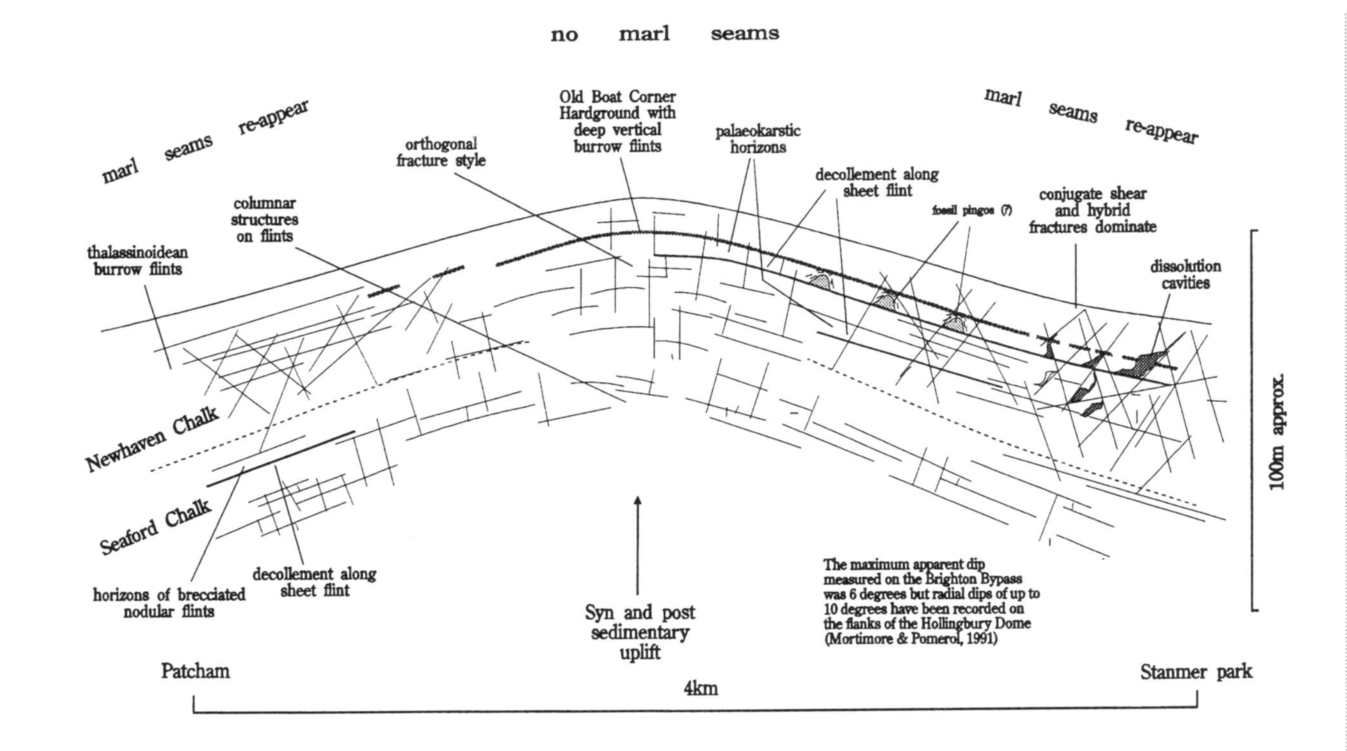

Fig. 7.14. Schematic cross-section through the northern flank of the Hollingbury Dome showing major changes in lithostratigraphy and rock mass characteristics.

Marl 1 (Lewes Nodular Chalk) in the Brighton region are reproduced in Fig. 7.13 and show four main structural features, these include:

1. the Sussex trough (Mortimore 1983) west of Round Hill;
2. Coney Hill syncline, a relatively small feature close to Red Hill (Mortimore 1979*b*);
3. the Hollingbury Dome (Henry & Drummond 1973), a syn- and post-depositional pericline near Old Boat Corner;
4. the Caburn Syncline (Henry & Drummond 1973).

The best example of structural control on sedimentology and lithostratigraphy on the Brighton Bypass is related to the Hollingbury Dome. This structure is an almost circular pericline with a diameter of approximately 4 m. Figure 7.14 shows a schematic cross-section through the northern flank of the Hollingbury Dome constructed using data from Patcham Court Farm, Braeside Avenue, Old Boat Corner, Great Wood and Marquee Brow cuttings. Several sedimentological effects observed in relation to this structure are described below.

1. Slight overall thinning of the sequence occurred over the Dome (Fig. 7.3).
2. Marl seams are usually abundant in the Newhaven Chalk but were weak on the flanks of the Dome and died out entirely near its crest.
3. The Old Nore Marl, usually the most prominent lithological marker in the Newhaven Chalk, was replaced by a glauconitic hardground (Old Boat Corner Hardground). This was accompanied by other less well developed, discontinuous hardgrounds replacing other, weaker marl seams.
4. Near the crest of the dome the Old Boat Corner Hardground comprised two well cemented layers with borings and stylolites. Descending the flanks of the dome, its character gradually changed to discontinuous hardground surfaces then nodular chalks with the eventual reappearance of the Old Nore Marl at approximately 2 km from Old Boat Corner.
5. The small 'finger flints' and nodular, thalassinoidean flints

which characterize the chalk around the Old Nore Marl were replaced over the dome by deep vertical burrow flints which extended over 1 m beneath the hardground. This probably represents a benthic response to a regressive event typified by the development of hardgrounds (Hancock 1989).

6. Between Old Boat Corner and Marquee Brow, bedding parallel and sub-parallel sheet flints were common. The most prominent of these occurred beneath the Old Boat Corner Hardground and was laterally persistent from Old Boat Corner to Marquee Brow.
7. The above sheet flint was interpreted to represent silicification (Clayton 1986) along a décollement surface. This is supported by the observed displacement of vertical burrow flints at Old Boat Corner across the sheet flint. Syn-sedimentary deformation in chalk sediments is well known (Kennedy & Juignet 1974; Mortimore 1979*b*; Gale 1980; Mortimore *et al.* 1990).
8. Horizons of brecciated nodular flint are indicative of syn-sedimentary reworking of chalk (e.g. Mortimore 1979*b*) and were found in the Cuckmere Beds of Patcham Court Farm cutting, on the northern flank of the dome.
9. In addition to changes in the dominant type of flint trace fossil, other faunal changes as the dome is approached included an abundance of non-burrowing fauna such as species of *Spondylus* and numerous small bioherms rich in bryozoa.

Effect on intact dry density

The sedimentological variations over the crest and down the flanks of the Hollingbury Dome were seen to broadly correlate with gradual changes in mean intact dry density. Table 7.2 indicates that mean intact dry densities for members of the Newhaven Chalk at Old Boat Corner, near the crest of the Hollingbury Dome, were consistently lower than for any of the other cuttings and that there was a gradual increase in intact dry density for these units away from the Hollingbury Dome; eastwards towards the Caburn Syncline and westwards towards the Sussex trough. At first sight therefore, it seems that mean intact dry density is affected by position relative to syn-sedimentary structures. However, it should be noted that the increase in intact dry density associated with the discrete Old Boat Corner Hardground (mean intact dry density 1.83 $Mg\,m^{-3}$) in the Old Nore Beds, is converse to the observed trend in the higher order stratigraphic units.

The above spatial distribution of intact dry density with respect to the Hollingbury Dome is only a correlation and any possible causal effects of sedimentology and tectonic setting on intact dry density are less clear. It is appreciated that larger overburdens give rise to a reduction in chalk porosity (Mimram 1977; Clayton 1983) though this is modified by lithological differences on a bed-to-bed basis, particularly because of bioturbation (Clayton & Matthews 1987). Tectonic deformation, most notably pressure solution effects, are very important in raising the intact dry density of chalk (Mimram 1977). For example, members of the Northern Province chalk in Yorkshire are much harder than the chalks in the south and contain many stylolites and have been shown to have experienced greater burial depths during diagenesis than their Southern Province counterparts.

Stylolites were only found in two places on the Brighton Bypass; between interpenetrating nodules in the Old Boat Corner Hardground (Fig. 7.3; Lamont-Black 1995) and above the Rottingdean marls in the Old Nore Beds of Benfield Hill cutting (i.e. towards the Sussex Trough). In Benfield Hill, the Old Nore Beds had a mean intact dry density of 1.70 $Mg\,m^{-3}$ but the stylolitized beds had an intact dry density of over 1.80 $Mg\,m^{-3}$. This differential was repeated at Old Boat Corner where these beds had an intact dry density of 1.65 $Mg\,m^{-3}$ compared to a mean intact dry density of 1.55 $Mg\,m^{-3}$ for the Old Nore Beds as a whole. As the beds of the Rottingdean Pair did not contain stylolites at Old Boat Corner, their higher intact dry density cannot be attributed to these features here. Examination of chalk from this horizon under a scanning electron microscope revealed the nannofossil *Micula* which had been found previously to correlate with harder chalks (Mortimore & Fielding 1990). This piece of evidence would suggest therefore, that the type of pelagic sediment (a basin-wide effect) can also influence the intact dry density.

The influences of sedimentology and tectonic setting on intact dry density from the Brighton Bypass therefore are diverse and include:

1. position with respect to zones of depression and uplift;
2. seafloor conditions, e.g the presence of hardgrounds and the type of benthos;
3. nature of the initial pelagic sediment input;
4. diagenetic effects, such as depth of burial and the development of stylolites.

Effect on rock mass characteristics

The association of conjugate shear and hybrid shear fractures with the Newhaven Chalk and an orthogonal fracture pattern with the Seaford Chalk was clear on the Brighton Bypass. Moreover, the absence of marl seams in the Splash Point, Old Nore and Peacehaven Beds near the crest of the Hollingbury Dome firmly established the association of marl seams in particular as one of the crucial elements in the development of inclined conjugate fractures (Fig. 7.14). Scan line fracture surveys of the Old Nore Beds made at Marquee Brow and Old Boat Corner showed that inclined conjugate fractures dominated the former compared to an orthogonal pattern at the latter (Fig. 7.15). Therefore, the absence of marl seams over the Hollingbury Dome resulted in a more 'Seaford Chalk type' fracture style in the Newhaven Chalk. Using the above sections and other sections from the Brighton Bypass and the Sussex coast, Lamont-Black (1995) interpreted these fractures as having an early syn-sedimentary origin related to the transient development of excess pore pressures in chalks with marl seams.

At 300 mm (fracture spacing Grade 2), the average fracture spacing of the Newhaven Chalk in the core of Old Boat Corner cutting was considerably less than the equivalent depth in Marquee Brow (Fig. 7.12) thus the fracture style has an effect on fracture spacing and therefore, block size. It seems likely, therefore, that the presence of these early fractures provided strain paths along which to dissipate tectonic stresses that would otherwise have resulted in the formation of the orthogonal fractures characteristic of the Seaford Chalk.

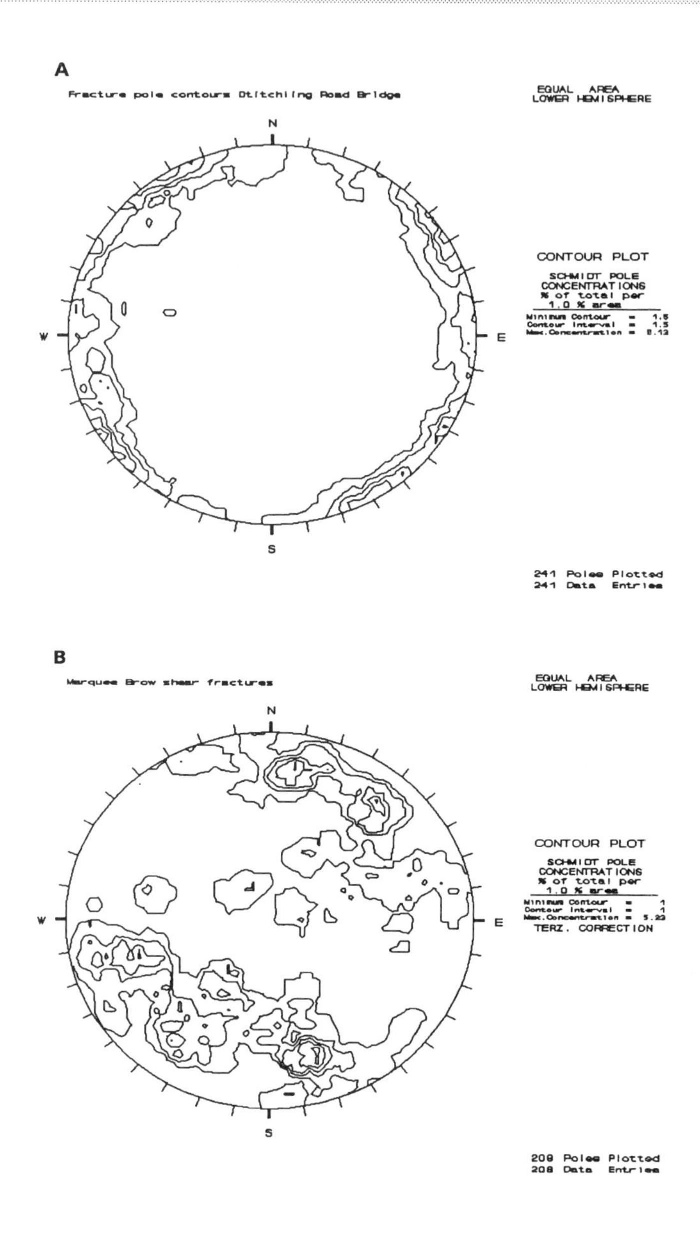

Fig. 7.15. Stereographic plots of fracture surveys in the Old Nore Beds of (A) Old Boat Corner and (B) Marquee Brow.

Geomorphology

During the main site investigations for the Brighton Bypass the terrain was classified into four basic types:

1. deeply weathered hill tops/dissolution pipe terrains;
2. dry valleys;
3. gentle slopes/'unweathered' hill tops;
4. steep valley sides.

The largest engineering operation on the Brighton Bypass was the earthworks which comprised approximately $4 \times 10^6\,m^3$ of earthmoving mostly in cuttings through hill tops and embankments across dry valleys. The condition of chalk excavated in cuttings proved critical to the earthworks, especially where hill tops were in dissolution pipe terrains. In this section, dry valleys are briefly described before a more detailed consideration of dissolution pipes and associated phenomena and their effects on engineering.

Gentle slopes and steep valley sides were relatively minor in their importance on the Brighton Bypass. Except for occasional valley side stress relief of fractures adjacent to deep valleys, chalk from these terrains was generally favourable for engineering and they will not be considered further here.

Dry valleys

The Brighton Bypass crossed a number of dry valleys, some of which contained up to 7 m of solifluction (coombe) deposits and preserved deeply weathered chalk to a depth of 20 m below ground surface. The construction of road bridges for interchanges in these settings required the use of deep piled foundations and so sections in these settings were unavailable for study.

The only dry valleys that were seen in section were on the eastern side of Old Boat Corner (Fig. 7.10) and in Patcham Court Farm cutting (Fig. 7.8). The main feature of these valleys was the downward displacement of rock mass grades owing to weathering effects. The dry valley in Patcham, in particular, showed that dry valleys need not have a large relief in order to significantly modify the weathering profile of the underlying chalk.

Physical characteristics and effects of dissolution features
Dissolution features can have a marked impact on design and construction with respect to earthworks foundations and tunnelling. The most commonly recognized dissolution feature in chalk is the dissolution pipe (Prestwich 1854; Jukes-Browne 1906; Davies 1929; Kirkaldy 1950; Higginbottom 1966; Higginbottom & Fookes 1970; Thorez *et al.* 1971; West & Dumbleton 1972; Walsh *et al.* 1973; DeBruijn 1983; Wilmot & Young 1985; Edmunds *et al.* 1987; Ford 1989; Mortimore *et al.* 1990).

Edmonds (1983, p.262) defined dissolution pipes as '. . . a cone or pipe-like cavity in vertical section, typically infilled with overlying deposits that have subsided into the cavity created by dissolution of the soluble chalk host rock.'

Using this definition with the added criterion of >1 m cross-sectional diameter, dissolution pipes were found to be concentrated in four major cuttings: Round Hill (200 per hectare), Red Hill (265 per hectare), Old Boat Corner (150 per hectare) and Great Wood (30 per hectare).

Dissolution pipes were described according to morphotype, cross-sectional shape, type of fill and effect on surrounding chalk material. Using this approach, two broad characters emerged.

1. The Red Hill character: mixture of circular and elliptical cross-sections, dominantly conical in shape (but with pinnacled and complex forms), filled with cohesive sediments, thick 'putty jackets', a broad zone of intact dry density reduction of proximal chalk and extensive filling of chalk fractures with clay and 'putty chalk'. This character describes the dissolution pipes in Red Hill and Round Hill.
2. The Old Boat Corner character: circular, dominantly cylindrical in shape, filled with sandy sediments and basal Palaeogene alteration minerals, thin putty jackets, short range reductive effect on chalk intact dry density and limited filling of chalk fractures with clay and 'putty chalk'. This character describes the dissolution pipes in Old Boat Corner and Great Wood.

In the Brighton area, Palaeogene sediments and Clay-with-flints are preserved in the axes of synclines which have preserved them from Quaternary erosion. The Coney Hill Syncline (Fig. 7.13) is responsible for the dissolution pipes in Red Hill and Round Hill, and the Caburn Syncline, which preserves over 5 m of Palaeogene sediments at Falmer, is responsible for the dissolution pipes at its westerly limit near Old Boat Corner and Great Wood. The two broad characters of dissolution pipes are therefore associated with two separate structures.

The correlation between the presence of dissolution pipes and overlying sediments (particularly towards the limits of outliers) is well known (e.g. Prestwich 1855; Dowker 1866; Edmonds 1983). The contact between the Chalk and the overlying Palaeogene, known as the sub-Palaeogene erosion surface (Jones 1980) varies in altitude and stratigraphic level in the Brighton area (Fig. 7.16). From Southwick Hill in the west to Falmer in the east the terrain along the Brighton Bypass can be divided into three tectonic domains: two broadly synclinal areas separated by the zone of uplift associated with the Hollingbury Dome.

In the west, the sub-Palaeogene erosion surface lies at approximately 125–135 m OD, slowly ascending the Chalk stratigraphy westwards. This stratigraphic ascension probably relates to pre-basal Palaeogene deepening of the Sussex Trough. Although not preserved on Foredown Hill (Brighton Bypass Contract 4), the few dissolution pipes there attest to the surface's former proximal position. To the east, the erosion surface descends from about 150 m OD near Old Boat Corner to about 91 m OD at Falmer. The preservation of a few small dissolution pipes in Marquee Brow cutting is analogous to the situation at Foredown Hill.

Accepting that marine erosion created the sub-Palaeogene erosion surface, its original character would likely have been near planar. Its absence on the crest of the Hollingbury Dome (178 m OD; TQ532 135–107 945) attests to a continued uplift of the Hollingbury Dome into Palaeogene times. Subsequent Quaternary erosion has left the projected erosion surface at considerable height above parts of the present land surface (Fig. 7.16). With respect to the degrading effect of dissolution pipes on chalk and from the viewpoint of material characteristics, areas well below the sub-Palaeogene erosion surface provided the best quality chalk for

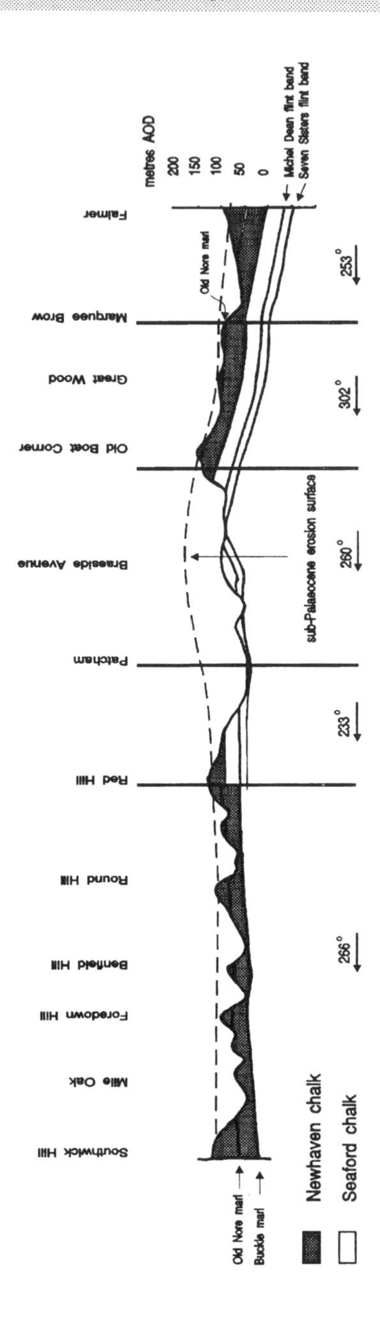

Fig. 7.16. Level of the sub-Palaeogene erosion surface along the Brighton Bypass.

earthworks and thus a knowledge of the level and shape of this surface prior to design and construction would be advantageous to engineering.

There are four main points of significance to engineering of dissolution features these include: (1) effect on intact dry density; (2) association of putty chalk; (3) palaeokarst horizons; and (4) caves.

Effect on intact dry density

Adjacent to individual pipes, chalk intact dry densities tended to increase with distance from the margin over a distance which varied from 0.2–3.0 m. Figure 7.17 shows this effect for equally sized dissolution pipes at Old Boat Corner and Red Hill. The chalk at Old Boat Corner and Red Hill was affected over a distance of approximately 1.5 and 3.0 m, respectively. Dissolution pipe concentrations of 265 per hectare and 150 per hectare for these respective areas indicates a greater intensity of dissolution at Red Hill and probably explains the wider zones of intact dry density reduction seen at this location.

The presence of dissolution pipes appeared to affect the intact dry density of chalk in cuttings as a whole (Table 7.2). For example, intact dry densities of the Meeching Beds of Benfield Hill and Round Hill cuttings were not significantly different (as determined by statistical T-tests for 95% certainty) but the same beds in Red Hill had a significantly lower intact dry density of 1.52 $Mg\,m^{-3}$. This is may be related to the fact that Benfield Hill had no dissolution pipes and the Meeching Beds in Round Hill were some 12 to 17 m beneath the contact with the overlying clay-with-flints (Fig. 7.6) whereas the Meeching Beds in Red Hill were directly beneath the clay-with-flints (Fig.7.7). A similar situation arises for the Old Nore Beds of these three cuttings. The intact dry densities of the Peacehaven Beds of these cuttings does not appear to fit this observation. However, the Peacehaven Beds at Red Hill occurred on the western flank of the hill where dissolution pipe intensity was much reduced.

Similarly, between Old Boat Corner and Marquee Brow the geomorphological domain grades from a deeply weathered hill top with a high concentration of dissolution pipes to a relatively

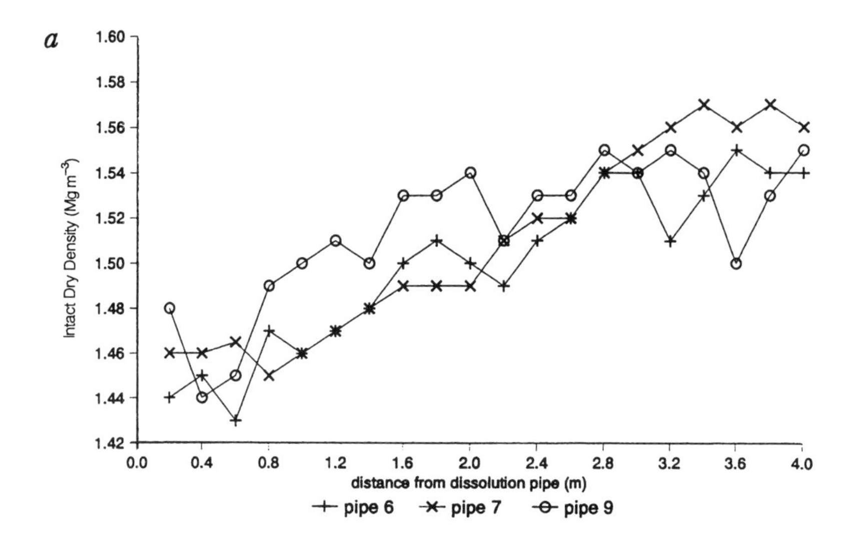

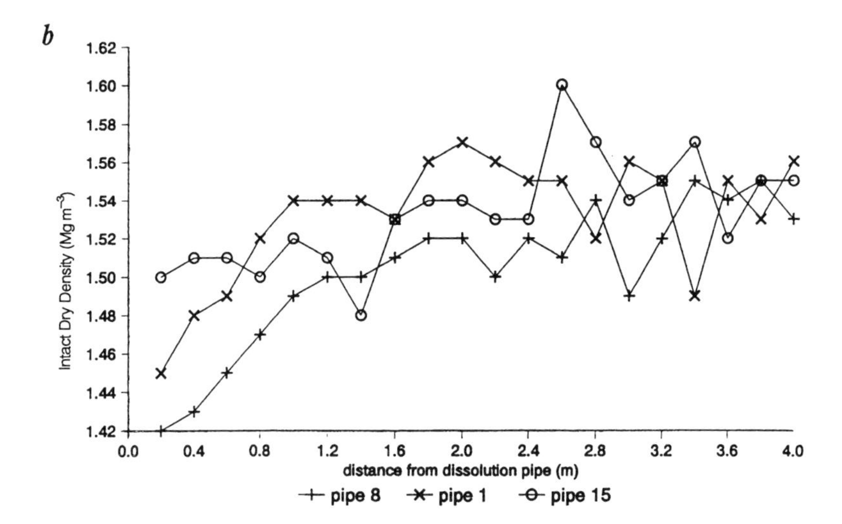

Fig. 7.17. Distribution of chalk intact dry density with distance from the edge of dissolution pipes at (A) Red Hill and (B) Old Boat Corner.

unweathered valley slope. The Peacehaven Beds at Old Boat Corner had a significantly lower mean intact dry density than those in Great Wood and Marquee Brow. The Old Nore Beds showed a significant decrease in intact dry density from Old Boat Corner to Marquee Brow correlating with the decrease in dissolution pipe concentration. Also, the Splash Point Beds exposed near the surface in the foundations of Coldean Lane bridge (no dissolution pipes) had a higher mean intact dry density than where they were exposed at a similar depth at Old Boat Corner.

Dissolution pipes were completely absent from Patcham Court Farm and Braeside Avenue cuttings (Figs 7.7 & 7.8) and the Belle Tout, Cuckmere and Haven Brow Beds all had similar intact dry densities and there was no reduction in intact dry density with stratigraphic level as seen in contrast to the Newhaven Chalk of Benfield Hill which also lacked dissolution pipes.

The above observations suggest that the presence of dissolution pipes has a reductive effect on intact dry density and this is related to the concentration of dissolution pipes and the proximity of the chalk in question to the pipes. It would seem, however, that this effect is superimposed upon pre-existing trends in intact dry density related to stratigraphy and structure. A knowledge of the expected distribution of dissolution pipes prior to design and construction is of great importance in engineering.

Association of putty chalk

The term 'putty chalk' was used by Higginbottom (1966) to describe mechanically disaggregated chalk silt and he inferred that it does not occur naturally *in situ*. Wakeling (1970) introduced Grade VI chalk to the Ward *et al.* (1968) Mundford grading system; the dominant chalky silt matrix in this material has been colloquially referred to as 'putty chalk'. On the Brighton Bypass, *in situ* putty chalk, occurred in several situations:

1. as very soft to firm off-white to light purplish grey chalk silt forming a 'jacket' around all dissolution pipes which varied in thickness from a few millimetres to 2 m;
2. as a weathering feature on fracture surfaces varying in thickness from a thin film to approximately 30 mm;

3. as off-white to cream granules up to 5 mm in diameter or as massive hemispherical nodes associated with palaeokarst surfaces developed on sheet flints (Lamont-Black 1995).

'Putty jackets' around dissolution pipes and 'putty' on fracture surfaces were usually found together but the distribution of palaeokarst horizons with associated 'putty chalk' and fracture surface 'putty' was independent from dissolution pipes.

The development of *in situ* putty chalk was greatest in Red Hill cutting where dissolution pipe development was most intense and both putty jackets and fracture surface putty reduced with increasing depth below ground surface. Figure 7.18 calculates the percentage of *in situ* putty chalk present in Red Hill cutting. Using a typical putty jacket thickness of 0.5 m, an average fracture spacing of 100 mm and taking an average of 5 mm of putty on fracture surfaces (typical values as seen in Red Hill cutting) a maximum figure of 19.4% *in situ* putty is derived. This had important implications for the suitability and volume of chalk available for earthworks.

Palaeokarst horizons

Horizons of palaeokarst developed above laterally extensive bedding parallel and sub-parallel sheet flints were encountered at Round Hill, Red Hill, Patcham Court Farm Old, Boat Corner, Great Wood and Marquee Brow cuttings. Their chief features are 'dissolution tubules' which take the form of irregular, anastomosing, bifurcating-upwards voids in the chalk; 1–50 mm in diameter and can be over 1500 mm in length which are choked with granules of putty chalk (Lamont-Black 1995).

The most prominent palaeokarst horizon on the Brighton Bypass extended over 2 km between Old Boat Corner and Marquee Brow, forming beneath and within the Old Boat Corner Hardground. Tests on this horizon revealed a mean reduction of 17% by mass dissolution of the chalk compared with unaffected parts of the hardground. Of the remaining material, up to 50% by mass was finer than 2 mm. This material was found to be unsuitable for earthworks. The engineering implications of fine grained and 'putty chalk' are considered in the next section. Several other palaeokarst

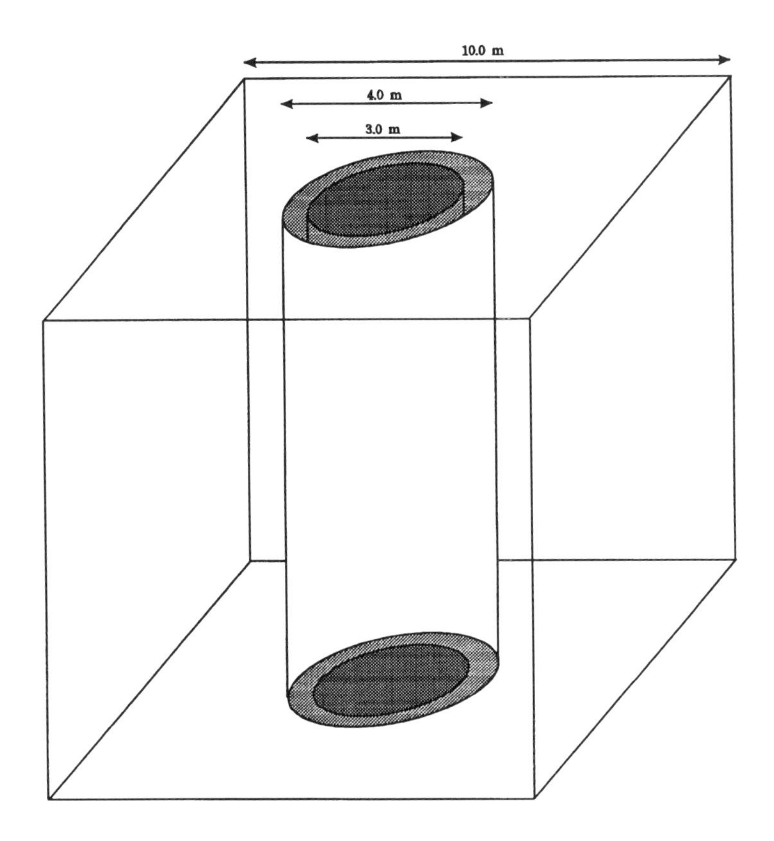

Volume of chalk mass = 1,000 m3
Volume of pipe infill = 71 m3
Volume of putty chalk jacket = 55 m3

If average fracture spacing = 0.10 m then average block volume = 0.001m3

Number of blocks = 1,000 - (55+71)/ 0.001 = 874,000

If putty thickness = 5 mm then volume of putty on each block = 1.43 x 10-4 m3
therefore total percentage by volume of putty in the chalk mass is

8.0% for 20% of all fractures affected
12.7% for 50% of all fractures affected
19.4% for 100% of all fractures affected

Fig. 7.18. Calculation of the volume of *in situ* putty chalk in a volume of rock in a dissolution pipe terrain (data taken from field observations at Red Hill).

horizons were developed always overlying sub-horizontal sheet flints which had developed on the flanks of the Hollingbury Dome.

In Old Boat Corner and Great Wood cuttings (Figs 7.10 & 7.11) several nodes of putty chalk were preserved overlying the sub-horizontal sheet flint upon which the main palaeokarst horizon was developed. These hemispheroidally shaped features were 2 to 5 m in diameter and 1 to 2.5 m in height and preserved relicts or ghosts of dissolution tubules developed in the palaeokarst horizon. The entire mass had been degraded to putty chalk. These nodes were associated with hemispheroidal/ellipsoidal concentric fracture sets and have been interpreted as fossil pingos (Lamont-Black 1995).

Caves

Dissolution cavities within the rock mass were encountered in Marquee Brow cutting, at the eastern end of Great Wood cutting and also at the far western end of the Brighton Bypass adjacent to the Kingston Interchange. Unlike dissolution pipes, these features were not obviously connected to the surface and were not formed by gradual subsidence of overlying sediments. Filled with silty clays to coarse sands, they showed an abundant variety of sedimentary bedforms such as parallel laminae, cross laminae, climbing ripples, scour features and imbricate bedding structures; indicative of deposition from rapid unidirectional currents.

These features were up to 6 m in length and occurred to a depth of 23 m below the ground surface. Their maximum development in size and number occurred above sub-horizontal sheet flints (Figs 7.12 & 7.14). They have been interpreted as filled caves of a system of palaeoswallow holes and are genetically linked to the development of palaeokarst horizons (Lamont-Black 1995). Similar features have been found in chalk with sheet flints adjacent to major dry valleys and river valleys in Sussex and in northern France. The presence of laterally persistent sub-horizontal sheet flint layers seems a pre-requisite for their development.

Being invisible from the surface and filled with sediments, chalk caves would prove difficult to detect by aerial photography or common geophysical methods. Further, their independence from dissolution pipes and their structural associations might mean their being overlooked at the site investigation stage.

Implications of dissolution features for rock mass characteristics
The thicknesses and distribution of weathering grades were closely influenced by the presence of dissolution pipes and clay-with-flints cover and rock mass quality was consistently poorer in dissolution pipe domains (Figs 7.6 & 7.7). In respect to individual dissolution pipes rock mass gradings were reduced compared to background profile to distances of up to 3 m from the edge of dissolution pipes. It has been suggested (Edmonds, pers. comm. 1997) that distortions of background weathering profiles could be identified by careful borehole drilling and used as a tool in the identification of dissolution pipes in specific locations.

Palaeokarst horizons also had an interruptive effect on weathering and rock mass grade profiles. Figures 7.9 and 7.10 show sharp boundaries between fracture spacing Grades 2 and 3 across the palaeokarst horizon with occurrences of fracture aperture Grades B and C present. This picture is, however, somewhat reversed in Marquee Brow as the remnants of the Old Boat Corner Hardground form a massive band of rock (spacing Grade 1) where the palaeokarst horizon is less well developed (Fig. 7.11)

Observed effects of rock mass and material variations

So far a brief summary has been presented of how geological components, interpreted within the scientific frameworks of stratigraphy, sedimentology, tectonics and geomorphology have interacted to create a variety of rock mass and rock material conditions on the Brighton Bypass. These conditions will be reviewed in the light of the engineering operations carried out as part of the construction process.

Earthworks

Variations in intact dry density, the presence of dissolution pipes, the occurrence of *in situ* putty chalk affected many different operations and especially affected earthworks.

The use of intact dry density in chalk classification and specification schemes for earthworks and foundations has been widespread (Masson 1973; Jenner & Burfitt 1975; Clayton 1977*a,b*; Ingoldby & Parsons 1977; Lord *et al.* 1994) and it is recognized as the most important index influencing the mechanical and engineering behaviour of chalk (Lamont-Black & Mortimore 1998). A detailed knowledge of intact dry density, prior to construction, is therefore crucial.

Chalk intact dry density was routinely measured on the Brighton Bypass and compared with field determined hardness using the scheme of Mortimore *et al.* (1990). The chalk varied in hardness from extremely soft to extremely hard (i.e. covering the full range of chalk hardnesses found anywhere in the country). The very hard and extremely hard chalks were only associated with reprecipitation of calcite associated with dissolutional processes and were rare. The geological factors that affected intact dry density include stratigraphy, sedimentology, tectonic setting and dissolutional weathering processes. Some of the issues arising from the earthworks operation are discussed below.

Trafficability and fill instability

There are many examples of instability or 'spongy' conditions occurring during the construction of chalk embankments, particularly when using soft chalks (Lewis & Croney 1966; Masson 1973; Clayton 1977*a*). Chalk often occurs at or near its saturation moisture content (SMC) (Cassel 1957; Higginbottom 1966; Lewis & Croney 1966; Jenner & Burfitt 1975) and for soft chalks (high SMC) the fines produced during excavation, transportation, deposition and compaction can be above their liquid limit (Clayton 1977*a*). The situation arises therefore, that softer and softer chalks produce more and more fines which are wetter and wetter thereby increasing the chance of instability during trafficking and compaction (i.e. a kind of negative feedback).

The percentage of fines in chalk fill has been identified as crucial to the development of instability (Rat & Schaeffner 1990) and Puig (1973) showed that above 20% the behaviour of fines controls the behaviour of the fill. Ingoldby & Parsons (1977) advised careful handling of chalk to reduce the chances of instability from the production of fines. Therefore, the occurrence of *in situ* putty chalk is particularly onerous for earthworks in soft chalks. There were numerous cases of instability in embankments and cuttings on the Brighton Bypass and many of these can be attributed to mechanical degradation of low intact dry density chalks. In some situations however, the presence of *in situ* putty chalk was directly responsible because although intact dry density tests on intact material showed it to be unlikely to cause instability on a large scale, instability still occurred.

Instability of cuttings and embankments created trafficability problems, temporary abandonment of haulage routes through cuttings and along embankments experiencing 'spongy conditions' and occasional removal of unstable fill from embankments. A detailed knowledge of intact dry density distributions and weathering effects such as *in situ* putty chalk (also clay fills of fractures) both within and between cuttings would be advantageous in order to predict and thus mitigate against instabilities. A calculation based on data from Red Hill (Fig. 7.18) indicated that even for fracture spacing Grade 3 chalk, the amount of *in situ* fines is very close to the level indicated by Puig (1973) at which the behaviour of the fines controls that of the fill as a whole. This is important because this is before the mechanically aggressive operations of excavation, transportation, deposition and compaction have taken place.

Extra excavation and haulage

The presence of unexpected dissolution pipes and in unexpectedly high concentrations (e.g. 265 features per hectare in Red Hill cutting, Fig. 7.7) required time-consuming selective excavation of sands, clays and clay-with-flints filling dissolution pipes and extra haulage of this material to designated sites. Selective excavation was adopted in certain situations because earthworks cut/fill volume balance meant that material could not be rejected without the potential for an overall shortfall in chalk available for structural fill. Extra haulage combined with an unpredicted pattern of instability and reductions in suitable material available from cuttings previously assessed to be acceptable meant that the construction critical paths developed by contractors prior to construction were either significantly interrupted or unworkable.

Volume considerations

When performing earthworks with chalk, degradation of the

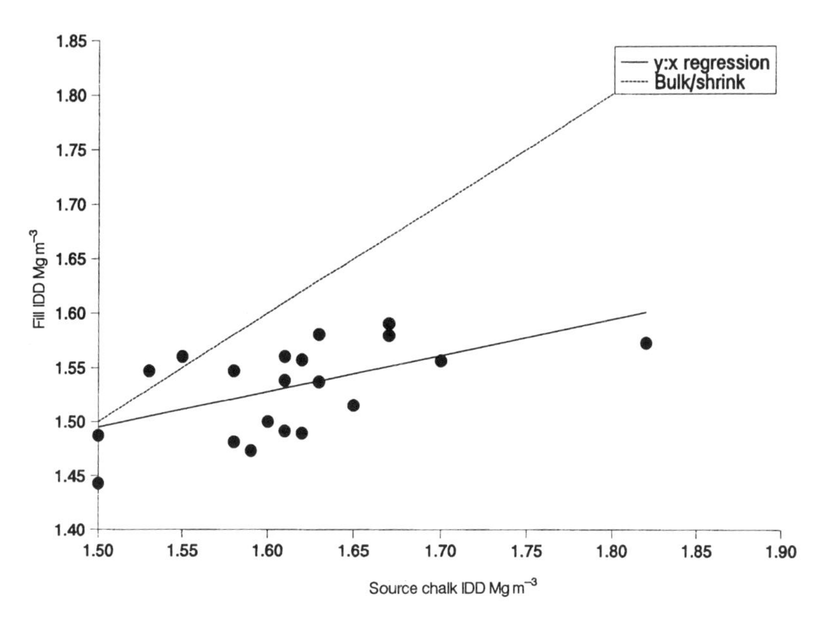

Fig. 7.19. Plot of *in situ* chalk intact dry density versus the density of fill achieved with this material. Note the convergence of the regression line and the 1:1 bulk shrink line at an *in situ* intact dry density of 1.50 Mg m^{-3}.

material and destruction of pore spaces can lead to a reduction in bulk volume. However, spaces created between intact lumps in a chalk fill lead to an increase in volume. When dealing with millions of cubic metres of chalk it is important to know which situation is dominant. Stratigraphic logging of cuttings and measurements of *in situ* chalk intact dry density and dry densities achieved in fill indicated that for the type of compaction used, a chalk intact dry density of about 1.50 Mg m^{-3} indicates the transition between bulking and shrinking (Fig. 7.19).

The palaeokarst horizon described above was associated with a large reduction in material volume owing to its unsuitability because of excessive chalk fines and also because the chalk had simply been dissolved. Between Old Boat Corner and Marquee Brow the Old Boat Corner Hardground had a mean intact dry density of 1.76 Mg m^{-3} which was reduced to a mean of 1.53 Mg m^{-3} for the solid material where karstified. This reduction in intact dry density combined with the loss of mass and the degradation to putty chalk lead to a maximum reduction in solid chalk of 64% associated with the palaeokarst horizon. Such a loss of material could, in some situations, prove critical for cut and fill volume balances.

Excavatability
Fracture style significantly influenced fracture spacing and tightness. Inclined, marl-smeared conjugate shear fractures, characteristic of the Newhaven Chalk, gave rise to large and often tight fractures especially in the cores of cuttings. Excavation beneath the level of the Old Nore Marl (i.e. the Old Nore Beds) in Benfield Hill required the rock to be loosened first by employing a bulldozer fitted with a ripping hook. Without this technique the excavating plant was unable to gain a purchase on blocks in the rock mass. Once excavated these blocks were unusually large and required further mechanical breakdown at the embankment site.

In the Old Nore Beds of Marquee Brow cutting average fracture spacings reached a maximum of 1820 mm (Fig. 7.11) and gave rise to numerous blocks well over 1 m in diameter. As compaction specifications required that the material was placed in 150 mm layers, considerable mechanical breakdown was required at the emplacement site. Contractors also found that with such large blocks transportation efficiency of dump trucks was reduced. Less than 600 m away in Great Wood cutting, contractors found that transportation efficiency was reduced by the soft and extremely soft chalks of the Peacehaven Beds associated with the palaeokarst horizon and dissolution pipes becoming stuck in dump truck bodies.

The Seaford Chalk was characterized by an orthogonal fracture pattern with spacings always <600 mm and provided ideal conditions for excavation. (Figs 7.8 & 7.9). Also a lack of dissolutional effects and a mean intact dry density 1.63 $Mg\,m^{-3}$ resulted in no instabilities and a very rapid rate of progress was achieved using this chalk. In some instances this chalk was mixed with poorer quality chalks from dissolution pipe terrains in order to reduce the occurrences of instability.

Slope instability

Occurrences of slope instabilities were quite rare on the Brighton Bypass. Dissolution pipes containing soft to stiff clays and loose to medium dense sands had to be selectively excavated and backfilled with chalk or concrete where they occurred in 1 : 1 cutting slopes.

Rock mass instabilities were associated with inclined conjugate shear fractures of the Newhaven Chalk in Benfield Hill and Marquee Brow cuttings. The failure of a temporary excavation in Marquee Brow (Fig. 7.11) occurred as a wedge failure along inclined conjugate shear fractures which were opened due to stress relief next to the steep valley slope of Coldean. Subsequent analysis of this failure showed that an appreciation of the rock mass conditions in the design of temporary works could be critical in preventing the failure of large volumes of unstable rock (Lamont-Black 1995).

The failure of a low cliff face along an inclined conjugate shear fracture in Benfield Hill resulted in the 'writing off' of an excavator.

Foundations

Foundation design on the Brighton Bypass involved either the use of piles in dry valleys with coombe deposits and deeply weathered chalk, or the use of spread foundations for structures in cuttings. An exception to this was the Dyke Road Bridge in Red Hill cutting (Fig. 7.7) which was designed with piled foundations owing the presence of dissolution pipes. All piled and spread foundations on the Brighton Bypass performed adequately.

A number of geophysical surveys were conducted to locate dissolution pipes at Old Boat Corner and Red Hill prior to the construction of bridges here. Comparison of ground conditions revealed after top soil stripping (Lamont-Black 1995) with the results of very detailed electromagnetic, seismic and galvanic resistivity surveys (Mortimore *et al.* 1990) made prior to construction showed unacceptable ambiguities and uncertainties in identifying dissolution pipes by geophysics in these settings. Uncertainties about dissolution pipes at Old Boat Corner at the time of writing the contract were catered for by the insertion of a clause to allow for further investigation during construction. In situations where risk from dissolution pipes is high but knowledge about their precise locations is low this approach, where possible, seems a sensible one.

Although not specifically a problem for foundations on the Brighton Bypass, the development of palaeokarst horizons could be potentially hazardous for foundations. Shallow foundations will often rely on trial pit excavations to assess the rock conditions. Palaeokarst horizons on the Brighton Bypass included large-scale dissolution of chalk and the nodes of putty/silt chalk. Such horizons were, in places, at depths of greater than 5 m with intact Grade B3 chalk above (Figs 7.10 & 7.11). Shallow trial pits could easily miss features such as these.

Conclusions

The ambition of adopting a 'whole rock approach' in studying engineering geology of the Chalk, was to demonstrate the insight that can be gained through considering many different aspects of geology which in combination produce the material and mass characteristics that are important to engineering. This approach has been of growing importance in recent years; Crighton *et al.* (1988), for example, observed that stratigraphic control on the Channel Tunnel '... has provided a more precise framework for correlating geological conditions and observed variations in geotechnical parameters...' and Mortimore (1993) suggested a resolution of stratigraphy to one metre to be both achievable and necessary for engineering operations.

Stratigraphy is the fundamental scientific framework. In the study presented above detailed litho- and biostratigraphic logging of chalk sections on the Brighton Bypass enabled correlation of chalk sections along the route (Fig. 7.4) and most of the relationships between geomorphology, intact dry density and earthworks performance could not have been reached without a detailed knowledge of stratigraphy. Likewise, identification of lateral and vertical variations in stratigraphy enabled the identification of tectonic structures and their influence on chalk sedimentology. These structures affected intact dry density, fracture style and the development of dissolution pipes in synclines.

It is appreciated that an index like intact dry density is controlled by a number of different factors and the isolation of one factor from another has not been specifically achieved in the present study. For example intact dry density was lowest at Old Boat Corner where dissolution pipes were preserved. It might be expected that intact dry density would have been reduced most at Red Hill where dissolution pipe concentration was approaching twice that of Old Boat Corner. However, it was observed that intact dry density reduced in all directions from Old Boat Corner, that is, passing from a site of a condensed sequence over a syn-sedimentary high into syn-sedimentary troughs.

In order to achieve a more statistically significant isolation of the factors affecting indices such as intact dry density, multivariate analyses of a much larger dataset should be considered. However, the identification of critical combinations of stratigraphy, tectonics, sedimentology, geomorphology and weathering is leading towards the development of ground models and given the enormous range of conditions that can be represented within each framework, the development of a 'whole rock approach' via ground modelling is appropriate for engineering.

The eastern half of the Brighton Bypass (Contract 3, Fig. 7.2) presented an interesting combination of factors upon which a ground model could be developed. The westerly extension of the Caburn syncline preserved Palaeocene sediments thus producing dissolution pipes at Old Boat Corner and Great Wood with their associated effects on intact dry density and *in situ* putty chalk. The Hollingbury Dome, in particular, was responsible for creating a condensed section, altering the lithostratigraphy of the Newhaven Chalk, thus changing the dominant style of fracturing and radically altering the rock mass conditions. Sub-horizontal sheet flints were crucial in the development of palaeokarst horizons and palaeos-wallow holes and their suggested origin as silicified décollement surfaces on the flanks of the Hollingbury Dome would genetically link basement tectonics to karstification. The presence of fossil pingos developed on this horizon can be similarly, if tentatively, linked to basement tectonics.

Therefore, the effects of Sub-hercynican block faulting relating to the Hollingbury Dome and Caburn syncline (Mortimore *et al.* 1990) can be traced through time affecting sedimentology, fracture development, dissolution pipes, palaeokarst horizons, caves and periglacial effects, all of which impacted on earthworks and foundations operations.

Acknowledgements

The research upon which the above is based was carried out with the support of L.G. Mouchel & Partners Ltd. and The Engineering and Physical Sciences Research Council and this is gratefully acknowledged. Permission to use results was given by The Highways Agency.

References

Bristow, C. R., Mortimore, R. N & Wood, C. J. 1997. Lithostratigraphy for mapping the Chalk of southern England. *Proceedings of the Geologists' Association*, **108**, 295–315.

Carter, P. G. & Mallard, D. J. 1974. A study of the strength, compressibility and density trends within the Chalk of south-east England. *Quarterly Journal of Engineering Geology*, **7**, 43–55.

Cassel, F. L. 1957. *In*: Discussion on soil properties and their measurement. *Proceedings of the 4th International Conference on Soil Mechanics and Foundation Engineering*, **3**, Butterworth Scientific Publications, London, 94–95.

Clayton, C. C. 1986. The chemical environment of flint formation in Upper Cretaceous chalks. *In*: Sieveking, De, G. & Hart, M. B., (eds.) *The Scientific study of Flint and Chert*, Proceedings of the 4th International Flint Symposium, Brighton, 1983. Cambridge University Press, 43–54.

Clayton, C. R. I. 1977*a*. Chalk in earthworks – Performance and Prediction. *Highway Engineer*, **24**, 14–20.

——— 1977*b*. Some properties of remoulded chalk. *In*: *Proceedings of the 9th International Conference on Soil Mechanics*, Tokyo, **1**, 65–68.

——— 1983. The influence of diagenesis on some index properties of chalk in England. *Géotechnique*, **33**, 225–241.

——— & Matthews, M. C. 1987. Deformation, diagenesis, and the mechanical behaviour of chalk. *In*: Jones, M. E. & Preston, R. M. F. (eds.). *Deformation of Sediments and Sedimentary Rocks.*

Geological Society, London, Special Publications, **29**, 55–62.
Davies, G. M. 1929. Field Meeting at Worms Heath (Report by the director). *Proceedings of the Geologists' Association*, **14**, 385–387.
De Bruijn, R. G. M. 1983. Some considerations on the factors that influence the formation of solution pipes in chalk rock. *Bulletin of the International Association of Engineering Geology*, **28**, 141–146.
Dowker, G. 1866. On the junction of the Chalk with the Tertiary Beds in East Kent. *Geological Magazine*, **3**, 210–213.
Edmonds, C. N. 1983. Towards the prediction of subsidence risk upon the Chalk outcrop. *Quarterly Journal of Engineering Geology*, **16**, 261–266.
———, Green, C. P. & Higginbottom, I. E. 1987. Subsidence hazard prediction for limestone terrains, as applied to the English Cretaceous Chalk. *In*: Culshaw, M. G., Bell, F. G., Cripps, J. C. & M. O'Hara, (eds) *Planning and Engineering Geology*. Geological Society, London, Engineering Geology Special Publications, **4**, 125–131.
Fletcher, T. P. 1978. *Lithostratigraphy of the Chalk (Ulster White Limestone) in Northern Ireland.* Institute of Geological Sciences Report, **77/24** HMSO, London.
Ford, T. D. 1989. Palaeokarst of Britain. *In:* Bosak, Ford, D. C., Gazek & Horafel, I. (eds) *Palaeokarst a Systematic and Regional Review*. Developments in Earth Surface Processes, 1. Elsevier, Amsterdam, 51–70.
Gale, A. S. 1980. Penecontemporaneous folding, sedimentation and erosion in Campanian Chalk near Portsmouth, England. *Sedimentology*, **27**, 137–151.
Gaster, C. T. A. 1951. The stratigraphy of the Chalk of Sussex. Part IV East Central area between the valley of the Adur and Seaford with zonal map. *Proceedings of the Geologists' Association*, **62**, 31–64.
Hancock, J. M. 1975. The petrology of the Chalk. *Proceedings of the Geologists' Association*, **86**, 499–535.
——— 1989. Sea level changes in the British region during the late Cretaceous. Presidential Address, May, 1987. *Proceedings of the Geologists' Association*, **100**, 565–594.
——— & Scholle, P. A. 1975. Chalk of the North Sea. *In:*

Woodland, A. W. (ed.). *Petroleum and the Continental Shelf of North-west Europe, 1 Geology*. Applied Science Publishers, London.

Hancock, P. L. 1985. Brittle microtectonics: principles and practice. *Journal of Structural Geology*, **7**, 437–457.

Henry, F. D. C. & Drummond, P. V. O. 1973. *The geology along two tentative routes for the proposed A27 trunk road between Kingston New Barn, Old Shoreham, and Lewes Road, Coldean in the county of Sussex and an appraisal of the problems therewith.* Unpublished, Report, Brighton Polytechnic.

Higginbottom, I. E. 1966. The engineering geology of chalk. *In: Proceedings of the Symposium on Chalk in Earthworks and Foundations*, April 1965. Institution of Civil Engineers, London.

——— & Fookes, P. G. 1970. Engineering aspects of periglacial features in Britain. *Quarterly Journal of Engineering Geology*, **3**, 85–117.

Ingoldby, H. C. & Parsons, A. W. 1977. The classification of chalk for use as fill material. *Transport and Road Research Laboratory, Laboratory Report* 806.

Jenner, H. N. & Burfitt, R. H. 1975. *Chalk: an engineering material.* ICE Southern Association Meeting, Brighton Polytechnic, 6th March, 1975. Paper with limited distribution.

Jones, D. K. C. 1980. The Tertiary evolution of south-east England with particular reference to the Weald. *In:* Jones, D. K. C. (ed.) *The Shaping of Southern England.* Institute of Geographers Special Publication, **11**, Academic Press, 13–48.

Jukes-Browne, A. J. 1906. The clay-with-flints: its origin and distribution. *Quarterly Journal of the Geological Society of London*, **62**, 132–164.

Kennedy, W. J. & Juignet, P. 1974. Carbonate banks and slump beds in the Upper Cretaceous (Upper Turonian – Santonian) of Haute Normandie, France. *Sedimentology*, **21**, 1–42.

Kirkaldy, J. F. 1950. Solution of the Chalk in the Mimms valley, Herts. *Proceedings of the Geologists' Association*, **61**, 219–224.

Lamont-Black, J. 1995. *The engineering classification of chalk with special reference to the origins of fracturing and dissolution.* PhD thesis, University of Brighton (2 volumes).

——— & Mortimore, R. N. 1998. A recommended method for the

determination of the intact dry density of irregular chalk lumps: Implications for natural variability. *Quarterly Journal of Engineering Geology*, **29**, 293–318.

Lewis, W. A. & Croney, D. 1966. The properties of chalk in relation to road foundations and pavements. In: *Symposium on Chalk in Earthworks and Foundations*, Institution of Civil Engineers, London, 27–41.

Lord, J. A., Twine, D. & Yeow, H. 1994. *Foundations in Chalk.* Funders Report 13, CIRIA Project Report, 11.

Masson, M. 1973. Petrophysique de la Craie. In *La Craie*, Bulletin des ponts et chaussées. Spécial **5**, 23–48.

Meigh, A.C. & Early, K. R. 1957. Some physical and engineering properties of Chalk. *Proceedings of the 4th International Conference on Soil Mechanics and Foundation Engineering* **1**, Butterworth Scientific Publications, London, 68–73.

Mimram, Y. 1977. Chalk deformation and large-scale migration of calcium carbonate. *Sedimentology*, **24**, 333–360.

Mortimore, R. N. 1979*a*. The engineering domains and classification of chalk in relation to Neolithic flint mining with special reference to Grimes Graves, England and Rijkholt-St. Geertruid, Holland. *Proceedings of the 3rd International Symposium on Flint*, **6**, 24–27 May, 1979, Maastricht, p. 30–35.

——— 1979*b*. *The relationship of Stratigraphy and Tectonofacies to the Physical Properties of the White Chalk of Sussex*. PhD. thesis (5 Volumes). CNAA, Brighton Polytechnic.

——— 1983 The stratigraphy and sedimentation of the Turonian–Campanian in the Southern Province of England. *Zitteliana*, **10**, 27–40.

——— 1986. Stratigraphy of the Upper Cretaceous White Chalk of Sussex. *Proceedings of the Geologists' Association*, **97**, 97–139.

——— 1990. Chalk or chalk. In: *Chalk, Proceedings of the International Chalk Symposium, Brighton 1989*. Thomas Telford, London, 15–45.

——— 1993. Chalk water and engineering geology. *In*: Downing, R., Price, A., M. & Jones, G. P. (eds) *The Hydrogeology of the Chalk of North-West Europe*. Clarendon, Oxford, 67–92.

———, Argent, K., Caillard, P., Snook., P.G., Smith, A. J., Tracey, N., Holliday, J. K. & Honeyman, W. N. 1990. Geophysical surveys over solution pipes and Neolithic mines in the Chalk of the South Downs, Sussex, England. *Cahiers du Quaternaire No. 17 - Le Silex de sa Genese a L'Outil.* Actes du V^0 Colloque international sur le silex.

——— & Fielding, P. M. 1990. The relationship between texture, density and strength of chalk. *In*: *Chalk, Proceedings of the International Chalk Symposium, Brighton, 1989.* Thomas Telford, London. 109–132.

——— & Pomerol, B. 1987. Correlation of the Upper Cretaceous White Chalk (Turonian–Campanian) in the Anglo-Paris Basin. *Proceedings of the Geologists' Association*, **98**, 97–143.

——— & ——— 1991. Upper Cretaceous tectonic disruptions in a placid Chalk sequence in the Anglo-Paris Basin. *Journal of the Geological Society, London*, **148**, 391–404.

———, ——— & Lamont-Black, J. 1996. Examples of structural sedimentological controls on chalk engineering behaviour. *In*: Harris, C. S., Hart, M.B., Varley, P. M. & Warren, C. D. (eds) *Engineering Geology and the Channel Tunnel*, Thomas Telford, London, 436–443.

Prestwich, J. 1854. On some swallow holes on the chalk hills near Canterbury. *Quarterly Journal of the Geological Society of London*, **10**, 222–224.

——— 1855. On the origin of sand and gravel pipes in the Chalk of the London Tertiary district. *Quarterly Journal of the Geological Society of London*, **11**, 64–68.

Puig, J. 1973. Problèmes de terrasement dans la Craie. In: *La Craie*, Bulletin de Liaison de Laboratoires des Ponts et Chaussées, Spécial **5**, 81–98.

Rat, M. & Schaeffner, M. 1990. Classification of Chalks and conditions of use in embankments. *In*: *Chalk, Proceedings of the International Chalk Symposium, Brighton, 1989.* Thomas Telford, London, 425–428.

Spink, T. W. & Norbury, D. R., 1990. The engineering geological description of chalk. *In*: *Chalk, Proceedings of the International Chalk Symposium, Brighton, 1989.* Thomas Telford, London, 153–160.

Thorez, J. Bullock, P. Catt, J. A. & Weir, A. H. 1971. The petrography and origin of deposits filling solution pipes in the chalk near South Mimms, Hertfordshire. *Geological Magazine*, **108**, 413–423.

Wakeling, T. R. M. 1970. A comparison of the results of a standard site investigation against a detailed geotechnical investigation in the Chalk at Mundford, Norfolk. *In*: *Proceedings of the Conference on In Situ Investigations in Soils and Rock*. British Geotechnical Society, London, 1969, 17–22.

Walsh, P. T, Elder, G. A., Edwards, B. R. Urbani, D. M., Valentine, K. & Soyer, J. 1973. Large scale surveys of solution subsidence deposits in the Carboniferous and Cretaceous limestones of Great Britain and Belgium and their contribution to an understanding of the mechanisms of karstic subsidence. *In: Proceedings of the International Association of Engineering Geology Symposium, Hanover, 1973*. T2.A1–T2.A10.

Ward, W. H., Burland, J. B. & Galloid, R. W. 1968. Geotechnical assessment of a site at Mundford, Norfolk, for a large proton accelerator. *Géotechnique*, **18**, 399–431.

West, G. & Dumbleton, M. J. 1972. Some observations on swallow holes and mines in the Chalk. *Quarterly Journal of Engineering Geology*, **5**, 171–178.

Wilmot, R. D. & Young, B. 1985. Aluminite and other aluminium minerals from Newhaven, Sussex: the first occurrence of Nordstrandite in Great Britain. *Proceedings of the Geologists' Association*, **96**. 47–52.

Woodcock, N. H. 1994. *Environmental Geology in Britain and Ireland.* UCL Press, London.

8 Hills of waste: a policy conflict in environmental geology

J. Murray Gray

Summary

- Landfilling currently accounts for *c*. 85–90% of household and commercial waste disposal in the UK.
- There is a shortage of landfill voids, particularly in SE England, and this has led to a large number of proposals to overfill and updome existing or approved landfill sites, or to create new waste hills on greenfield sites.
- While 'landraising' is supported by the Environment Agency since it allows greater flexibility in the location of waste sites, places them further from groundwater resources and permits easier site design and monitoring, some Local Planning Authorities are attempting to adopt policies to discourage landraising because of the visual impact both during construction and following restoration.
- 22 recent planning applications for major landraise sites in England are discussed and a more geomorphologically sensitive approach and the development of model planning policies are proposed.

Household and commercial waste in the UK has traditionally been disposed of into disused pits and quarries, thus playing the dual role of disposing of waste and restoring the pits. Landfilling currently accounts for around 85–90% of waste disposal in England and Wales (Department of Environment & Welsh Office 1995) and although this is likely to decline as local authorities and others adopt and implement the waste hierarchy in which waste reduction, reuse and recovery (recycling, composting and energy) are all given priority over disposal, there will inevitably be at least a medium-term requirement for significant landfill void space.

There is, however a growing shortage of landfill sites. A recent questionnaire survey of English shire counties indicated that 60%

Fig. 8.1 The Packington landraising site in Warwickshire as seen from the Forest of Arden golf course in summer 1996, rising in the distance to over 50 m above the surrounding countryside.

of the authorities have landfill void shortages in all or part of their areas (Gray 1997). There is a particular problem in SE England, where two recent reports have warned that the area could run out of landfill sites within 15 years (SERPLAN 1994; SEWRAC 1994). One of the main problems is London's waste which traditionally has been taken to landfill sites in the surrounding counties, particularly Essex.

This national shortage of landfill void space, together with the desire of waste disposal companies to maximize the capacity of their landfill sites, has led to a large number of recent proposals to overfill and updome landfill sites, or to create new waste hills on greenfield sites. These developments which raise levels above those previously existing are referred to as 'landraising'. The resultant hills of waste, such as that currently rising to 50 m above the surrounding countryside at Packington in Warwickshire (Fig. 8.1), are probably the most significant constructive landforms being created by humans today, yet geomorphologists have scarcely been involved in their planning or design, nor in the framing of policy aimed at controlling their development. The issue of landraising is also giving rise to an important policy conflict in environmental geology.

Policy issues

Government guidance on landraising is contained in *Planning Policy Guidance Note 23 (PPG 23)* (1994), Para 5.10 which states that: 'Landraising may... provide an appropriate method of disposal, particularly in areas where insufficient landfill opportunities exist, provided it can be designed to blend in with the surrounding landscape, and the impact of disposal operations on the environment are acceptable'.

This requirement for landraising schemes to 'blend in' with the landscape has been a crucial one in deciding whether landraising proposals are acceptable, and is discussed later in this chapter. The government's recent Waste White Paper states that landraising schemes will become 'more prevalent. Landraising also offers advantages in improved leachate control, but gas collection is more difficult than with conventional, below-ground sites' (Department of the Environment & Welsh Office 1995, p. 64).

The Environment Agency (by implication through its predecessor the National Rivers Authority (1994)) favours landraising since it allows wastes to be placed further from the groundwater, is easier to design and monitor, and allows greater flexibility in the location of waste disposal sites, since they can be sited away from vulnerable groundwater supplies (i.e. they are more 'footloose'). This is in contrast to landfill sites utilizing quarries whose locations are predetermined and whose floors are often close to, or even below, the local water table.

On the other hand, several Local Planning Authorities have been attempting to introduce policies to restrict or prevent the development of landraising within their areas. Hampshire's Minerals and Waste Local Plan (Hampshire County Council 1993, Para 6.37), for example, states that there is a 'presumption against landraising': 'The County Council considers landraising generally to be the most unacceptable form of waste disposal because of the adverse environmental impact it has on otherwise undisturbed land and the permanent change to the landscape that it causes.' It therefore included Policy 47: 'The County Council will not grant permission for the disposal of waste by land raising unless it can be clearly demonstrated that there is no satisfactory alternative way of meeting the need for the disposal of that waste.'

However, following the Public Inquiry into the plan, the Inspector, in his report, considered the presumption against landraising to be contrary to government policy and recommended the deletion of the policy. In response, the County Council Members were 'particularly concerned that the Inspector considered there was no justification for a presumption against landraising in Hampshire' and 'considered that a substitute policy should be drawn up to ensure adequate control over land-raising' (Hampshire County Council 1996, Para. 31).

Similarly, Kent County Council's *Waste Local Plan* (Kent County Council 1993) states that: '...in rural areas landraising leads to permanent change in the character of the countryside and so works against the strategic objective of protecting the countryside for its own sake'.

Essex County Council's Draft *Waste Management Strategy* (Essex County Council 1994) is unequivocal: 'Deposition of waste above original ground level, known as landraising, either by increasing heights on existing tips or on virgin land is unacceptable'.

There is therefore a serious policy conflict between the Local Authority planning policies which are seeking to limit or prevent the use of landraising, and the Environment Agency's groundwater protection policies which are seeking to encourage it. This research project was therefore initiated with the aim of exploring the policy conflict and seeking a balanced assessment of the possible ways forward.

Methodology

This study of waste disposal by landraising, whose preliminary results are described in this chapter, has four stages.

1. Full review of current national, regional, structure and local plan policies and those of other organizations related to landraising.
2. Detailed assessment of the issues involved in all recent (post-1990) planning applications for major (over 10 m post-settlement) landraising schemes in England.
3. Review of the issues of landform 'blending'/'incongruity' in relation to natural landscapes, and recommendations on landraise siting and landform design in different landscapes.
4. Development of model policies on landraising for incorporation in development plans, waste local plans and informal policy guidance.

Work so far has concentrated on stage 2 and up to December 1996, 22 major landraising planning applications submitted to English Local Authorities since 1990 had been studied. The sites included in the study have been identified from notifications in the national waste technical or planning press or by contact with individual planning authorities.

Information about each site was collected on visits to the Local Authority offices, including study of committee reports, environmental statements, inspectors' reports and files of correspondence. A field visit has been undertaken to each site. Details of the sites studied are given in Table 8.1, while Fig. 8.2 shows the location of the sites. The study has also included attendance at 2 Public Inquiries (Hardwick Airfield, Norfolk, January/February 1993, and Rivenhall Airfield, Essex, March 1995). Informal discussions about attitudes to landraising have also taken place with Council planners and waste disposal companies.

The study has suggested that at least three types of landraising proposals can be identified. These are defined as follows, and each site is so classified in Table 8.1.

- *Primary landraising* (p) involves waste disposal on unexcavated (greenfield or brownfield) land, but usually including an initial shallow excavation to provide capping materials and soils.
- *Secondary landraising* (s) involves the filling of a disused or planned mineral void to levels above those reasonably needed for settlement and drainage purposes.

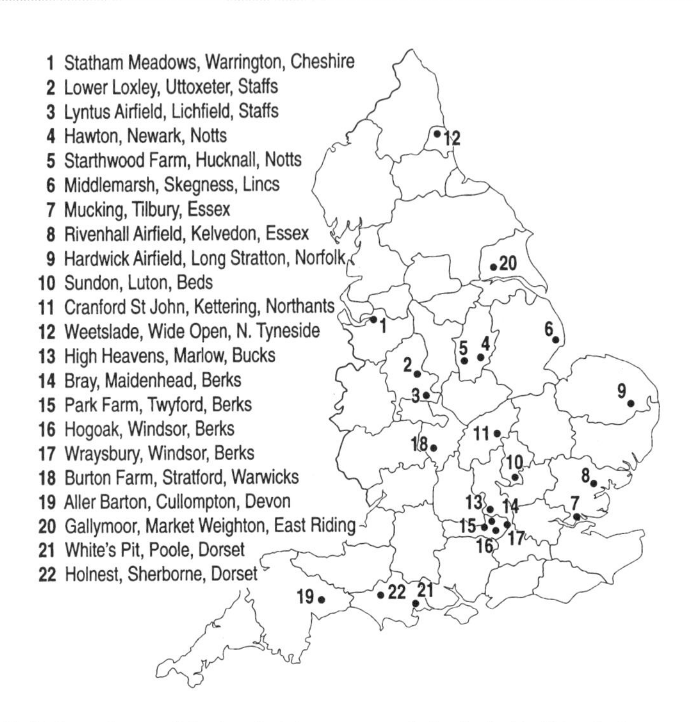

Fig. 8.2. Locations of the landraising proposals included in this study.

- *Tertiary landraising* (t) involves further raising of levels on already built or approved landfill sites.

It should be noted that the use of waste to rebuild hills or ridges that have been excavated for minerals (e.g. eskers for aggregate) is not regarded as landraising, but as restoration of the previously existing contours. This approach has been used in Stockholm (Morfeldt 1993) to reconstruct esker morphology using inert demolition waste and could be used to beneficial effect at several sites in the UK.

Outcome of the planning applications

As can be seen from Table 8.1, most of the planning applications have been refused permission, often at least partly due to visual

Table 8.1 Summary data for the 22 landraising site proposals included in this study

No.	Site	WPA	Applicant	Date	Area (ha)	Max. Ht (m)	Max. Grad.	Topography	Type*	Capacity **	Life (yrs)	Decision	Appeal?	Decision	Date
1	Statham Meadows	Cheshire CC	Waste Management Ltd	1992	58	22	1:5	Flat, floodplain	P	5	10	Refuse	Yes	Refuse	1993
2	Lower Loxley	Staffs CC	Leigh Environmental	1993	25	17	1:8	Shallow valley head	P	1	7	Refuse	Yes	Refuse	1994
3	Lyntus Airfield	Staffs CC	HJ Banks & Co Ltd	1994	61	15	1:10	Very shallow valley	P	2	10	Refuse	?		
4	Hawton	Notts CC	Leigh Environmental	1991	48	35	1:6	Flat, floodplain	T	5.3	20	Refuse	Yes	Refuse	1994
5	Starthwood Farm	Notts CC	UK Waste	1992	52	17	1:10	Gently rolling	P	3.6 mte	14	Refuse	Withdrawn		
6	Middlemarsh	Lincs CC	Lincs CC	1993	9	10.5	1:10	Flat, coastal marsh	P	0.3 mte	8	Approve			
7	Mucking	Essex CC	Cory Environmental	1992	287	+15	1:7	Flat, floodplain	T	16	25–33	Refuse	Yes	Refuse	1994
8	Rivenhall Airfield	Essex CC	Blackwater Aggregates	1993	210	18	1:15	Flat, till plain	S	14 mte	23	Non-determined	Yes	Refuse	1995
9	Hardwick Airfield	Norfolk CC	Norfolk CC	1991	57	10	1:23	Flat, till plain	P	1.5	22	Called in	Refuse	1993	
10	Sundon	Beds CC	Beds CC	1990	41	+11	?	Chalk scarp slope	T	?	5	Approve			
11	Cranford St John	Northants CC	Northants CC	1992	21	12.5	1:12	Moderately rolling	S	0.9	10	Approve			
12	Weetslade	N. Tyneside CC	Biffa Waste	1992	60	25	1:3	Gently rolling	P	3	10	Refuse	Yes	Refuse	1994
13	High Heavens	Bucks CC	Bucks CC & Grundon Waste	1991	63	14	1:5	Dry valley head	P	3.3	13.5	Called in	Refuse	1994	
14	Bray	Berks CC	Biffa Waste	1990	89	+8	1:5	Flat, river terrace	T	2.7	12	Refuse	Yes	Refuse	1992
15	Park Farm	Berks CC	Terry Adams Ltd	1990	48	12	1:7	Flat, river terrace	P	1.9	15	Refuse	Yes	Refuse	1991
16	Hogoak	Berks CC	Terramicus/MRM	1990	25	17.5	?	N slope of ridgeline	P	1.5	10	Refuse	No		
17	Wraysbury	Berks CC	ARC Ltd	1991	36	6	?	Flat, floodplain	S	3	12	Refuse	No		
18	Burton Farm	Warwks CC	Burton Farms Ltd	1992	30	17.5	1:5	Shallow valley	P	1.3	16	Refuse	Yes	Refuse	1994
19	Aller Barton Farm	Devon CC	Devon LAWDC	1994	26	13.5	1:12	Gently rolling	P	0.9	9	Refuse	?		
20	Gallymoor	Humberside CC	Humberside CC	1991	27	10	1:12	Flat	P	1.5	20	Approve			
21	White's Pit	Dorset CC	WH White plc	1993	38	+35	1:3	Flat, heathland	T	2.2	9	Refuse	Yes	Refuse	1995
22	Holnest	Dorset CC	Dorset CC	1991	12	12.5	?	Moderately rolling	P	0.4	15–25	Approve			

*P = Primary landraising, S = Secondary, T = Tertiary (see text for definitions)
**Capacity $\times 10^6$ m^3 except where followed by mte which denotes $\times 10^6$ tonnes

impact of the proposed landform. All applications that have gone to an appeal hearing or public inquiry have been refused, and again the landform issue has often been significant in the decision notices.

Only five of the sites in Table 8.1 have so far been approved, and all five have involved situations in which Local Authorities have been responsible for approving their own schemes or those of their 'arms length' Waste Disposal Companies (LAWDC). A sixth site (Site 9) would have been approved by the Local Authority had the Secretary of State not called in the application. This raises the whole issue of deemed consent whereby Councils can grant themselves planning permission, and the Nolan Committee has stated that this issue is included in its current deliberations on Local Authority procedures (Nolan Committee 1996, Para 54). However, the author is aware of several sites where private operators have been granted consent for landraising (e.g. Port Clarence in Teesside, Southleigh in Hampshire and Bedington in south London) and these sites will be included in the completed study.

The main objection to the landraising proposals studied has been that they would be 'alien', 'incongruous' or 'unnatural' in the landscape, or that the final landform would not 'blend in' with the existing topography. Such comments have been made by local objectors, landscape consultants, local authorities, inquiry inspectors and the Secretary of State for the Environment.

This is a particularly important issue in areas of flat topography such as floodplains, river terraces and till plains where objectors have frequently objected to the visual impact of the completed hills or the closure of views. Impact during construction has also been a significant objection. Since the waste is being deposited above ground, the operations are inevitably more visible and in addition the more exposed nature of the sites may mean greater problems with litter, dust, noise and lighting. Where bunds have been proposed to help screen the operations, in turn these have been criticized as also being incongruous.

In support of landraising, applicants have cited need, a particularly potent argument for tertiary landraising as defined above. If there is shortage of voids in an area, but a landfill site already exists, the arguments for building that site higher are quite powerful. In these circumstances, landraising has been difficult to

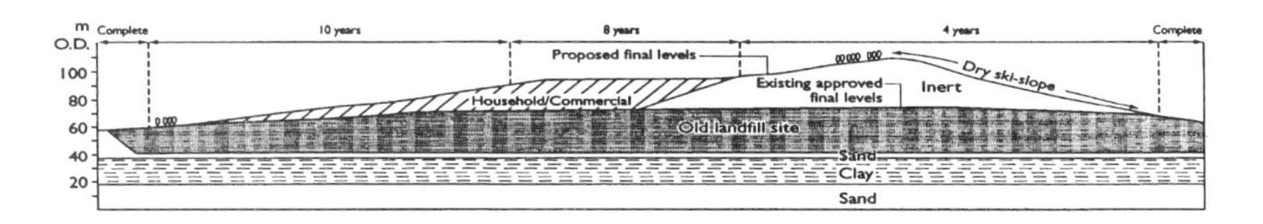

Fig. 8.3. Cross-section through the proposed tertiary landraising site at White's Pit, Poole, Dorset, showing the area of inert waste underlying the proposed dry ski-slope and related infrastructure, and the degradable waste areas forming the remainder of the hill.

argue against because most landfill sites are overfilled anyway to provide a gradient for surface drainage (which reduces rainwater infiltration and thus the generation of leachate), and are built to a suitable height to allow for settlement as the waste degrades. This can amount to 25% of the waste depth even when initially compacted. The argument as to whether higher waste hills are necessary for settlement and drainage purposes or are simply being used to get more waste into the site has been the subject of considerable debate at several public inquiries.

Other arguments used to support landraising have included improved public access in the long term, and provision of recreational uses including summit viewpoints, golf courses and even dry ski-slopes. The latter is illustrated by the planning application at White's Pit, Dorset (Site 21) where inert waste was to form the ski slope section of the hill where it was important that waste settlement should be minimized, and degradable waste was used to form the rest of the hill (Fig. 8.3). Some applicants have also argued that the restored waste hills will improve the existing landscape or screen less attractive landscape elements such as industrial plants or motorways.

These arguments for and against landraising are illustrated below by four case studies.

Statham Meadows, Cheshire

The most vivid language encountered to describe the incongruity of a landraise site, was by an Inspector on the proposal to extend the

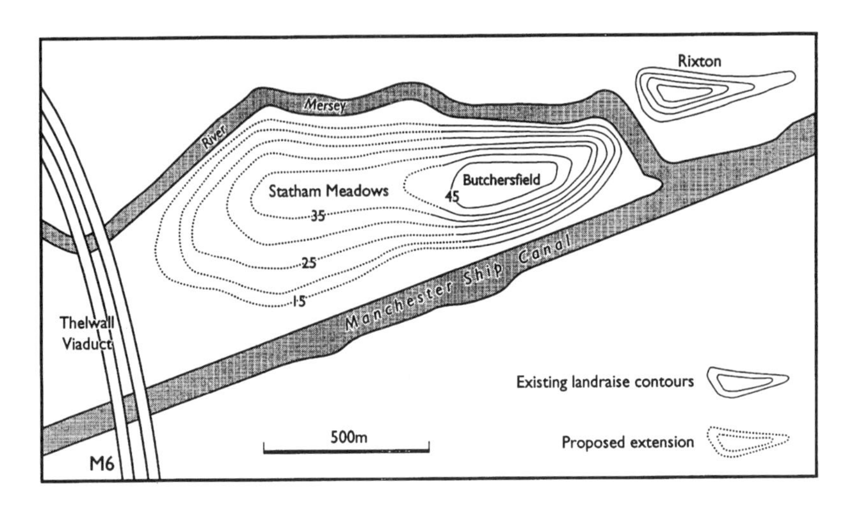

Fig. 8.4. Contours on the proposed Statham Meadows extension to the Butchersfield landraise site.

Butchersfield landfill site onto Statham Meadows (Site 1) near the Thelwall Viaduct in Cheshire. The Inspector wrote as follows about the existing landfill site: 'Prior to the inquiry I had not visited this locality for some years. The impact of the Butchersfield site, even though the senses were prepared by the photographic evidence, was such as to cause involuntary exclamation on first seeing it on crossing the A57 bridge, such as might happen on first crossing a mountain pass and seeing, unexpectedly a wide panorama. In this instance, however the sensation of amazement was not mixed with delight'. The site is part of the Mersey floodplain, and although it has been used for the dumping of canal dredgings in the past, the Planning Authority, Inquiry Inspector and Secretary of State all agreed that the proposed extension (Fig. 8.4) was inappropriate in this landscape and would not blend with it. Cheshire County Council has a policy that landraise proposals should blend with the local topography. The applicants argued that the extension would blend in with the existing landfill site and its neighbour and mitigate the effects of the existing site. This was rejected, but it is not the only example of this type of argument.

The applicants proposed to open the site to the public after restoration and argued that the elevated landscape would provide a vantage point from which the public can view the surrounding countryside and observe the wildlife habitats of the site.

Part of Cheshire County Council's objection to this extension to the Butchersfield site was that the site lies in a wide river valley where open views were already impaired by the unnatural forms of the Thelwall viaduct and the existing Butchersfield site. The proposal would close the remaining open views across the valley, extending one intrusive landform with one of greater mass. The five local Parish Councils in their objection agreed, stating that for those whose views now included Butchersfield, a significant proportion of the horizon would be composed of dominating, alien features.

The Cheshire Mineral and Waste Local Plan (Cheshire County Council 1987) contains a policy which states that proposals for new waste sites should not be approved where they would not be capable of being well screened from public view. In the view of the Inspector at the Statham Meadows appeal this policy 'is largely incapable of satisfaction if applied to finished landraisings'.

Bray, Berkshire

At Bray (Site 14) near Maidenhead, an application to raise approved levels (tertiary landraising) was attempted (Fig. 8.5). The site lies on a Thames terrace and was objected to by the local District Council as being 'an alien element' and by Berkshire County Council as: 'incongruous in this generally flat and open area . . . out of character as an isolated peak . . . disassociated from the general grain of the topography, and would form, an outlier within the floodplain'.

The applicants disagreed, arguing that the feature was not a simple dome but had two higher areas separated by a col: 'This creates a subtle, naturally sculptured quality to the proposed landform. This type of whaleback landform occurs frequently in natural landscapes, often as outliers of higher ground on the edge of floodplains'.

They also argued that the proposed landraising would screen the M4 motorway traffic: 'The proposed landform . . . by screening

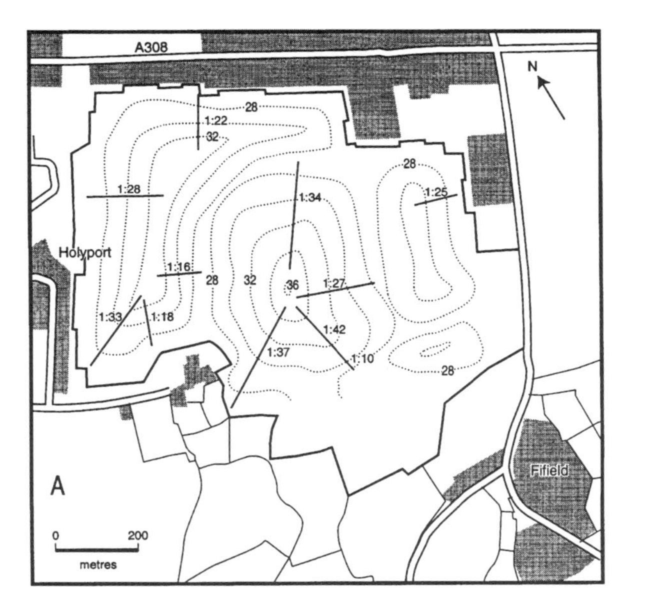

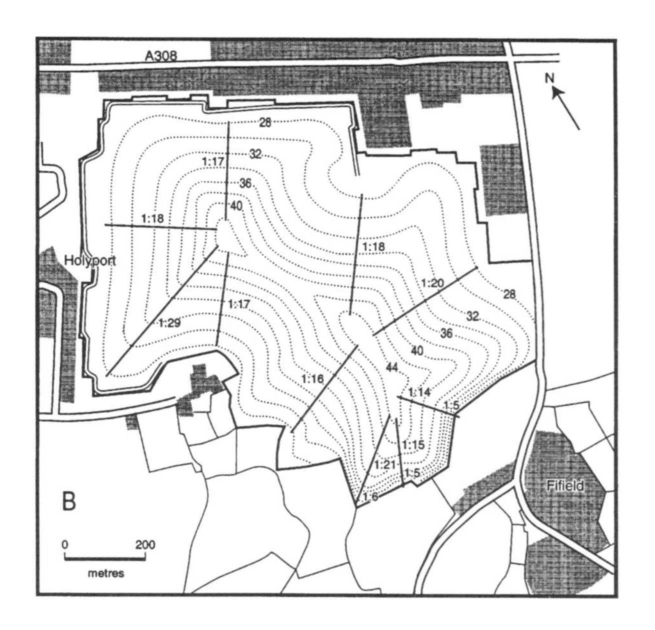

Fig. 8.5. (**A**) Contours on the approved landraise site at Bray, Berkshire; (**B**) contours on the proposed further landraising of the site, 1990/91.

moving vehicles on the M4 and house rooftops would in fact enhance this view'.

The Inspector again disagreed in refusing the application. According to him the proposals would create 'an isolated protuberant landform completely out of character with the rest of the floodplain'.

Rivenhall Airfield, Essex

At Rivenhall Airfield (Site 8) in Essex, a waste disposal-led mineral extraction was planned (secondary landraising), with an overfilling to 16 m above existing till plain levels. The applicants claimed that they were improving an old airfield landscape which was described as flat, bland and even 'boring'! They would create: 'a more undulating topography to replace the flat featureless airfield site at levels and gradients which are sympathetic to the surrounding terrain'. The design strategy included the following: 'The creation of a new landform within the site with a topographical trend sympathetic to the local easterly/south easterly topographical grain. The creation of generally asymmetrical profiles in the new topography, to add interest and diversity and avoid regularity of landform shape'.

A valley tributary to the Blackwater would be created and would appear as a natural feature between a former continuous ridgeline. The historic field pattern would be recreated as part of the restoration scheme, which also included picnic sites, country footpaths, visitor centre and summit viewpoints (Fig. 8.6). Essex County Council objected on the grounds that: 'The result is a landform which is out of context to the neighbouring levels and inappropriate to the wider topographical context... the gradients bear no relationship to the existing plateau location. The proposals are therefore quite alien and do not respect the existing location. The proposals create two mounds within the existing topography. Each mound has a multifaceted form with slopes in all directions, contrary to the existing grain'.

Their landscape witness produced various contour maps for the public inquiry showing the impact. She also described a 7.5 m high screening bund with gradients of 1 : 3 to 1 : 5.5 as an alien element in

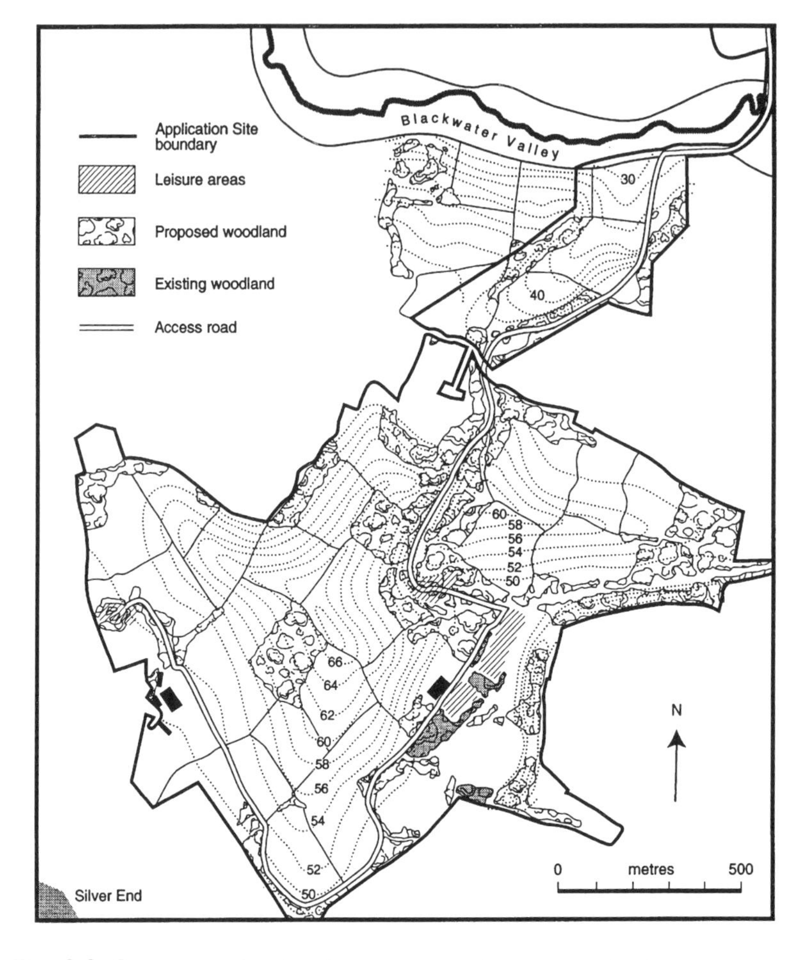

Fig. 8.6. Contours and restoration scheme on the proposed landraising scheme at Rivenhall Airfield, Essex.

the landscape with a new and totally different landform being created which does not relate to the immediate or wider landscape context, or to the grain of the landform.

The Inspector's conclusion was that: 'the Appellants' restoration strategy would be visually unsuccessful. If the creation of a side valley to the Blackwater is desirable, then it is more likely to be achieved by connecting up the individual low level restorations than by raising the plateau even higher. In my judgement, the proposed landform would not relate well visually to any existing feature, and when completed would not appear to be an authentic part of the local scene'.

He also predicted a visual impact during the construction phase: 'the landfill would probably be seen above the perimeter mounds as a hotchpotch of restored land, newly seeded or planted ground, bare soil and spoil, fencing and tops of machinery. Thus for over 20 years there would be strange and unattractive heaps on the old Airfield'.

Hardwick Airfield, Norfolk

This primary landraising scheme in South Norfolk (Site 9) was proposed by Norfolk County Council in 1991 as a replacement for an existing landfill site in disused gravel pits at Morningthorpe. Leachate from the latter site had been leaking into a local stream and the applicants felt that a landraising scheme sited on the Lowestoft Till would give greater groundwater protection. While this is certainly true, the engineering of the proposal proved to be insufficient to satisfy the Inquiry Inspector and the site was rejected partly because of an 'inadequate knowledge of components of the geology and land drainage system' (Department of Environment 1993, p. 112).

The major flaw in the site design was the absence of an adequate containment lining system. Instead, the till itself was argued to be of sufficiently low permeability to meet the required specification of $10^{-9}\,m\,s^{-1}$. However, three of the five field permeability measurements were well above the specification. The applicants' consultant geologist (Norfolk County Council 1993) argued that the high permeabilities were the result of sand lenses in the till, but this was not confirmed by the borehole logs. In fact, the highest permeabilities were recorded where no sand lenses were identified in the logs. Instead it appears that the till itself in its upper few metres is more permeable in the field than in the laboratory. The most likely explanation is fissure flow. All 26 trial pits on the site describe the till as fissured and some show roots, iron staining and

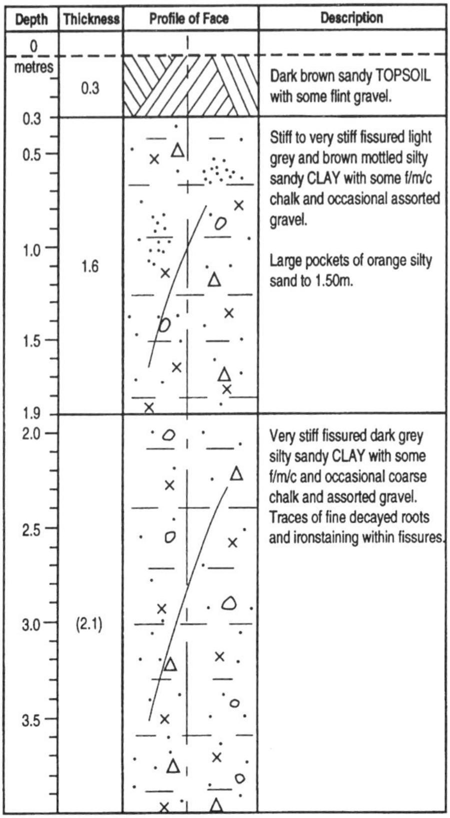

Depth	Thickness	Profile of Face	Description
0 metres			
0.3	0.3		Dark brown sandy TOPSOIL with some flint gravel.
0.5 1.0 1.5 1.9	1.6		Stiff to very stiff fissured light grey and brown mottled silty sandy CLAY with some f/m/c chalk and occasional assorted gravel. Large pockets of orange silty sand to 1.50m.
2.0 2.5 3.0 3.5	(2.1)		Very stiff fissured dark grey silty sandy CLAY with some f/m/c and occasional coarse chalk and assorted gravel. Traces of fine decayed roots and ironstaining within fissures.

Fig. 8.7. Borehole log showing fissuring in glacial till at Hardwick Airfield, Norfolk (modified from Gray 1993).

silt within the fissures which imply that the fissures are hydraulically active (Fig. 8.7). It is known from studies in Scandinavia and Canada that fissuring in till can increase permeabilities by 1–2 orders of magnitude above laboratory values (e.g. Fredericia 1990). The till fissuring was not mentioned in the Environmental Statement or site investigation report. Both before and during the Public Inquiry objectors argued that the site would not conform to the permeability specification and believed that a composite lining system should be installed (Gray 1993, 1996).

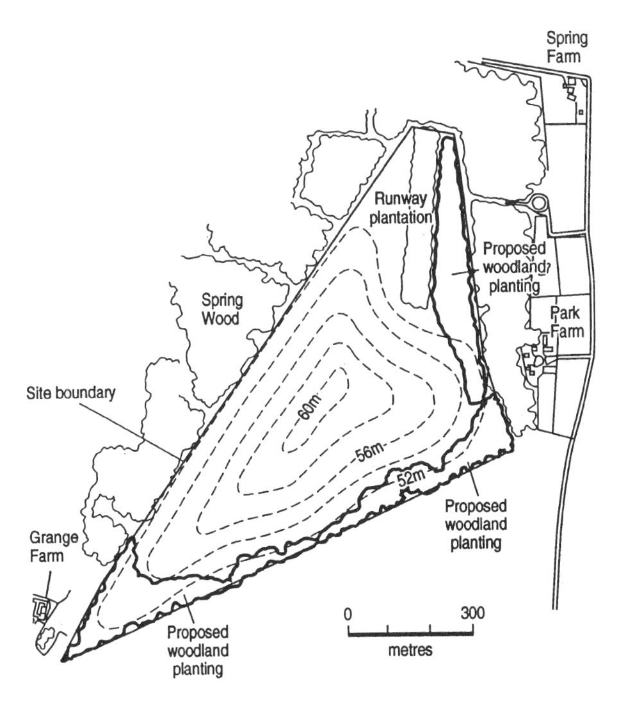

Fig. 8.8. Contours on the proposed landraising scheme at Hardwick Airfield, showing the existing and planned woodland screening.

Having rejected this argument throughout this period, at the Public Inquiry the County Council's consultant geologist accepted the essence of these arguments and subsequently the Inquiry Inspector concluded that 'the evidence is clear that the original Norfolk County Council assumption about the geological advantage of the Hardwick site was misplaced'. This is perhaps an overstatement, but what is clear is that the more 'footloose' nature of landraise siting should be used to increase groundwater protection, not as a means of saving money by skimping on site engineering.

The landraise hill was to rise 10 m above the surrounding till plain, but in this case the Inspector did not believe that this would be an unacceptable visual intrusion since the triangular site was already bordered on two sides by woodland and was to be screened by extensive planting on the third side (Fig. 8.8).

Discussion

The above case studies illustrate some of the arguments that have been used to object to, or support, landraising proposals. Supporting arguments have included the need for the sites, perceived landscape improvements, public access and recreational uses. Arguments against have mainly comprised the impact of the operations during construction and landscape impact of the restored hills.

As indicated, the issue of what constitutes unnatural, incongruous or alien landforms has been one of the most important and is a debate to which geomorphologists ought to be able to contribute. After all, some of the world's most famous landforms are arguably alien in not blending in with the surrounding landscape, for example Ayers Rock, the Matterhorn and Sugar Loaf Mountain, or nearer to home, The Wrekin in Shropshire or Arthur's Seat in Edinburgh. Indeed these landforms are famous because they are incongruous. Some have argued, therefore, that it is wrong to reject landraise proposals simply because they result in incongruous landforms whose size, steepness or shape are out of keeping with the surrounding landscape. At an appeal against Nottinghamshire County Council's refusal of a planning application for a landraise site at Hawton near Newark (Site 4), the applicants' landscape consultants argued that there are examples of relatively isolated natural hills which project noticeably above the surrounding landscape in which they are located, and gave as examples several Gloucestershire Hills. However, these arguments miss the crucial point that whereas natural hills are timeless, distinctive and valued parts of local landscapes, creating a new landraise hills alters familiar landscapes and closes views for a non-benign purpose as far as the public is concerned. Hills of waste are likely to be viewed as just that, not as enduring and valued elements of the physical, historical and cultural, local landscape.

In considering the impact of landform change it is appropriate to discuss what elements of the morphology should be analysed. The author's review of the major recent landraise proposals in England suggests that the following aspects need to be included.

1. Size: a landform may be out of place with its surroundings because of its plan scale.
2. Height: similarly, a landform may have an amplitude out of keeping with the surrounding relief.
3. Gradients: may be too steep, a characteristic particularly noticeable in flat terrain.
4. Shape: ought to be authentic, for example thin, straight, sharp-crested ridges and intricate hollows are rare in nature.
5. Detailed variation: natural slopes are rarely regular or evenly graded, but will be subject to small-scale variation.
6. Geomorphological context: the design ought to be sensitive to its surroundings, for example. It is not valid to design gradients on a waste hill to match those of a valley side, as proposed at Rivenhall Airfield (Site 8).

In addition to these geomorphological parameters, landform acceptability will depend on other factors including impact on the skyline/views, future landuse and restoration planting schemes.

Fig. 8.9. The landraising site at Milton (not included in this study), near Cambridge, rises over 10 m above the flat Cambridgeshire fenland landscape (foreground).

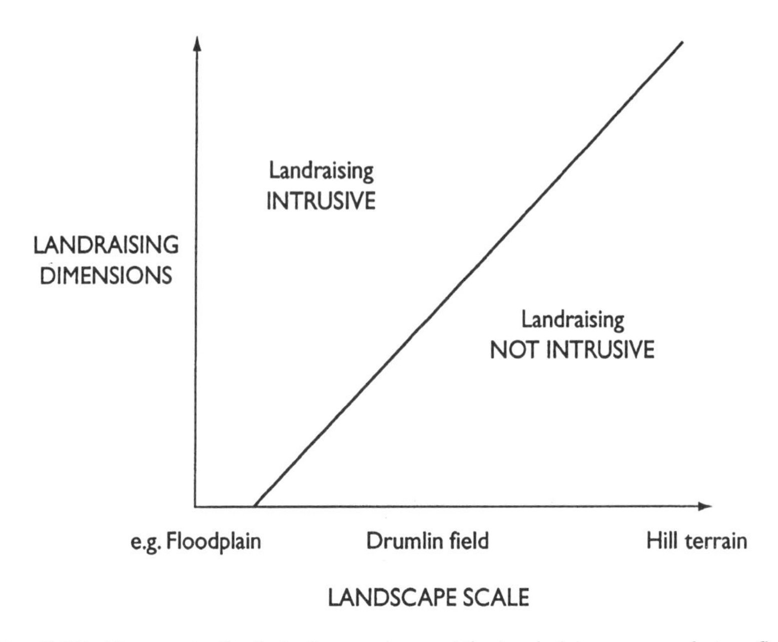

Fig. 8.10. Diagram to illustrate the requirement for landraising proposals to reflect the scale of the surrounding landscape.

The importance of some of these factors will vary depending on the ability of the landscape to absorb morphological change. A flat landscape, such as a river terrace or coastal fenland, can accommodate almost no change (Fig. 8.9), whereas high relief topography is much less sensitive to landform remodelling. There is a limited recognition of this point by English Heritage *et al.* (1996, p. 75) who note that landraising 'should be avoided in river floodplains where it results in unnatural landforms and may exacerbate flooding problems. In other areas its suitability should be judged against local landscape character'. This idea is illustrated in Fig. 8.10, which simply illustrates the point that the three-dimensional scale and design of landraising sites ought to be in keeping with the natural landscape and should be sensitive to its geomorphological evolution.

Figure 8.11 indicates the points at which a geomorphologist might contribute to landraising site selection and design. This will

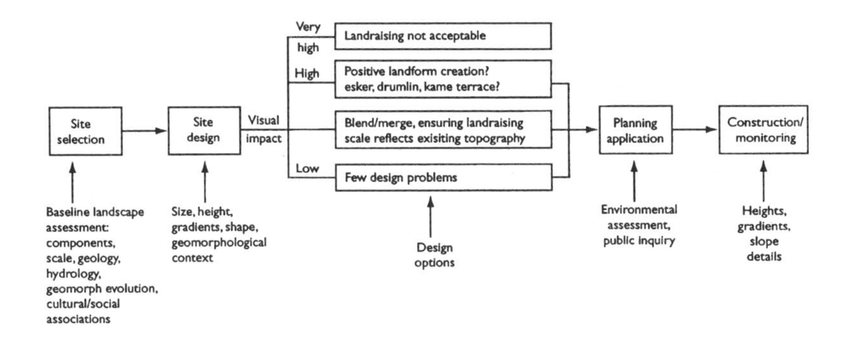

Fig. 8.11. The potential geomorphological input to the site selection, design and construction of landraising sites.

include desk studies of landscape topography, geology, geomorphological evolution and hydrology, through site selection, landform design and the planning application/public inquiry stages, to construction monitoring and restoration. Of course the geomorphological input will be only one of a range of inputs required during this process, but the point of this chapter is to indicate that it is an input not being undertaken at present.

One of the most appropriate ways of controlling morphological change and ensuring geomorphological conservation in the wider landscape is through the planning system. An example of this approach is included in *South Norfolk Local Plan, Deposit version* (South Norfolk County Coucil, 1997): 'The low-gradient open landscape character of South Norfolk makes it particularly vulnerable and sensitive to topographical change. It is important that submitted landscaping schemes reflect the local landscape character and distinctiveness'.

The policy (BEN 5) that follows reads: 'The District Council will require where appropriate a site survey of all existing natural or semi-natural features and a comprehensive landscaping scheme to be submitted with detailed planning applications... The landscaping scheme to be submitted should... (b) reflect the local landscape character and distinctiveness... (f) seek to ensure that any land modelling proposed, as associated with uses such as golf courses, landfilling etc. is sensitive to the local topographical character in terms of height, gradient, scale and shape...'

If suitable planning policies such as the above are in place, it will be much clearer to applicants what is required and much easier for Local Planning Authorities to encourage geomorphologically authentic landraising design.

Acknowledgements

I am very grateful to the Planning Authorities who have assisted in giving access to public files used in this research project. Alan McKirdy and Matthew Bennett made some very useful comments which have improved the paper. The figures were kindly drawn by Ed Oliver at the Department of Geography, Queen Mary & Westfield College.

References

Cheshire County Council 1987. *Cheshire waste disposal Local Plan.* Cheshire County Council, Chester.

Department of the Environment. 1993. *Hardwick landfill site and haul road: inspector's report.* File no. E1/X2600, E1/X2600/3/2, Y/DN/5066, HMSO, London.

——— & Welsh Office 1995. *Making Waste Work: a strategy for sustainable waste management in England and Wales.* Cm 3040. HMSO, London.

English Heritage, Countryside Commission & English Nature. 1996. *Conservation Issues in Local Plans.* English Heritage, London.

Essex County Council. 1994. *Waste management strategy: consultation draft.* Essex County Council, Chelmsford.

Fredericia, J. 1990. Saturated hydraulic conductivity of clayey tills and the role of fractures. *Nordic Hydrology*, **21**, 119–132.

Gray, J. M. 1993. Quaternary geology and waste disposal in South Norfolk, England. *Quaternary Science Reviews,* **12**, 899–912.

——— 1996. The containment properties of glacial till: a case study from Hardwick Airfield, Norfolk. *In:* Bentley, S. (ed.) *Engineering Geology of Waste Disposal.* Geological Society, London, Engineering Geology Special Publications, **11**, 299–307.

——— 1997. Environment, policy and municipal waste management in the UK. *Transactions of the Institute of British Geographers,* New Series **22**, 69–90.

Hampshire County Council. 1993. *Minerals and waste Local Plan.* Hampshire County Council, Winchester.

——— 1996. *Minutes of a Meeting of the Minerals Policy Panel of the Planning and Transportation Committee, 15 July 1996.* Hampshire County Council, Winchester.

Kent County Council. 1993. *Waste Local Plan: consultation draft.* Kent County Council, Maidstone.

Morfeldt, C. O. 1993. Landscaping with waste. *Engineering Geology*, **34**, 135–143.

Nolan Committee. 1996. *Aspects of conduct in local government in England, Scotland and Wales: issues and questions.* HMSO, London.

Norfolk County Council. 1993. *Hardwick landfill site: with and without haul road. Public Inquiry Proof of Evidence on geology and engineering design.* W. S. Atkins Environment, Epsom.

National Rivers Authority. 1994. *Landfill and the water environment: NRA position statement.* National Rivers Authority, Bristol.

SERPLAN. 1994. *Advice on planning for waste reduction, treatment and disposal in the South-East 1994-2005.* The London and South East Regional Planning Conference, London.

SEWRAC. 1994. *Waste disposal in the south east region: results of the 1993 waste monitoring survey.* South East Waste Regulation Advisory Committee, London.

South Norfolk County Council. 1997. *South Norfolk Local Plan: deposit version.* South Norfolk Council, Long Stratton.

9 Contaminated land: problems of liability

David Cuckson

Summary

- The Environment Act 1995 sets out a new legal regime for dealing with historically contaminated land. When the new provisions are brought into force, 'the appropriate person' will be made liable for the remediation of contaminated land, but what will fall within the definition of 'contaminated land?' and what will be the standard of remediation required?
- An attempt has been made to avoid the amount of litigation experienced in relation to the US Superfund regime, but the proposed exclusion tests to narrow down the scope of liability, and exclude those 'less responsible', create their own problems.
- The extent to which the new measures will succeed in their objective of cleaning up large areas of historically contaminated land which exist will depend on the resources which the government makes available to operate the new procedures and to meet the costs of remediation where these fall on the public purse.

A new statutory regime for the control of the environmental problems of contaminated land is due to be implemented in 1998. The framework is contained in the Environment Act 1995, in Section 57 of that Act. When brought into force this will have the effect of amending the Environmental Protection Act 1990 (here referred to as 'the Act'), by introducing a new Part II A which will be made up of Sections 78A to 78YC. Implementation of these provisions is dependent on the making of regulations and the issue of statutory guidance, as referred to in a number of places in the Act. The statutory guidance is particularly important to an understanding of how the new regime will operate. This chapter relies on the formal consultation draft of the statutory guidance which was published in September 1996. A large number of representations have been submitted in response to the consultation exercise and the House of Commons Environment Committee has also reviewed the government's proposals, particularly as contained in the draft statutory guidance, and produced its own report. The government's response to this report and the final form of the statutory guidance are currently awaited. Only then will a date be set for the new provisions to come into force.

After a brief reference to the policy background to the current measures, this chapter will then summarize some of the key features of the new regime, focusing particularly on issues of liability for the remediation of contaminated land and what this entails, before turning to consider some of the problems which are likely to arise in the practical application of the new controls.

The 'suitable for use' approach

The current measures are the outcome of a long period of gestation. The government's first attempt to deal generally with the problem presented by contaminated land was the proposal of local authority registers of land which had been subjected to contaminative uses, contained in Section 143 of the Act. This met such an adverse reaction from those involved in property ownership, development and investment that the measure was repealed before it was implemented. The government then asked those who objected, and others, what kind of provisions might work, and this process of consultation has continued, both on a formal and informal basis, ever since.

Throughout its consideration of how to deal with the problems caused by contaminated land the government has adopted a pragmatic approach, which was described in *Paying for our Past* in 1994 as 'working better with the grain of the market' (Department of Environment & Welsh Office 1994*a*, p.9). This policy was well summed up in an earlier document, namely the 1990 White Paper issued under the title, *The Common Inheritance: Britain's Environmental Strategy*, which included the statements: '"Action" on the environment has to be proportionate to the costs involved and to the ability of those affected to pay them. So it is particularly important for Governments to adopt the most cost effective instruments for controlling pollution and tackling environmental problems. And we need to ensure that we have a sensible order of priorities, acting first to tackle problems that could cause most damage to human life or health and could do most damage to the environment now or in the future.' (Department of Environment 1990, p.13). 'These new approaches have been described loosely as the market based approach to the environment, since they involve integrating economic and environmental concerns and applying market economics more broadly.' (Department of Environment 1990, p.14)

The aims of what is characterized as the 'suitable for use' approach are described (Department of Environment & Welsh Office 1994*a*, p.10) as follows:

1. to deal with actual or perceived threats to health, safety or the environment;
2. where practicable, to keep or to bring back such land into beneficial use;
3. to minimize avoidable pressures on greenfield sites.

There is a clear concern not to jeopardize wealth creation within the UK economy by imposing excessive burdens on industry and commerce. Where circumstances justify it thorough remedial works will be required. It is also open to an owner, occupier or developer to undertake earlier or more thorough action if they wish to do so. The intention is to avoid insisting on disproportionate or unnecessarily early steps to treat land and to consider the costs and benefits of particular schemes and to balance the need for economic growth and environmental protection. This is reflected in a policy 'that the works, if any, required to be undertaken for any contaminated site should deal with any unacceptable risks to health or the environment, taking into account its actual or intended use' (Department of Environment & Welsh Office 1994*a*, p.9).

This approach is then repeated and reinforced in *Framework for Contaminated Land* which presents the outcome of the Government's Policy Review initiated by *Paying for our Past*. The 'suitable for use' approach is described as supporting 'sustainable development both by reducing the damage from past activities and by permitting contaminated land to be kept in, or returned to, beneficial use wherever practicable – minimising avoidable pressures for new development to take place on greenfield sites.' (Department of Environment & Welsh Office 1994*b*, p.4). Remedial action will only be required where (Department of Environment & Welsh Office, 1994b, p.4):

1. the contamination poses unacceptable actual or potential risks to health or the environment.
2. there are appropriate and cost-effective means available to do so, taking into account the actual or intended use of the site.

Definition of 'contaminated land'

'Contaminated Land' is defined in Section 78A(2) of the Act as any land which appears to the local authority in whose area it is situated to be in such a condition, by reason of substances in, on or under the land, that: (1) significant harm is being caused or there is a significant possibility of such harm being caused; or (2) pollution of controlled waters is being, or is likely to be caused; and, in determining whether any land appears to be such land, a local authority shall 'act in accordance with guidance issued by the Secretary of State ... with respect to the manner in which that determination is to be made'.

The government's underlying intention is that action will only be required where there are 'unacceptable actual or potential risks to health or the environment' and where there are 'appropriate and cost effective means to do so taking into account the actual or intended use of the site' (Department of Environment *et al.* 1996, p.28).

The statutory guidance embodies the concept of risk assessment. 'Risk' here means the combination of: (1) the probability, or frequency, of occurrence of a defined hazard; and (2) the magnitude (including the seriousness) of the consequences to a specified receptor. This involves the analysis of the three elements of source, pathway and receptor, as follows:

1. the source is a contaminant or potential pollutant which has the potential to cause harm or to cause pollution of controlled waters;
2. the receptor (or target) is either a living organism, a group of living organisms, an ecological system or some piece of property, as listed in the statutory guidance, or controlled waters;
3. the pathway is the route or routes or means by or through which the receptor is being, or could be, exposed to, or affected by, the source.

The relationship between a contaminant, a pathway and a receptor is termed 'a pollutant linkage'. The next, and important, question is whether you have 'a significant pollution linkage'. Remediation action is likely to be required where there is a 'significant pollutant linkage'.

'Significant pollutant linkage'

Guidance is given as to what harm is to be regarded as 'significant'. For human beings this includes death, serious injury, cancer or other disease, genetic mutation, birth defects, or the impairment of reproductive functions. For protected habitats it is to be harm which results in an irreversible or other substantial adverse change in the functioning of the habitat or site. Relevant harm to livestock includes death, disease, or other physical damage such that there is a substantial loss in their value (for which a loss of 10% of value is suggested as a bench mark). For buildings, it is envisaged that the definition would only include structural failure or substantial damage.

Guidance is also given as to whether the possibility of significant harm being caused is significant, by reference to: (1) the nature and degree of harm; (2) the timescale within which the harm might occur; and (3) the vulnerability of the receptors to which the harm might be caused.

The example of an explosion helps clarify this. If an explosion could cause extensive damage, even though the chances of an explosion occurring are low, but not minuscule, then the possibility of significant harm being caused would be regarded as significant. The threat of pollution of controlled waters is not qualified by the expression, 'significant', in the same way. This is to avoid any conflict with the existing water legislation, contained primarily in the Water Resources Act 1991.

Inspections and determinations

The local authority will be required to carry out a systematic inspection of its area to identify land which merits detailed individual inspection (normally the Borough or District Council). It will be required to publish a formal written strategy within 15 months of the issue of the guidance in its final form. In preparing its strategy it will take into account information from other statutory bodies and also information provided by businesses, voluntary organizations and members of the public.

Where there are reasonable grounds for believing that any land may be contaminated, the local authority is required to carry out further investigations and inspections, including, where appropriate, the taking of soil samples. If the contamination in question is such as would lead the site to be categorized as a 'special site', then responsibility passes to the appropriate national agency, for example, the Environment Agency or the Nuclear Installations Inspectorate.

Regulations will spell out the criteria for land to be designated as a special site, by reference to the presence of named substances or by reference to their use or occupancy. In particular, it is proposed that contaminated land which has been used for a prescribed process designated for central control under the Integrated Pollution Control provisions of Part I of the Act should be designated as a special site. The impact, or potential impact on controlled waters, particularly those which are used for the supply of drinking water for human consumption, will also be a factor.

In determining whether 'significant harm is being caused', the local authority should be satisfied, on the balance of probability and in the light of all the relevant evidence and an appropriate scientific assessment, that significant harm is being caused by reason of substances in, on or under the land. Similar principles apply to the identification of 'a significant possibility of significant harm being caused', and, in addition, the local authority needs to check that there are no adequate risk management arrangements in place to prevent such harm. In relation to the pollution of controlled waters, the local authority should follow an approach consistent with that of the Environment Agency in applying other statutory provisions concerning pollution of controlled waters, and also should consult the Agency before making its determination.

The standard of remediation

The aim of remediation is not to achieve a total clean-up (although it is, of course, open to anyone to do this) but to ensure that the land in question no longer falls within the definition of 'contaminated land'. This is the 'suitable for use' approach.

Section 78E of the Act provides that the only things by way of remediation which the enforcing authority may do, or require to be done, 'are things which it considers reasonable, having regard to: (1) the cost which is likely to be involved; and (2) the seriousness of the harm, or pollution of controlled waters, in question'.

The authority has a measure of discretion here and, indeed, it is only required to 'have regard to' the statutory guidance. The authority is expressly advised to have regard to the extent that an assessment of the costs likely to be involved and the resulting benefits shows that those benefits are worth incurring those costs. The authority should also have regard to the practicability of any remediation scheme, for example, whether relevant technologies are commercially available on the necessary scale or whether the presence of buildings or similar structures, on the land would permit the relevant remediation actions to be carried out in practice. Similarly the effectiveness of any remediation scheme or its durability should be taken into account.

Remediation can include further assessment action, remedial treatment action and/or subsequent monitoring action. Remediation should not be required to be carried out for the purpose of making demands for any uses other than its current use.

Who is to bear the responsibility for remediation?

Section 78F of the Act refers to a person who is to bear responsibility for a remediation action as an 'appropriate person'. The section refers to two broad categories of person, namely: (1) Class A: persons who 'caused or knowingly permitted'; and (2) Class B: owners and occupiers of the land.

Only if no Class A person can be found, after reasonable enquiry by the authority, will a Class B person (i.e. the owner or occupier) be responsible for remediation by virtue solely of that ownership or occupation.

In many cases more than one appropriate person will be identified. On the basis that some can reasonably be considered to be 'more responsible' than others, the authority is required to apply a series of tests to exclude those less responsible from liability. Where there are still a number within the liability group then the responsibility is to be apportioned between them. In applying these tests the authority should pay no regard to the financial circumstances of those concerned, nor as to whether there may be any insurance cover in place. Where two or more appropriate persons have agreed between themselves as to the basis on which they wish to share the costs of a remediation action and have given notice of that agreement in writing to the authority, the authority shall normally give effect to this agreement in making its determination on exclusion and/or apportionment.

This approach contrasts with that of the so-called 'Superfund' regime in the United States of America where, essentially, anyone who has anything to do with the contaminated land, either as a causer of the contamination or by virtue of having an interest in the land, can be targeted by the enforcing body. Apart from some limitation on the liability on those whose only involvement in the polluting activities has been as lenders of financial resources, anyone can be held liable for the whole of the cost of remediation. It is up to that person to claim in turn for contributions from other of the 'potentially responsible parties'. This system has inevitably led in practice to extensive inter-party litigation which, in addition to being very costly, has delayed actual clean-up operations.

The exclusion tests in the UK regime are to be applied in

sequence, but only up to the point where applying the test will still leave at least one person liable. If there is more than one 'significant pollutant linkage' affecting a particular site, then the tests are applied in respect of each such linkage. The tests for Class A persons are.

- Test 1 'Excluded Activities': where a person's only involvement is by reason of specified activities, including providing financial assistance, carrying out any action necessary for the purpose of underwriting an insurance policy, providing legal, financial, engineering, scientific or technical advice, or certain kinds of contract where the other person knowingly took responsibility.
- Test 2 'Payments Made for Remediation': where a person has already paid someone else to carry out adequate remediation.
- Test 3 'Sold with Information': where a person has sold land where there are significant pollutants, or let it on a long lease, and has ensured that the purchaser or lessee had information as to the presence of those pollutants and thus had the opportunity to take that into account in agreeing a price.
- Test 4 'Changes to Substances': where a significant pollutant linkage has been created only because another substance, which interacted with substances already present, was later introduced to the land.
- Test 5 'Escaped Substances': where the contamination is caused as a result of the escape of substances from other land, for which another Class A person was responsible.
- Test 6 'Introduction of Pathways or Receptors': where the contamination issue has arisen solely because of subsequent development on the land, or a change of use, by another liable person.

There is a test for Class B persons so as to exclude from liability those who do not have an interest in the capital value of the land in question (i.e. licensees or tenants at a rack rent).

Apportionments and mitigation

If, after the application of these exclusion tests, there remains more than one person within the liability group, the authority is to determine the apportionment of liability among the members of the liability group in proportion to its assessment of the relative degree of responsibility that it attributes to each member of that group. If, for example, pollution has occurred over a long period of time, with various parties involved, and it is possible reasonably to estimate relative quantities for which each party has been responsible, then the cost of remediation should be borne pro rata.

In the case of Class B persons, where different areas of land are affected by the same contamination, or there are different interests in the same land, then the authority is to apportion liability in proportion to capital values. Where there are two or more significant pollutant linkages which are best treated by a combined remediation scheme, then the cost is to be apportioned between the liability groups. In cases where there is what is termed an 'orphan linkage', where there is no appropriate person on whom the authority can serve notice, then the cost of that share has to fall on the public purse.

Section 78P(2) of the Act requires the authority to have regard, in deciding whether to recover the cost of remediation, and, if so, how much of the cost, to any hardship which the recovery may cause to the person from whom the cost is recoverable. In the case of a small- or medium-sized enterprise the authority is advised to give sympathetic consideration to reducing the recovery of remediation costs where it is clear that enterprise would otherwise become insolvent. The cost to the local economy may be greater if that enterprise ceases to exist than if the authority meets some of the costs of remediation. There are mitigating factors for Class A persons where other potentially appropriate persons have not been found or where the damage was not reasonably foreseeable. In relation to Class B persons, the authority should take into account any evidence that normal enquiries before purchase did not reveal the presence of the significant pollutants. Authorities are advised to waive or reduce recovery of remediation costs where the appropriate person in such circumstances is an owner/occupier of a dwelling 'unless the financial circumstances of the owner/occupier would enable him to meet easily the remediation costs in question'. Authorities should also give sympathetic consideration to circumstances where the costs of remediation are likely to exceed the value of the land.

What constitutes 'knowingly permitting'?

In commenting on the test of 'causing or knowingly permitting', the draft statutory guidance starts from the point that the test has been used as a basis for establishing liability for environmental legislation for more than one hundred years, with the implication that there should not be a problem of interpretation for us today. It then quotes Lord Wilberforce in the leading case of *Alphacell v. Woodward ([1972] 2 All ER 475)* (in relation to sub-section 2(1) of the Rivers (Prevention of Pollution) Act 1951): The sub-section clearly contemplates two things – causing, which must involve some active operation or chain of operations involving as a result the pollution of the stream; knowingly permitting, which involves a failure to prevent the pollution, which failure, however, must be accompanied by knowledge.

The draft statutory guidance then goes on to comment that there is a substantial body of case law relating to 'causing', but without actually referring to any other cases (Department of Environment *et al.* (Anon 1995) 1996, p.53). Perhaps it is worth just noting here the comment from the decision of the Court of Appeal in *Attorney General's Reference (No. 1 of 1994) [1995] 1 WLR 599*, to the effect that the word 'causes' should be given its plain common sense meaning.

With regard to the term 'knowingly permitting', the draft statutory guidance quotes the Minister of State for the Environment, the Earl Ferrers, who stated in a House of Lords debate on behalf of the government that (Department of Environment *et al.* 1996, p.53): 'The test of 'knowingly permitting' would require both knowledge that the substances in question were in, on or under the land and the possession of the power to prevent such a substance being there.'

He also addressed the question as to the extent to which this test might apply with respect to banks or other lenders, where their clients have themselves caused or knowingly permitted the presence of pollutants, commenting that: 'I am advised that there is no judicial decision which supports the contention that a lender, by virtue of the act of lending the money only, could be said to have 'knowingly permitted' the substances to be in, on or under the land such that it is contaminated land. This would be the case if for no other reason that the lender, irrespective of any covenants it may have required from the polluter as to its environmental behaviour, would have no permissive rights over the land in question to prevent contamination occurring or continuing.' (Department of Environment *et al.* 1996, p.53).

The point was raised in the course of the informal consultation process as to whether an owner or occupier would automatically become a 'knowing permitter' on being informed by the authority of the presence of pollutants such that the land was regarded as being contaminated. The logical result of such an interpretation is that you would never have a Class B person. This does appear to be logically absurd, in view of the way in which the legislation is framed, and the government's view is stated in paragraph 18 that a person who merely owns or occupies the land in question does not become a 'knowing permitter' merely as a consequence of having been consulted by the authority. Presumably the inference is that the key point at which the analysis is made is at the start of the statutory process, i.e. when the land is identified as contaminated.

There are, however, some outstanding issues to be clarified. In the case of a corporate person, it will be material who within the company has the knowledge. Following the decision of the Privy Council in *Meridian Global Funds Management Asia Limited v. Securities Commission (1995) 3 WLR 413*, in order to decide who in a company must have the requisite knowledge, it will be necessary to consider whose knowledge or state of mind was for the purpose concerned intended to count as that of the company. A director's knowledge no doubt will count, whereas that of a casual worker probably will not. It will be helpful to have some further guidance on how this question should be addressed, to ensure, as far as possible, uniform interpretation throughout the country. It is also open to argument how far a person can avoid liability by shutting his/her eyes to the obvious or refraining from enquiry, largely to avoid the risk of becoming a 'knowing permitter'. Further guidance on this question of constructive knowledge would be welcome.

Interaction with the water pollution control system

Reference was made earlier, in relation to the definition of 'contaminated land', that the threat of pollution of controlled waters is not qualified by the expression, 'significant', so as to avoid any conflict with the existing water legislation, contained primarily in the Water Resources Act 1991. There is an acknowledged overlap here between the two separate systems of control and there is a concern at present that there is no guarantee that they will operate in the same way, so as to make the same person liable for any remediation which may be required. The water pollution control system is administered by the Environment Agency, and their powers are due to be enhanced by amendments to the Water Resources Act 1991 contained in the Environment Act 1995 (not yet brought into force). Whereas at present the statutory powers of the Environment Agency in relation to clean-up works are limited to carrying out any works considered necessary and then seeking to recover the cost from the person or person responsible for the pollution, under the new arrangements the Agency will be able to serve 'works notices' which will operate in a similar way to the remediation notices under the contaminated land regime. However, at present, there is no obligation on the Agency to follow the same procedure for identifying 'the appropriate person' or for apportionment and mitigation.

It is understood that a protocol is in course of preparation to deal with the overlap between the water pollution control system and the new contaminated land regime. It is hoped that this may reduce the scope for confusion and uncertainty, and it is desirable that such arrangements should be binding on both the Environment Agency and the local authorities involved as enforcing bodies. Where the major issue is of the pollution of controlled waters, but there is some residual contamination of land, guidance is needed as to how serious the land contamination should be for the contaminated land powers to be used in preference to those under the Water Resources Act. In addition, there needs to be certainty as to the operation of the works notices procedure and, in particular, as to the identification of the person(s) on whom a works notice should be served. It is not sufficient that the Environment Agency should voluntarily choose to follow the procedures and the tests set out in the guidance on contaminated land. Parties to a property transaction, whether a seller and buyer or borrower and lender, need to be able to predict where the risk of future liability lies. Accordingly, it is important that the same basis of liability is applied under both regimes.

Interaction with the planning regime

At present liability for the remediation of contaminated land commonly arises in practice where the land is being redeveloped. The local planning authority has considerable powers to impose requirements to deal with risks from land contamination where application is made for planning permission for development or change of use. Department of the Environment Circular No. 11/95 (Department of Environment 1995) advises on the use of conditions in planning permissions relating to the development of contaminated sites and sets out some model conditions dealing with contaminated land and soil decontamination (Model Conditions Nos. 56 to 59). Account is taken of both the actual and intended uses of the land and it is up to the local planning authority to ensure that land is not contaminated land when used for any purpose permitted by the decision.

There is already governmental guidance on the interaction between planning and pollution control, in *Planning Policy Guidance Note 23 (PPG 23)* (Department of Environment 1954). It is understood that *PPG 23* will be revised in due course to take account of the interaction between the new contaminated land regime and the existing planning controls. There are a number of specific issues which, it is hoped, will be addressed in this revision. For example, how far will a local planning authority be bound by the guidance on the standard of remediation of contaminated land in deciding on what is to be required by way of condition for planning permission for the development or change of use of land? Where a landowner has submitted his/her own proposal for remediation in a remediation statement, which has been accepted by the enforcing body, and then submits a planning application for development or change of use such as would not change the basis of risk assessment applicable, will the local planning authority be bound by the terms of the remediation statement in relation to any conditions it may consider applying to the planning permission? Will it be possible for a landowner to deal with a contamination problem by an application for planning permission for change of use, for example from residential to industrial? Such a situation could arise where the assessment of the land in its current use led to a finding that it was 'contaminated land' but that an assessment in the light of the proposed future use would not.

Resources available to the enforcing bodies

The effectiveness of the new contaminated land regime will depend to a considerable extent on the resources available to the enforcing bodies. At the present time the position is not clear. If systematic inspections of their areas and the identification of contaminated land are to be carried out as envisaged by the statutory provisions and the client guidance, then significant additional resources will need to be made available to the enforcing bodies. Otherwise, local authorities, in particular, will end up only responding to specific complaints. Even then, there will be a temptation to delay taking action where there is a significant possibility that any finding of liability will be resisted by the person or persons identified, or that much or all of the cost of remediation will fall ultimately on the public purse. Such a half-hearted implementation would result in an uneven application of the new provisions across the country and the new system of control may fall generally into disrepute.

If financial resources are unduly restricted, then local authorities will be discouraged in practice from applying the hardship provisions as envisaged by the draft guidance. Where persons identified as liable for remediation consider that they are being treated unduly harshly in the application of the new provisions, they will be more inclined to contest the decisions of the enforcing bodies. This runs counter to the overall objective of minimizing the scope for dispute and the reduction of expenditure on professional costs not directed primarily to the actual clean up of contamination.

Conclusion

The proposed new contaminated land regime is already having an impact on the property market, even before it has been brought into force. Because the new provisions will be retrospective in effect, they are already beginning to be taken into account in relation to current transactions. This means that assumptions are being made on the basis of the present wording of the draft statutory guidance. Accordingly, although there has been critical comment on some of the details of the proposals, including from the House of Commons Environment Select Committee (1996), there is a general view that the new regime should be implemented as soon as possible to provide greater certainty and consistency of approach.

In the immediate aftermath of the enactment of Section 143 of the Act, in relation to the proposed registers of land subject to the contaminative uses, there was strong anecdotal evidence that some sales and purchases failed to be completed because of fear of the potential impact of an entry on the proposed register, and some lenders adopted a very cautious line where there was perceived to be a threat of 'blight'. People will continue to be unsure about how to deal appropriately with contamination issues until the legislative position is more settled. It is hoped, therefore, that the current proposals will be made as good as reasonably possible by the obvious points of difficulty being identified and improved, but that then the new contaminated land regime be implemented without further delay. The application of the new provisions can be kept under review and the controls subsequently refined as necessary in the light of experience. To this end the House of Commons Environment Select Committee specifically recommended that the Environment Agency should be charged with this task, and that it should report its findings to the Department of the Environment within four years of the new regime coming into force, recommending where appropriate further modifications to the guidance or the statutory regime (House of Commons Environment Select Committee 1996, p.xxxii).

References

Alphacell v. Woodward [1972] 2 All E R 475. Butterworths, London.

Attorney General's Reference (No. 1 of 1994) [1995] 1 WLR 599. Incorporated Council of Law Reporting for England and Wales, London.

Department of Environment. 1990. *The Common Inheritance: Britain's environmental strategy*. HMSO, London.

———. 1994. *Planning Policy Guidance: Planning and Pollution Control (PPG23)*. HMSO, London.

———. 1995. *Department of Environment Circular No. 11/95*. HMSO, London.

——— & The Welsh Office. 1994*a*. *Paying for Our Past – A consultation paper*. HMSO, London.

——— & ———. 1994*b*. *Framework for Contaminated Land – Outcome of the Government's Policy Review and Conclusions from the Consultation Paper Paying for Our Past*. HMSO, London.

———, ——— & The Scottish Office. 1996. *Consultation on Draft Statutory Guidance on Contaminated Land*. HMSO, London.

House of Commons Environment Select Committee. 1996. *Second Report: Contaminated Land*. HMSO, London..

Meridian Global Funds Management Asia Limited v. Securities Commission [1995] 3 WLR 413. Incorporated Council of Law Reporting for England and Wales, London.

10 The generation of geological data from environmental and construction projects in the Penarth area of southeast Wales

Chris Lee

Summary

- South Wales has long been known for its association with the exploitation of raw materials such as coal, ironstones, limestone for building stone and cement manufacture, and clays in the manufacture of bricks.
- The extractive industry and its transport infrastructure associated with many of these minerals is now in decline and has generated large areas of so called 'brownland' which are contaminated or derelict in some way.
- Development of contaminated land in the Penarth area has required and generated much geological information providing a better insight into the regional geology of the area.

The area examined in this paper occupies the eastern part of the Vale of Glamorgan and Cardiff Unitary Authorities in southeast Wales (Fig. 10.1). It comprises a dissected lowland, which rarely reaches above 100 m, together with much of the coastline bordering the Bristol Channel which exposes a Mesozoic succession in a range of cliffs. These cliffs are broken by wide coastal plains associated with the rivers Taff, Ely and Rhymney. Inland the best exposures are seen in quarries as well as escarpments. The geology of this region has greatly influenced the development of the area. For example, Fig. 10.2 lists the environmentally sensitive sites with an underlying geological control or influence which can be recognized in an area of some 50 km^2 within this region.

Many, if not all, of the features listed in Fig. 10.2 were geologically or geomorphologically controlled or influenced during their inception and in turn have generated more geological information during subsequent redevelopment and reappraisal.

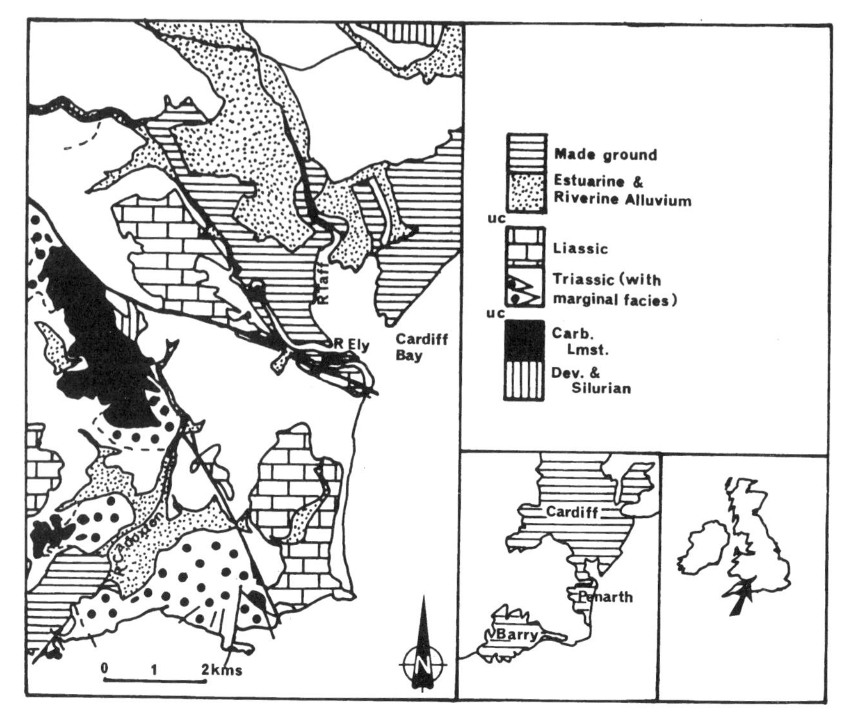

Fig. 10.1. Location map.

This geological information includes numerous site reports and a large number of boreholes. These data are usually site or job specific and consequently tend to be treated in isolation. However, by integrating this information, often from multiple sources, an important resource for regional planners and decision-makers can be obtained.

This chapter presents integrated information from a range of sources gathered during development and redevelopment in the Penarth region. The principal projects from which data were gathered are given in Fig. 10.2. Perhaps the most important information is that derived from cores. Cores were obtained from both onshore and offshore sources and were commonly cut into the superficial deposits, although some reached the Mercia Mudstone Group at depth. Coring in the Mercia Mudstone was often difficult

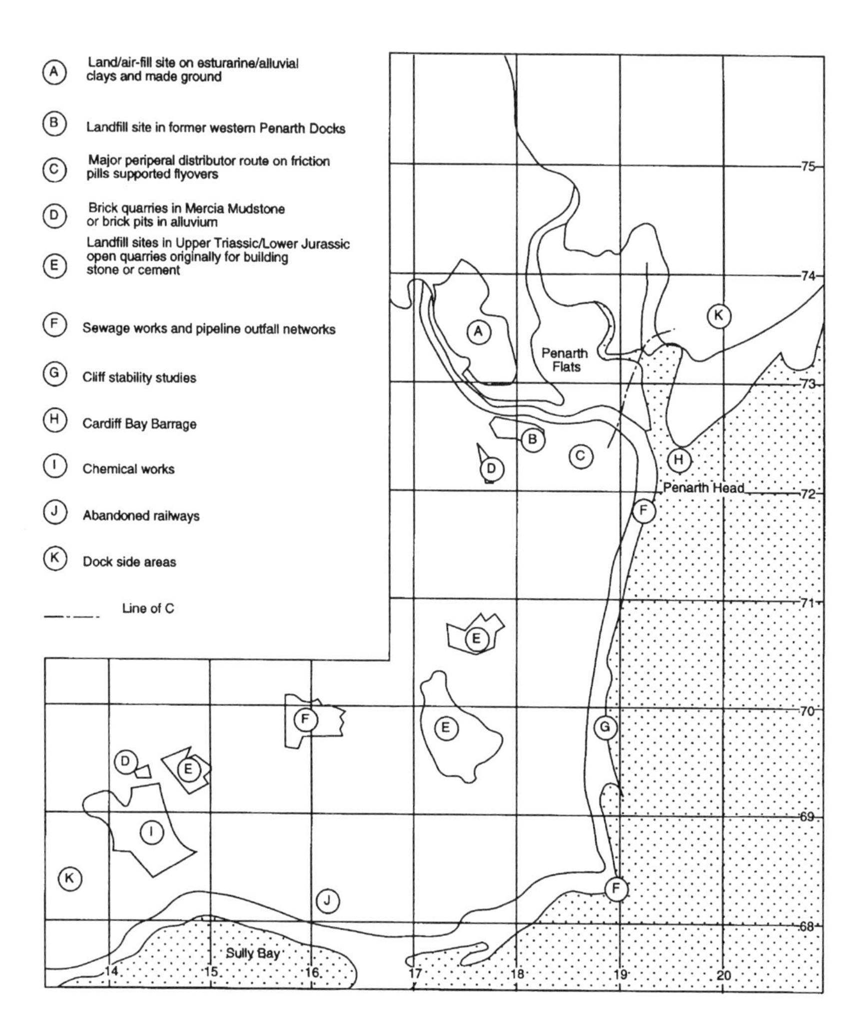

Fig. 10.2. Environmentally sensitive sites with an underlying geological control or influence whose intial development or redevelopment provides geological data.

as the strength of the material, owing to weathering, was variable and hard to predict. Borehole stability and voids at depth could also be a problem. Some correlateable 'weak' zones were distinguished across the route of the peripheral distributor road and were similarly identified by geophysical logs along the line of the Cardiff Bay Barrage. To the south and west of Cardiff Bay many boreholes were drilled for various site investigations in Penarth Marina, Penarth Dock Landfill Site and at Penarth Moors. Welsh Water/Dwr Cymru have also collected considerable borehole data as part of their plan to improve the sewage and general drainage systems of the area at Cog Moors and Lavernock Point.

The aim here is to show how an overview of the geology of this region of southeast Wales and its importance in influencing environmentally sensitive sites can be constructed from this diverse data resource.

Geology of the Penarth region

The geology of the Penarth region can be determined from the limited surface exposures, mostly in scarp faces and coastal cliffs, as well as from published sources. Most importantly, new data are derived from environmental projects, particularly in the development of landfill sites and in the exploration for groundwater. The following outline is taken from these studies.

Bedrock

Mesozoic rocks are predominant in this region (Fig. 10.1) but inliers of Silurian, Devonian and especially Carboniferous (Dinantian) rocks occur locally. During much of the Permian and early Triassic periods, these Palaeozoic rocks were deeply dissected by erosion, and in the late Triassic, sediments began to accumulate on the Palaeozoic erosional remnants. These sediments consist of red, green and grey mudstones with evaporites and some carbonates of the Mercia Mudstone Group which successively onlapped onto the irregular topography beneath. Stratigraphically, the argillaceous ‘normal’ Mercia Mudstone Group passes laterally and/or vertically into a marginal facies, which tends to be much coarser grained. This marginal facies, with a good porosity, tends to be the principal aquifer of the area. The ‘normal’ Mercia Mudstone Group may be divided into four zones in the Cardiff Bay area on the basis of its weathering characteristics (Chandler & Davis 1973). These zones are independent of stratigraphy and as one would expect, tend to decrease in effect downwards from the surface. However, considerable variation with depth occurs and the zones may be intermixed and may not follow a general pattern of reduction of weathering effects with depth (Fig. 10.3). The five zones identified are as follows.

1. Zone IVb. Matrix only.
2. Zone IVa. Matrix of clay/silt with coarse sand and occasional lithorelics up to 3 mm diameter.
3. Zone III. Matrix of silt with unweathered mudstone lithorelics up to 25 mm diameter.
4. Zone II. Mudstones weathered along fissure surfaces, weathered soil forming a thin veneer of silty clay.
5. Zone I. Unweathered mudstone.

The unweathered mudstones are generally compact, hard and break with a conchoidal or sub-conchoidal fracture. Leigh (1976) records a cyclical pattern in the engineering strength characteristics of these mudstones which has been loosely linked to variations in the depositional environments. However, the often enigmatic relationships between the zones in closely spaced boreholes (Fig. 10.3) may indicate a variety of reasons for the weathering differences, including depositional, diagenetic, structural, deep periglacial weathering or pressure/overpressure release (Lee 1996).

The 'normal' Mercia Mudstone is overlain by marine carbonates, grey shales, sandstones of the Rhaetian Penarth Group, followed by early Jurassic calci-lutites and mudstones of the Blue Lias. The Blue Lias is made up of three formations, subdivided on the basis of limestone–mudstone ratios. The lowest, the St Mary's Well Bay Formation, comprises mudstones and significant but just subordinate limestones, and was often used for building stone and cement, while the Lavernock Shales above are mostly mudstones. The overlying Porthkerry Formation in which limestones dominate has yielded materials for the manufacture of Portland Cement.

There are no solid rocks younger than the Early Jurassic in this area, although the presence of later Mesozoic and Tertiary strata in the Bristol Channel Basin suggest that such deposits were probably laid down in this area and have now been eroded.

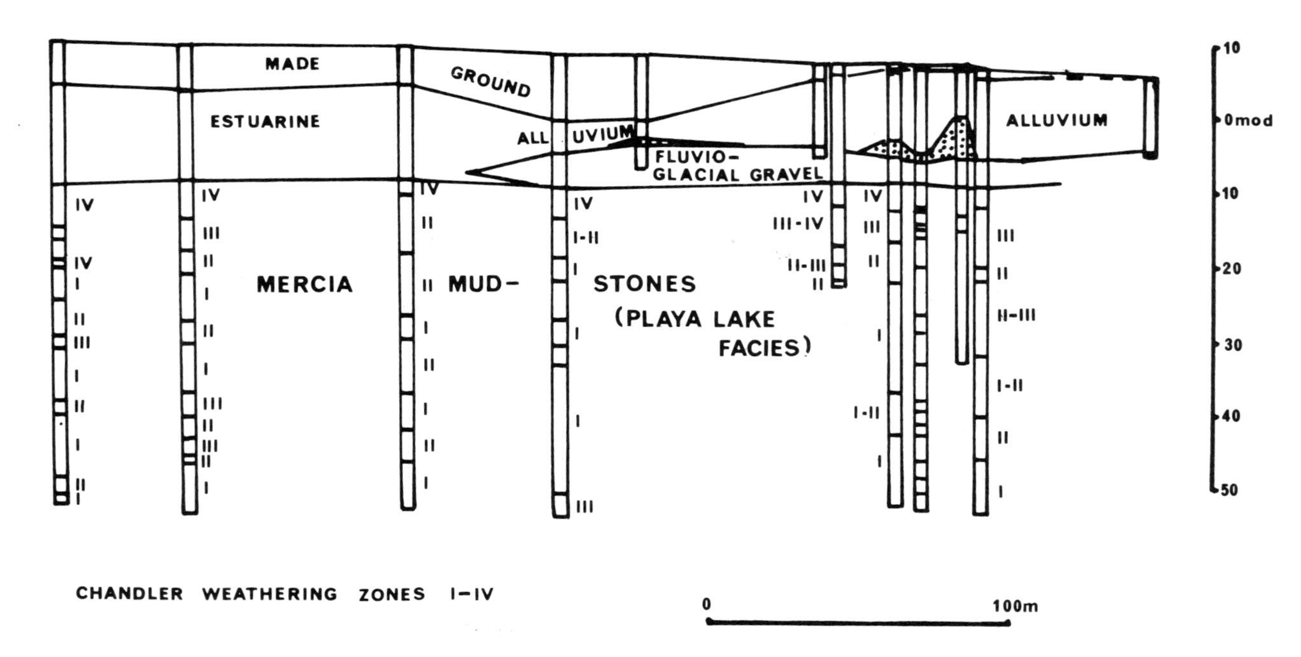

Fig. 10.3. Cross-section of part of the Butetown Link, South Glamorgan Distributor Road.

Superficial deposits
Five main superficial deposits, of Quaternary to Recent age, can be identified in the Penarth region. These are given below.

1. Made ground. This is generally soft to firm, red, brown, blue-grey sandy, gravely clay with abundant evidence of human artefacts, such as brick, clinker, ash and wood. This type of deposit varies considerably and can rapidly change from one location to the next both laterally and vertically. It may consist of any combination of reworked alluvium, peat, top soil, wood, domestic refuse, slag, ash, demolition debris, colliery spoil and various rock fill. It may also include landfill material of domestic or commercial waste, although some industrial/chemical waste may be included.
2. Estuarine alluvium. These deposits are post-Devensian and are typically found in the estuaries of the Cadoxton, Ely, Taff and Rhymney rivers, along the low-lying coastal plain of the Bristol Channel. This is typically very soft silty clay which may be stiff at depth. It may also contain some subordinate silts, sands and gravels. Human artefacts have been found in the upper portions. In places, discontinuous peat beds with associated methane gas zones occur. The deposit tends to coarsen at the base and may show a diachronous peat layer (Kidson 1977).
3. River gravels. These are clean, mature, medium to coarse gravels with rounded to sub-rounded cobbles and occasional small boulders, generally of a sandstone (quartzitic) lithology.
4. Glacio-fluvial sand and gravel. These deposits are of Devensian age and are deposited in river valleys (Anderson 1968). Generally these are only distinguished from the river gravels by a greater sand and fines content. Cobbles and occasional small boulders with a sandy matrix dominate and may grade down into glacial till. These beds constitute the principal aquifer of the superficial deposits.
5. Glacial till/Periglacial Head. This comprises predominantly silty clay but with variable proportions of glacio-fluvial gravel. These deposits tend to be structureless and unpredictable in their proportion of clay to granular material at any one point but where it is cohesive it may show failure along discrete surfaces.

Geological information from landfill sites

Principal landfill sites which have provided geological information are given in Fig. 10.2. The history of the landfilled areas in this region is closely linked with its development as a commercial and industrial centre. Landfilling began in the 1700s, and increased after 1850 when it expanded into areas of marshland and saltings on the margin of the estuary (Figs 10.1 & 10.2). Increasingly, commercial quarries, now exhausted of raw materials, have became available for landfill operations. Landfills vary in age from pre-1945 to the present day. Pre-1945 tips are usually considered to be inert and all post-1945 tips in the area have now reached their capacity and are under active waste management schemes or being redeveloped.

Location of landfill sites

Pre-1945 landfill is mostly found in Cardiff Docks, but can also be found between Sloper Road and Saltmead (Cardiff Moor) and between Roath Dock and Pengam Moors (Fig. 10.2).

Post-1945, the principal landfill site of the region is at Penarth Moors, although tipping stopped here in 1995/96. This site was first tipped in 1969, and is bounded to the east by Ferry Road and the Penarth Flats and the River Ely to the west. Two major meanders of the River Ely which originally ran through the site were plugged, drained and filled before tipping over the whole area began. These old channel routes may have locally cut down into the glacio-fluvial gravels that occur extensively in this area, and as such there could present a pollutant pathway into the main aquifer of the Cardiff Bay Region. Extensive post-1945 landfill is found at Leckwith Moors. Tipping was completed in 1974 and like Penarth Moors, this site contains relict channels of the River Ely that were channelized prior to its development. Other areas of relatively small post-1945 landfill tips are seen in the Timber Ponds at either end of Queen Alexander Dock and in Butetown.

To the south of the River Ely the Penarth Dock Tip occupies the western end of the old docks. Tipping began in the 1960s and was completed in 1995/96 with 750 000 m^3 of waste. The site is also underlain by the gravel beds, and consequently as part of the development programme this 23 ha site has been contained by 1100 m of slurry wall and diaphragm wall up to 23 m deep (Privett 1996).

In north Penarth scattered abandoned quarries occur both in the Blue Lias where building stone and cement were extracted, and in the red mudstones of the Mercia Mudstone Group, where materials were exploited for brick making. Further south larger limestone quarries occur in the St Mary's Well Bay Formation such as at Cwrt-Y-Vil which was landfilled and developed on in the 1960s (Fig. 10.2).

There are several disused limestone quarries in Cosmeston Country Park that were initiated at the turn of the century. These occur on both sides of the main road and the quarried material was transported via conveyors and tramways to the South Wales Cement and Lime Works where it was processed for cement (Fig. 10.2). Subsequent to the closure of the quarry in the late 1960s infilling of the northwestern end of the quarry with domestic waste was undertaken while the northeastern area was used as a tipping site for a chemical firm. Upon closure the cement works site was subsequently 'cleared' and used for housing. A large area of landfill is also present to the northeast of Sully Moors Road. It was used for the tipping of domestic and building refuse during 1976 and 1977 to a depth of 7 m. This site was founded on the alluvium of the Cadoxton River where there had been extensive excavations in the alluvium and underlying Mercia Mudstone Red Marls for brick making in the past.

All of these landfill sites provide important geological information – particularly relevant in determining the pollution pathways from leachates and landfill gasses. The gathering of these data, from both exposed rocks and boreholes, is therefore an important task for an environmental geologist.

Geological information from groundwater investigations

With the increase in environmentally sensitive sites or projects, regional groundwater studies have become increasingly important, especially in those aquifers close to the surface which could be affected by pollutants from above (Edwards 1997; Heathcote *et al.* 1997; Thomas 1997). These studies have provided much information about the nature of subsurface geology.

Numerous potential aquifers exist in this region due to the suitability of the geology and since the average annual rainfall of between 950 and 1 300 mm exceeds evapotranspiration (530–535 mm). The Carboniferous (Dinantian) Limestones form the most important aquifer in South Wales as a whole but are confined by the mudstones of the Mercia Mudstone Group. However, the limestone aquifer may be exploited when in hydraulic contact with the more permeable Triassic marginal facies. A well at Biglis [ST 147 698] is used for public supply close to the landfill site (F) in Fig. 10.2. Elsewhere across the area the marginal facies of the Mercia Mudstone Group is the main aquifer and produces good but very hard water, which can rise close to the surface under piezometric pressure. Stratigraphically higher, in the Mercia Mudstone Group, local beds of sandstone and dissolved gypsum bands occur that may produce water. Near the coast both the lower and upper Mercia Mudstone reservoirs become brackish and may exhibit tidal effects. Tidal variations can also be seen in the Carboniferous Limestone aquifers below.

The superficial deposits, especially the glacio-fluvial gravels were extensively used in the past for public supply. Several abstraction licences are held, especially where there is an association of the gravel bed with the underlying Mercia Mudstones that is confined by the estuarine alluvium above. This composite aquifer may comprise the upper weathered parts of the Mercia Mudstones, the glacial tills (if present) and the glacio-fluvial and river gravels depending upon its geographical position in Cardiff Bay. The permeability of this horizon is variable but approximates 5×10^{-4} $m\,s^{-1}$ with a regional water table sloping very shallowly to the southeast. The overlying alluvium is semi-confining with an average permeability of $10^{-8}\,m\,s^{-1}$. This will allow some fluid flow and exceeds the $10^{-9}\,m\,s^{-1}$ value that is recommended for clay barriers in containment sites.

The use of geological data derived from landfill and groundwater studies

Three landfill sites – Ferry Road (A), Penarth Dock (B) and Sully Moors Road (G) (Fig. 10.2) are sited on alluvial clays. All three date from the 1960s, and were probably designed as attenuation (dilute and disperse) sites, with the mobile contaminants being ameliorated as they passed down through the alluvial clays. However, if the leachate gains access to the underlying gravels, this may cause its delayed release into the environment. These gravels in the Cardiff Bay area could, therefore, be polluted by landfill leachate unless physical barriers are constructed. As the pattern of any groundwater or associated pollution is dominated by the local geology/hydrogeology and the localized nature of the potential pollution source an integrated approach will link the many localized sources and evaluate them as part of the regional system.

The gravels in the Cardiff Bay area are also important as they form a foundation level for many of the major structures of the area such as the Cardiff Bay Barrage and the Bute Town Road link. As groundwater from these gravels may rise several metres by peizometric pressure and show tidal fluctuations in wells, hydraulic heave on structures may be significant. In the Cardiff Bay area extensive groundwater monitoring has provided a model for the consequences of a rising, post-barrage, groundwater table (Edwards 1997) and at Cosmeston Country Park the possibility of leachate flowing into the lakes.

These examples show how the geological data derived from these studies may be combined to provide valuable environmental information. An integrated, mainly project-based, approach provides a historical and environmentally important database formulated from engineering, geological and hydrogeological information. This gives the characteristics, extent and location of the aquifers, acquicludes and aquitards of the area and enables groundwater flow charts to be constructed to provide control on the siting or management of new landfills. The increased borehole information strengthens the stratigraphical control of the area and establishes the often extreme variability in the strength characteristics of the Mercia Mudstone Group. The distribution and nature of the anthropogenic fill material is also now better known and this will undoubtedly influence the type of remediation that may have to take place before development.

Conclusions

This paper has described some of the various projects from which new geological and engineering data have been generated. Within the area redevelopment has produced an extensive database, which in the case of the Cardiff Bay area may number in excess of 2000 borehole logs (Edwards 1997). The linking together of these data across the many individual projects from which they were obtained can give an essential overview and will allow the characteristics and extent of regionally significant beds, such as the glacio-fluvial gravels to be examined. The information may then be projected into areas of unknown geology to provide decision-makers with an extra tier of information. This enables the decision-maker to at least anticipate some of the problems that may need to be addressed before any new developments can take place. A project when viewed in isolation may look favourable but when viewed from a holistic standpoint may place the natural system under pressure.

Acknowledgements

The author wishes to acknowledge the geological and engineering staff of Cardiff Unitary Authority, Cardiff Bay Development Corporation, Dwr Cymru/Welsh Water and the Vale of Glamorgan Unitary Authority without whom this paper could not have been written. The manuscript was formatted by Pam Rees.

References

Anderson, J. G. C. 1968. The Concealed Rock Surface and Overlying Deposits of the Severn Valley and Estuaries from Upton to Neath. *Proceedings of the South Wales Institute of Engineers,* **83**, 27–47.

Chandler, R. J. & Davis, A. G. 1973. *Further Work on the*

Engineering Properties of Keuper Marl, CIRIA Report 47.
Edwards, R. J. G. 1979. A review of the hydrogeological studies for the Cardiff Bay Barrage. *Quarterly Journal of Engineering Geology*, **30**, 49–62.
Heathcote, J. A., Lewis, R. T., Russell, D. I. & Soley, R. W. N. 1997. Cardiff Bay Barrage: investigating groundwater control in a tidal aquifer. *Quarterly Journal of Engineering Geology*, **30**, 63–77.
Kidson, C. 1977. The Coast of South West England. *In:* Kidson, C. & Tooley, M.J. (eds) *The Quaternary History of the Irish Sea.* Seel House Press, Liverpool. 257–298.
Lee C. W. 1996. *The Geological Setting of the Upper Triassic Red Mudstones Piling Seminar.* South Glamorgan County Engineering Consultancy.
Leigh, A. C. 1976. The Triassic Rocks with particular reference to predicted and observed performance of some major foundations. *Géotechnique*, **26**, 391–452.
Privett, K. 1996. Physical containment of contaminated land: vertical slurry wall barriers. *In: Assessment and Treatment of Contaminated Land.* Speakers Notes, CIRIA Seminar Series.
Thomas, B. R. 1997. Possible effects of rising groundwater levels on a gasworks site: at case study from Cardiff Bay UK. *Quarterly Journal of Engineering Geology*, **30**, 79–93.

11 Environmental impacts of lead mining in the Ullswater catchment (English Lake District): dam failures and flooding

Jane Anderton, Elizabeth Y. Haworth, David J. Horne and David S. Wray

Summary

- Ullswater, in the English Lake District, is a lake with a history of mining in its catchment, the most important being the Green Side lead mine which closed in the early 1960s.
- Reservoirs facilitated the provision of hydraulic and hydro-electric power for the mine, but dam failure is known to have been a problem.
- Contamination of lake sediments with lead and other metals, together with the potential for slope failures in the spoil heaps around the mine, continues to be a cause of concern 35 years after the Green Side mine closed.
- The mine's history is reviewed, with emphasis on the causes and impacts of the dam failures, and correlated with the lake sediment record.

Ullswater is the second largest lake in the Lake District National Park (11.8 km long, 145 m above sea-level, maximum depth 63 m; Ramsbottom 1976), occupying a glacially-deepened valley which opens northeastwards into the Vale of Eden near Penrith (Fig. 11.1). The boundary of its 145.5 km^2 catchment includes several summits above 800 m, including Helvellyn and High Street, and encloses an area of varied geology ranging from Ordovician Borrowdale Volcanics in the south, to Ordovician Skiddaw Slates, Devonian Mell Fell Conglomerate and Carboniferous Limestone in the north. The southern part of the catchment, composed mainly of Borrowdale Volcanics tuffs and lavas variously overlain by Quaternary sediments, contains several abandoned mines, the most

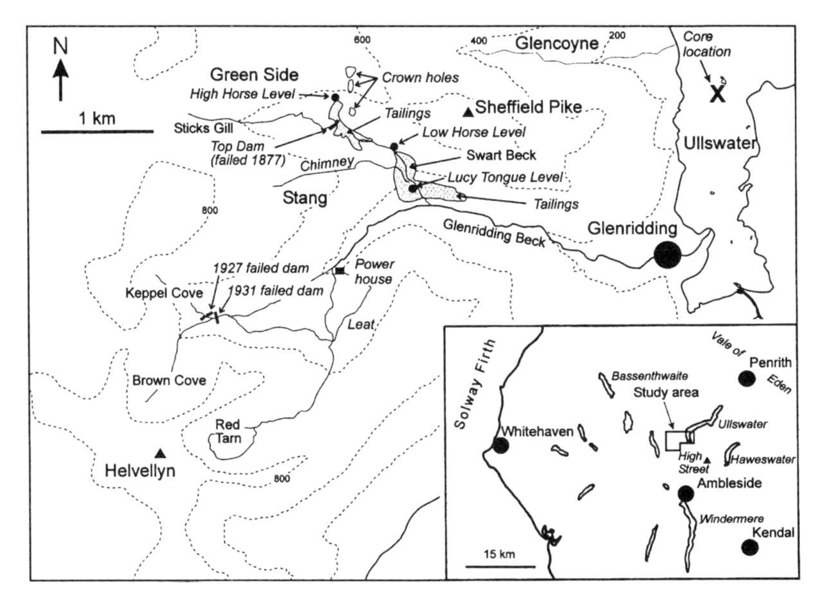

Fig. 11.1. Map showing the location of the Ullswater catchment in the English Lake District and the area around the Green Side lead mine, illustrating features referred to in the text.

important of which is the Green Side lead mine (Fig. 11.2) which closed in the early 1960s. As part of a research programme on the environmental impacts of lead mining in the Ullswater catchment we have focused on the effects of a series of dam failures and flood events associated with the Green Side mine which are thought to have flushed lead-contaminated sediment into Ullswater. The purpose of this chapter is to review the historical and physical evidence of the dam failures and to briefly consider the preliminary results of palaeolimnological investigations aimed at determining the record of these events as preserved in the lake sediments of Ullswater.

The Green Side lead mine

Working of the Green Side lode, which extends approximately south to north from the vicinity of the junction of Swart Beck and Glenridding Beck to the head of Glencoyne (Fig. 11.1), may have

Fig. 11.2. Green Side lead mine: (**a**) view down Glenridding valley from the vicinity of Swart Beck/Lucy Tongue Level (Fig. 11.1), showing spoil heaps planted with grass in the foreground. The south end of Ullswater may be seen in the distance; (**b**) view northwards towards Green Side showing crown holes (arrowed) and tailings in the foreground (Fig. 11.1).

begun as early as the late seventeenth century (Borlase 1908; Tyler 1992). The mine exploited fault-controlled lead/zinc mineralization in Borrowdale Volcanics. The vein contains galena (with a high silver content), chalcopyrite, sphalerite, quartz, calcite and barytes. In the early nineteenth century, water supply for the water wheels, which powered the crushers and other processing machinery for the mine (then concentrated at high levels between Green Side and Stang End), was provided by a leat (channel) constructed from the head of Sticks Gill. Waste from the higher levels was either backfilled in abandoned workings or tipped in a pattern of low ridges fanning out across the upper valley of Swart Beck (Fig. 11.1). The establishment of the Green Side Mining Company in 1827 was followed by the construction of a road up to Swart Beck, ore extraction from lower levels, the tipping of waste along the northern slopes of the Glenridding valley (Fig. 11.2a) and the construction of an elaborate water supply system involving dams and leats. Lucy Tongue Level was driven in the 1850s. Top Dam, on Sticks Gill (Fig. 11.1), was probably constructed in the 1830s to improve the water supply to the mine; the remains of the dam now show it to be an earth wall, faced with stone and linking moraine ridges, but it may have been enlarged at least twice during its history. Top Dam is now breached, but this is probably the result of a deliberate act when the mine closed, and is not related to the partial failure of Top Dam in the 1870s. The purpose of a small dam lower down on Swart Beck above Low Horse Level seems to have been to divert the beck and alleviate flooding in the lower levels, rather than to ensure water supply. The first edition 1 : 10 560 Ordnance Survey map (surveyed 1860) shows Keppelcove Tarn as already possessing a sluice and being supplemented by influx from two artificial watercourses, although there is no indication of any outgoing leats; other mine-related features shown on this map include the Top Dam reservoir on Swart Beck and a chimney extending nearly 3 km up the fell side from the smelter near the confluence of Swart Beck and Glenridding Beck to a position high up on Stang (Fig. 11.1).

Three large irregular craters which now exist in a line on the southern slope of Green Side (Fig. 11.2b) are crown holes, probably related to a major collapse of material through several levels in 1862

(Tyler 1992). Some leats in this area may have been cut subsequent to the collapse, to intercept and divert surface water away from the crown holes and thus prevent flooding of the mine (Royal Commission on Historical Monuments 1993).

The main water management infrastructure took shape in the 1860s and 1870s, incorporating Red Tarn, Brown Cove Tarn and Keppelcove Tarn (Fig. 11.1). According to Tyler (1992) the Keppelcove earth dam was constructed around 1867, but it is clear that this 'construction' was actually only the modification of an existing arcuate end-moraine formed over 10 000 years ago by a Younger Dryas (Loch Lomond Readvance) glacier (Carling & Glaister 1987; Binney 1992). The inner, water face of the dam was lined and strengthened with stone paving, peat and soft bricks, a 40 m long iron pipe (*c*. 40 cm diameter) was driven through near the base of the 10 m high moraine, and leats were constructed to carry water in from Brown Cove Tarn and away to Red Tarn Beck. We agree with Carling & Glaister (1987) that there is no evidence to support Hay's (1928) contention that the Keppelcove dam was artificially raised to increase reservoir volume. The narrow spillway, still preserved, seems to have been built at the natural outflow point of the orginal tarn, in the top of the moraine. A small (uncompleted) dam on Brown Cove Tarn served to maintain water levels to top up Keppelcove Tarn; another small dam was built to raise water levels at Red Tarn, from which a leat (augmented by the incoming leat from Keppelcove) carried water to a position high up on the southern side of the Glenridding valley, where it was dropped through a pipe to a hydroelectric power house at the foot of the slope near the junction of Red Tarn and Glenridding becks (Borlase 1908). The Green Side mine was reliant on this power source until the late 1930s when it became connected to the National Grid. The mine closed in 1962 and, in 1977, most of its remains became the responsibility of the Lake District Special Planning Board, whose stewardship of this contaminated and dangerous site is made even more challenging by its designation as a Scheduled Ancient Monument. The growing spoil heaps, particularly those composed of fines from the crushers, became increasingly problematic throughout the working life of the mine and continue to be a cause of concern today. A serious mudflow occurred during the 1940s as a result of a misplaced drainage pipe introducing water directly into the Lucy Level spoil heaps, and another major slope failure in the late 1980s engulfed the old smelter building (Tyler 1992). Measures aimed at stabilizing them began in the 1940s (Connor 1955) and have included the planting of grass and trees (Fig. 11.2a), the application of sewage sludge and improvements to the drainage (Lake District Special Planning Board 1978, 1992). Remedial drainage work was carried out in 1996.

Dam failures and their consequences

Top Dam failure, 1877

A catastrophic flood down Swart Beck after heavy rain one night in 1877 is attributed to the failure of Top Dam, although as Tyler (1992) points out, it is difficult to see how the water could then have bypassed extensive spoil heaps in its path without disturbing them (Fig. 11.1). In fact the outflow from Top Dam tarn was apparently culverted under spoil heaps which originated from the High Horse Level in the 1830s. Possibly water escaping through the culvert accumulated behind a smaller dam that had been built a little way downstream near the dressing floor, and it was this dam which failed. Whatever its cause, the flood was noteworthy for the destruction of a silver refining house with the loss of 1000 ounces of silver, the present-day value of which Tyler (1992) estimates at over £50 000. Wainwright (1955) shows the reservoir as still in existence but little more than marshy ground now remains; presumably it was drained as a safety measure following the mine's closure in 1962.

Keppelcove Tarn dam failure, 1927

Although the Keppelcove dam failure and flood of 1927 is much-better documented than the Top Dam event, there are nevertheless discrepancies between reports. Surprisingly there is even disagreement about the date of the event. Tyler (1992) puts it at 2.00 am on the morning of Saturday 5 November, but Carling & Glaister (1987) give 29 October, with the flood wave from the breached dam passing through Glenridding village (4 km downstream) between 1.30 and 1.40 am. According to Heaton Cooper (1983) the moraine

Fig. 11.3. Kepple Cove Dam failures: (**a**) view southwards showing the breached moraine (arrowed) of Keppelcove Tarn; the dark area is the marshy floor of the drained tarn; (**b**) the concrete dam at Keppel Cove, looking upstream, showing the breach at its north end (arrowed).

gave way in the early hours of Saturday 30 October. The correct date appears to be Saturday 29 October, corroborated by local newspaper reports of the disaster. The report in the *Cumberland and Westmorland Herald*, for example, was published a week after the event on Saturday 5 November, perhaps giving rise to Tyler's error. It is clear from contemporary accounts that the dam failed after prolonged heavy rainfall and the resultant flood wave caused extensive damage to property on its way through Glenridding to Ullswater, although no human lives were lost. The moraine dam failed east of the spillway, in the vicinity of the iron pipe through which water was normally taken from the tarn, resulting in a v-shaped breach about 35 m wide and about 10 m deep (Fig. 11.3a). Keppelcove Tarn, containing an estimated 124 000 m^3 of water (Carling & Glaister 1987), was completely emptied and remains so to this day. About 1 m of material was removed by scouring immediately below the breach, while further downstream, entrained sediment from the moraine or from sediments in Glenridding beck was deposited in fans and berms. The stream-bed was raised by as much as 1 m by fine gravels in its lower course, and a considerable amount of sediment was added to the delta on the shore of Ullswater (Carling & Glaister 1987). In Glenridding houses and bridges were severely damaged by the flood and the main street and shops were filled with mud and debris (Tyler 1992).

The precise cause of the breach is uncertain. A likely scenario is that water flowing along the outside of the pipe that had been inserted near the base of the moraine had gradually washed out quantities of fine sediment (sapping or piping). The heavy rainfall prior to the failure would have increased the hydrostatic pressure in the tarn and consequently the rate of sapping along the outside of the pipe. This process is self-accelerating: as fines are removed the flow increases, leading to greater sediment loss. This could have resulted in sagging of the crest of the dam above the pipe, with consequent overtopping leading to a catastrophic washout of the moraine. As noted by Carling & Glaister (1987), the gravelly moraine contains very little cohesive silt or clay and would have been rapidly eroded once the breach started. Another possibility is that quarrying of gravels on the outer, air face of the dam (as shown on the second edition 1 : 10 560 OS map) weakened the moraine and resulted in its collapse during the storm (Royal Commission on Historical Monuments 1993).

A further consequence of the dam failure was the loss of adequate power supply for the mine. It was decided not to try to repair the breached moraine, however, but to construct a new concrete dam immediately downstream of the failure site (Fig. 11.1).

The concrete dam failure, 1931

The concrete dam was completed in 1929 and still stands today. It is 80 m long and 5 m thick at the base, narrowing to less than 2 m at the top; its maximum height is approximately 12 m. An iron sluice valve still *in situ* at the base of the dam on the downstream side bears the date 1869; Tyler (1992) suggests that it was rescued from the original Keppelcove dam, although how it could have survived the breaching of that dam is hard to imagine. The concrete dam is built across, and partly obscures, an earlier stone spillway and low dam; possibly this functioned as a stilling-pond, in which sediment could settle out of water extracted from the original Kepplecove Tarn, before it flowed on into the leat leading to the power station. However, although still standing, the concrete dam retains no water due to a ragged 10 m wide hole at its base on the north side of the stream (Fig. 11.3b). This breach occurred in 1931 according to Tyler (1992), although Heaton Cooper (1983) gives it as 1932. The dam failure was almost certainly a consequence of the north side of the dam having been footed, not on bedrock, but on moraine. The hydraulic gradient through the moraine would have been large, leading to natural piping, washout of the moraine and failure of the consequently unsupported part of the dam. Compared to the Keppelcove moraine collapse, this failure appears to have been relatively slow and gentle; that it is the least-documented of the three may be simply because no catastrophic flooding occurred. Nevertheless it added to the accumulating problems of the mine company (compensation payements for the Keppelcove disaster, collapse and flooding in some of the mine levels and a depression in the lead market) which went into liquidation (Tyler 1992). In 1936 it was taken over by the Basinghall Mining Company.

Palaeolimnological investigations

It has been known for some time that former mining activity around Ullswater has resulted in elevated lead levels in the lake. The presence of a lead-rich layer in the lake sediments was familiar enough for George Thompson (Laboratory Steward at the Freshwater Biological Association) to write in his notebook in 1961 that a certain core from Ullswater was unusual in that it 'did not contain the usual thickness of white (lead) deposit'. Stockner & Lund (1970) found that live algae only occurred in the upper 8 cm of Ullswater sediment underlain by a layer of silt and clay with a high lead content, as analysed by F. J. H. Mackereth. Some aquatic macrophytes were found to contain high lead levels (Welsh & Denny 1980), and it was rumoured (but unproven) that swan deaths might be linked to former mining activity. Denney (1981) showed that although metal concentrations were highest around the southern end of Ullswater, near the source of the contamination, elevated levels could be detected along the entire length of the lake, a fact which he attributed to possible cycling of contaminants through macrophytes and phytoplankton. Our initial investigations, using a Jenkin Surface Mud Sampler to take short cores (10–30 cm long) of lake sediments, showed that a distinctive light grey, lead-rich mud layer could frequently be encountered at core depths of 10–20 cm in the southern part of the lake, contrasting with the usual brown, organic lake muds. We assumed, at first, that this grey layer represented the well-known Keppelcove Tarn dam failure flood event. The evidence gathered so far, however, suggests that the story is more complex. First, analysis of cladoceran remains from a 1 m core provided evidence of changes in zooplankton assemblages associated with a bundle of five discrete grey layers at 18–35 cm depth in the sediment: chydorids were dominant below the grey layers, bosminids above, with both groups showing abundance minima within the grey-layer sequence itself (Burrows 1994; Field 1994). While it is tempting to attribute this change to elevated lead levels, other causes such as high amounts of suspended sediment in the water column or changes in the trophic status of the lake must also be considered. Second, a more recently-obtained 1 m core from Ullswater shows at least five well-defined

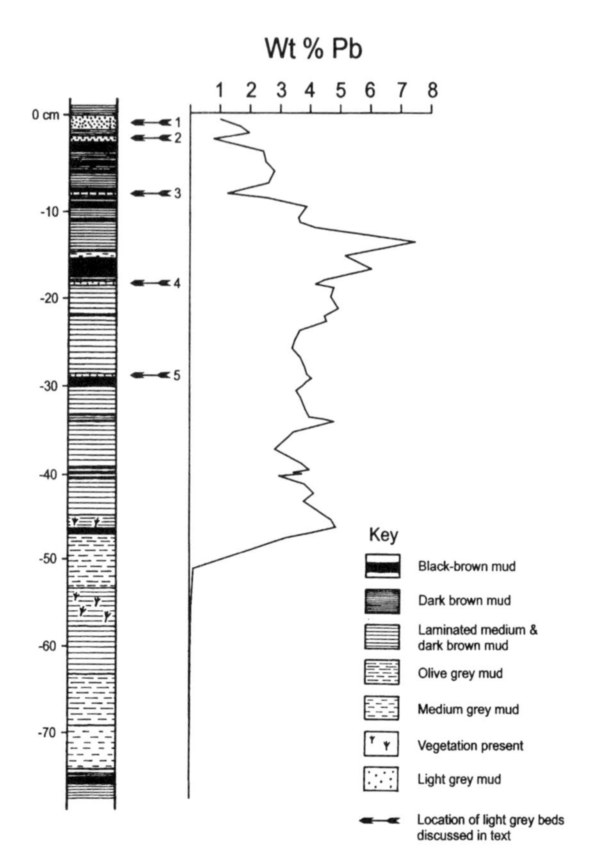

Fig. 11.4. Lead levels in a lake sediment core from Ullswater (see Fig. 11.1 for location). Note that the top 20 cm of sediment is omitted. Five grey clay/silt layers are indicated.

grey layers, not just one. This core may also have penetrated a grey layer within the top 20 cm of sediment, similar to that obtained with the Jenkin Surface Mud Sampler, but since this section of the core became disturbed during transport to the laboratory, it has been disregarded. Third, preliminary geochemical analyses of this core show elevated lead levels through a considerable thickness, not just confined to the first-recognized grey layer (Fig. 11.4). None of the peaks in lead concentration seem to correspond exactly with any of the grey layers. In view of our knowledge of at least three flood events in the mine's history, we cannot confidently attribute any single layer to any single event without dating the core, a project currently in hand. In fact we are no longer convinced that the best-documented dam failure, that of Keppelcove Tarn, resulted in any recognizable lead-rich layer in the lake sediments. Given the route taken by the flood, it is by no means certain that any of the spoil heaps were directly affected by it, although it is probable that the heavy rain at the time would have increased the general influx of contaminated sediment to the lake.

Geochemistry

In this section we present results and a provisional interpretation of lead data derived from a 77 cm sediment core collected from Ullswater at a site to the north of the Glenridding Beck delta (Fig. 11.1).

The core was collected using a 1 m Mackereth corer. Bedding parallel samples were taken at approximately 1 cm intervals. Beds less than 1 cm in thickness were sampled individually. Unfortunately, approximately the top 20 cm was disturbed during transit and hence was not sampled. The datum line for the core was established at the top of the first prominent light grey layer that could be successfully sampled. Depths discussed below are measured relative to this datum. Prior to preparation and analysis for lead, samples were stored in sealed plastic bags in a refrigerator. Samples were dried, and ground using an agate ball mill. Before weighing, ground samples were dried at 105°C overnight. A sample of 1 g was placed in an acid washed test tube and leached using AnalaR HNO_3 for three hours at 70°C. The resulting solution was filtered into a 100 ml volumetric flask, made to the mark with deionized water and stored in plastic bottles. Analysis was undertaken by ICP-OES, calibration was achieved using synthetic standards derived from a 1000 mg/l BDH SpectrosoL standard. Leach efficiency, based on results from two certified reference materials (marine sediment NRC PACS-1 and stream sediment NRCCRM C73-11) which were prepared in the same manner as the unknowns was better than 80%. Precision, based on analysis of an unknown sample prepared and analysed in duplicate, was better than 3%.

Data are plotted in Fig. 11.4. Results below −70 cm range from 20 to 28 $\mu g\, g^{-1}$. Lead concentration then rises gradually to a level of 1214 $\mu g\, g^{-1}$ at −51 cm before climbing sharply to a level of around 4 wt %. Although somewhat irregular, lead levels remain broadly constant until −16 cm where a further climb is observed to a maximum of 7.5 wt % at −13.5 cm. Above this point lead levels decline, with two noticeable drops coinciding with light grey beds.

Lead levels in the lowest part of the core are comparable to values from the lower portion of a core sampled from nearby Brotherswater (Rae & Parker 1996) and to the value for average shale (Alloway 1995), and hence probably represent the catchment's background concentration. Above −70 cm all values are significantly above catchment background and abnormally high for shales, implying that mining activity in the catchment was contributing lead to the lake sediments. The lead may have been derived from primary processing activities such as the crushing and washing of the ore, or from erosion of tailings. The relatively gradual increase in lead observed to −51 cm implies that mining was very localized and small scale. Lead values above −51 cm are comparable to, or higher, than those reported from soils in the proximity of other old mining areas (Alloway & Davies 1971; Fuge *et al.* 1989; Xiangdong & Thornton 1993) and imply a rapid expansion of mining activity at and above this level. Results are orders of magnitude greater than the highest values reported by Rae & Parker (1996) from Brotherswater, in whose catchment mining activity occurred on a much smaller scale. Irregular fluctuations in lead concentration above −51 cm may reflect slight variations in mining activity although changes in lake sedimentation rates and tailings erosion rates could also account for this pattern. The peak at −13.5 cm is considerably higher than the normal fluctuation and hence it seems likely that this point represents a true increase in mining activity. Likewise, the gradual fall in lead levels above −13.5 cm probably represents at least a reduction in mining activity from this point onwards. It is possible that the fall represents a complete cessation of activity with the lead contribution above −13.5 cm derived solely from erosion of tailings and material already contained within the stream bed. We hope to resolve this and other questions in the near future through ^{210}Pb and ^{137}Cs dating of the core. The sharp drop in lead levels associated with light grey layers 2 and 3 is not repeated in layers 4 and 5 and not clearly shown to occur in layer 1. It is therefore difficult to directly relate these layers to mining activity and the rapid erosion of tailings, as had been proposed previously, and further work is required to establish their significance.

Conclusions

It seems likely that the lake sediments of Ullswater contain records of both natural flood events and anthropogenic impacts, which we hope will be elucidated by work on a further suite of cores. Furthermore, the impacts of other (often less well-known) lead mines in the catchment still have to be considered; for example, the onset of lead mining in one sub-catchment has been dated at around 800 BP on the basis of a 4.5 m sediment core from Brotherswater (Rae & Parker 1996). Possible sequestering of contaminants in soils and stream sediments will also have to be investigated.

Ullswater is a lake with a history of lead mining in its catchment and which is known to contain elevated levels of lead and other metals. The Green Side lead mine had an extensive water management scheme for the provision of hydraulic and hydro-electric power; three dam failures occurred during the working life of the mine, one of which caused extensive damage in Glenridding village. The potential for slope failures in spoil heaps is another hazard which continues to give cause for concern 35 years after the mine closed. The prospect of correlating the palaeolimnological record with the historical record of lead mining in the catchment is impeded by several factors, including the possibility of biological cycling of lead within the lake, as well as uncertainty about the influence of flood events related to dam failures. It is clear that the abandoned mines continue to be a source of pollution and a potential hazard; successful assessment, management and mitiga-tion of such hazards are ultimately dependent on a full and detailed understanding of the nature and history of the problem. It is towards such an understanding that our investigations are aimed.

Acknowledgements

Peter Allen (Institute of Freshwater Ecology) is thanked for his assistance with preparations for coring on Ullswater. Our understanding of the impacts of lead mining on Ullswater has been furthered by discussions with Glen George, Ed Tipping and many others at the IFE, to whom we are indebted. We are grateful to the numerous undergraduate students undertaking project work in the area who have added to our knowledge, in particular Shirley Wheatley, Rachel Paice, Becky Hardman, Justine Stanghan, Marie Lawlor, Kelly Watson and Joanne Parry. This paper forms part of a PhD research programme by Jane Anderton, supported by the University of Greenwich and the Institute of Freshwater Ecology. David Horne wishes to thank Norman Bell of Windermere and Roland Wood for accompanying him on his first research visit to Keppelcove Tarn.

References

Alloway, B. J. (ed.) 1995. The origins of heavy metals in soils. *In*: Alloway, B. J. (ed.) *Heavy Metals in Soils*. Blackie and Son, Glasgow.

——— & Davies, B. E. 1971. Trace element content of soils affected by base metal mining in Wales. *Geoderma*, **5**, 197–208.

Binney, H. A. 1992. *A palaeoenvironmental investigation of Keppel Cove in the English Lake District*. MSc dissertation, City of London Polytechnic/Polytechnic of North London/Thames Polytechnic.

Borlase, E. T. 1908. The Green Side lead mines, Cumberland. *The Engineering & Mining Journal*, **85**, 297–301, New York, February 8.

Burrows, Y. M. 1994. *An examination of the cladoceran remains in a short core from Ullswater*. MSc dissertation, University of North London/London Guildhall University/University of Greenwich.

Carling, P. A. & Glaister, M. S. 1987 Reconstruction of a flood resulting from a moraine-dam failure using geomorphological evidence and dam-break modeling. *In*: Mayer, L. & Nash, D.

(eds) *Catastrophic Flooding*. Allen & Unwin, Boston, 181–200.
Connor, C. 1955. Green Side mine. *Mine & Quarry Engineering*, June, 222–231.
Denny, P. 1981. Limnological studies on the relocation of lead in Ullswater, Cumbria. *In*: Say, P. J. & Whitton, B. A. (eds) *Heavy Metals in Northern England: environmental and biological aspects*. University of Durham, Dept of Botany, 93–98.
Field, C. R. 1994. *An investigation into the effects of lead pollution on the Order Cladocera by the preparation and subsequent analysis of a sediment core taken from the bed of Lake Ullswater in the Cumbrian Lake District, England.* MSc dissertation, University of North London/London Guildhall University/University of Greenwich.
Fuge, R., Paveley, C. F. & Holdham, M. T. 1989. Heavy metal contamination in the Tanat Valley, North Wales. *Environmental Geochemistry and Health*, **11**, 127–135.
Hay, T. 1928. Glenridding flood of 1927. *Geographical Journal*, **73**, 90–91.
Heaton Cooper, W. 1983. *The Tarns of Lakeland* (3rd edn), Frank Peters Publishing, Kendal, Cumbria.
Lake District Special Planning Board. 1978. *The management of Glenridding Common and Green Side Mine. A report for consultations*. Lake District National Park, Kendal.
——— 1992. *Green Side Mines. Five year plan*. Lake District National Park, Kendal.
Rae, J. E. & Parker, A. 1996. Techniques for validating the historic record of lake sediments. A demonstration of their use in the English Lake District. *Applied Geochemistry*, **11**, 211–215.
Ramsbottom, A. E. 1976. *Depth charts of the Cumbrian lakes*. Freshwater Biological Association Scientific Publication no.33, Titus Wilson, Kendal.
Royal Commission on the Historical Monuments of England, 1993. *The Green Side lead mines*. Unpublished report.
Stockner, J. G. & Lund, J. W. G. 1970. Live algae in post-glacial lake deposits. *Limnology & Oceanography*, **15**, 41–58.
Tyler, I. 1992. *Green Side. A tale of lakeland miners*. Red Earth Publications, Ulverston.

Wainwright, A. 1955. *A Pictorial Guide to the Lakeland Fells. Book 1, The Eastern Fells*. Westmorland Gazette, Kendal.
Welsh, R. P. H. & Denny, P. 1980. The uptake of lead and copper by submerged aquatic macrophytes in two English lakes. *Journal of Ecology*, **68**, 443–425.
Xiangdong, L. & Thornton, I. 1993. Multi-element contamination of soils and plants in old mining areas, U.K. *Applied Geochemistry*, Supplementary Issue **2**, 51–56.

12 The role of academic environmental geoscientists in radioactive waste disposal assessment

G. D. Couples, C. McKeown, R. S. Haszeldine & D. K. Smythe

Summary

- Academic environmental geoscientists can provide an independent scientific view on the issues involved in the disposal of nuclear waste, as illustrated by the proposed site at Sellafield, Cumbria.
- UK NIREX Ltd had identified Sellafield as a possible site for the disposal of low and intermediate level nuclear waste, but following an extensive Public Inquiry, it was judged to be unsuitable.
- Here we discuss some of the contentious issues which played a role in the Inquiry, and use computer simulations to better understand the geosphere processes involving hydrogeology and geochemical interactions.
- An open forum needs to be established within which the academic community, together with commercial geoscientists, can reach an understanding on the best approaches, and on the sites which appear best suited, to solve the pressing environmental problem of radioactive waste disposal.

During the last half-century of the 'nuclear age', significant quantities of radioactive waste materials have been generated (in all the major nuclear countries); nearly all of these wastes are presently in storage awaiting a decision on how best to achieve their permanent disposal. The volumes of this waste are not large in comparison with domestic or industrial waste: 257 000 m^3 of Intermediate Level Waste (ILW), forming 26% of the total radioactive waste volume, will have been produced in the UK by AD 2030 (NIREX 1992*a*). As an aid to visualizing the quantity of waste, this volume is similar to that occupied by the 1×10^6 tonnes of metal in a small, commercial mine although a mine will have a substantial amount of gangue materials which will contribute to the void space created during its extraction.

Deep underground burial is presently considered to offer the greatest chance for long-term isolation of the waste (Bredehoeft & Maini 1981; Chapman & McKinley 1987; Billington *et al.* 1989; Nuclear Energy Agency 1989; Chapman 1994; Royal Society 1994) and is currently favoured in several countries (Hooper 1995; Karlsson 1995; Langmuir 1995; Horseman 1996). Assessment of the failure potential of such a disposal scheme is usually divided into two parts: the engineered system (waste form, containers and their surrounding packing); and that of the geosphere, which includes the hydrogeological system and the host rock. The disposal industry refers to containment in terms of an engineered barrier, which is surrounded by a geosphere which results in dilution and dispersion of escaped wastes. Although the failure probabilities of the engineered systems are relevant, most safety debates focus on the geosphere. In this chapter, we consider the role of the geosphere in retarding the return of radionuclides to the surface of the Earth. In doing so we emphasize the behaviour of the hydrogeological system, including the interaction of groundwaters with their natural and engineered settings.

Our purpose is to describe the ways in which environmental geoscientists can be involved in evaluating the expected performance of a nuclear waste disposal scheme – that is, the performance of the geosphere in slowing or preventing the migration of escaped wastes. Geoscientists are uniquely qualified to participate in such an assessment because of our breadth of knowledge concerning a wide range of interacting earth processes, and because of our appreciation of the complexity of earth processes across a spectrum of scales. Central to the involvement of environmental geoscientists is a recognition of the ways that mass can be transported through a body of rocks. Such concerns arise in all areas of geoscience (Wood 1997). Groundwaters are a readily-identified potential agent of transport, but gas-phase or additional liquid-phase mechanisms cannot be ignored. Within the groundwater domain, there is also a need to consider if the materials being transported (e.g. radionuclides) are retarded by physical and chemical interactions with the rock mass, or by changes in the chemical and/or electro-chemical state of the waters as they migrate to other crustal locations. Each of these concerns is relevant to some of the radioactive species which might escape from a disposal site. Many of the topics addressed in this chapter are applicable to other forms of waste which could be disposed of in the geosphere. We use the issue of nuclear waste disposal to illustrate general points that we seek to make concerning the involvement of academic environmental geoscientists in the broader topic of geosphere pollution assessment, prevention and remediation.

Context of geoscience involvement

We consider it useful to contrast two situations which arise in the effort to develop a safe disposal scheme for nuclear waste materials: (1) the exploration for potential sites; and (2) the evaluation of proposed sites. In the exploration phase, it is possible that academic and commercial geoscientists act similarly, but once a potential site has been identified and the safety of the site is being evaluated, the two groups certainly have differing agendas, as we explain below.

The content of this chapter refers specifically to the disposal site for radioactive waste that was proposed to be constructed at Sellafield, in Cumbria, UK, since the scientific and political issues raised prove particularly instructive. Other views on the geology and hydrogeology of the Sellafield site are summarized by Bath *et al.* (1996), Black & Brightman (1996), Heathcote *et al.* (1996) and Michie (1996). The disposal site at Sellafield was planned to be located in the Borrowdale Volcanic Group (BVG) at 650 m depth, at a location between the Cumbrian mountains and the coastline. These rocks are overlain (from 450 m up to the present land surface) by Carboniferous–Triassic marine deposits, terrestrial sandstones, and mudrocks, along with a veneer of Recent deposits.

It is important to point out that on 17 March 1997, the UK Secretary of State for the Environment upheld the decision by Cumbria County Council to reject a planning application to construct an underground Rock Characterization Facility (RCF) at the Sellafield site. His decision was based on the recommendations of the Planning Inquiry Inspector, who stated that: 'The indications are, in my judgement, still overwhelmingly that this site is not suitable for the proposed repository, and that investigations should now be moved to one of the more promising sites elsewhere' (McDonald *et al.* 1996). Since NIREX have stated that they will not challenge this decision the Sellafield repository site has, in effect, been abandoned.

Exploration

In common with disposal organizations overseas (Karlsson 1995; Horseman 1996), it is proposed that the entire UK inventory of ILW should be disposed of in a deep, geological repository (NIREX 1993*b*). Such repositories are subject to regulatory guidelines (Holmes 1995; Hooper 1995) such that for the period after closure of the disposal facility the radiological risk to the human population will be 1×10^{-6} (i.e. 1 death in 1 million per year). Given that ILW contains a significant inventory of very long-lived radioactive constituents (NIREX 1992*c*; Royal Society 1994), it is UK policy that the disposal concept provides adequate containment for greater than 10^8 years.

A sub-surface disposal site needs to meet certain basic characteristics. In terms of geosphere criteria, these have historically been paraphrased as:

(1) stable and workable rock types, to permit the safe and easy excavation of an open cavern;
(2) suitable hydrogeology;
(3) suitable geochemical environment.

Additionally, for the long timescales of radioactive waste duration, consideration must be given to:

(4) past and future tectonic activity;
(5) possible future intervention by humans seeking to retrieve the material, or to exploit a resource;
(6) climate change;
(7) meeting the present and future regulations on permitted dose rates (Chapman & McKinley 1987).

There has been an emphasis around the world to identify potential sites in 'homogeneous' crystalline (hard) rock masses, although other rock types (such as rock salt or mudrocks), and other situations (such as former mine workings) are also considered. Crucial to meeting the first characteristic is a lack of, or minimal number of, discontinuities (joints, faults) – since these have a major impact on rockmass behaviour. The matter of 'suitable hydrogeology' is less clear. The fact that groundwater is at all important results from the possibility that contaminants could be transported by the movement of groundwaters (in the worst case, to the Earth's surface, or perhaps into the shallow subsurface). The pattern and rate of undergroundwater flow must, therefore, be both predictable and safe for geologically long times into the future.

The diverse geological factors affecting such predictions, and some of their difficulties of measurement and forecasting, have been reviewed elsewhere (e.g. Chapman 1994). Fundamentally, the groundwater flow prediction must be most concerned with the path taken by any moving water after it leaves a repository and the rate of its movement. If the motion is such that any escaped waste gets carried to depth, and that the water mixes there (i.e. is diluted), then these conditions are favourable, and even better is the case where these movements are slow. Upwards flow paths, and the possibility of mixing with and polluting shallow aquifers – or even of discharge to the surface – are much less favourable, and rapid flows are undesired. It would obviously be ideal if groundwaters were static and any dissolved waste would not move by advective flows, but only by extremely slow diffusion.

Such static groundwaters might conceivably occur in situations with no hydraulic energy, although we are doubtful as to the existence of such circumstances. A variant on this theme is the case at Yucca Mountain, in the USA, where the intended disposal site is located well above a deep water table; here the hydrogeological concerns focus on transient phenomena (such as seismic pumping, climate change or hydrothermal activity) which conceivably might cause a drastic change in the groundwater system. A more typical view of hydrogeology is one in which groundwaters are present in a dynamic, connected system with an ever-present potential for movement depending on a range of parameters (Neuzil 1995; Toth 1995). Accepting this paradigm, the exploration exercise becomes one of locating sites where, as groundwaters move away from the disposal site, they flow either to deeper parts of the crust, or flow only very slowly back to the Earth's surface. In simple terms, it is essential that a repository be sited in an area of very slow local and regional groundwater movements (i.e. in an area with low regional hydraulic gradients; Chapman & McEwen 1986). This could mean that an ideal site would be in the recharge zone of a major groundwater system, whose flow paths away from the site were long in both time and distance, and directed to greater depth (Bredehoeft & Maini 1981; Toth & Sheng 1996).

Even though a groundwater system may be sluggish, consideration needs to be given to perturbations which could be induced as a consequence of the creation of a repository. For example, although ILW does not, by definition, produce much heat, this energy, and that from exothermic cement reactions (combined with slow, diffusive heat dispersion), could result in the temperature of a repository volume reaching 80°C (NIREX 1995*d*). Consequently, even a naturally static water system may be induced to circulation or convection following the emplacement of waste. This point is revisited below in the context of permeability changes brought about by construction works.

Exploration for a potential disposal site needs to consider the hydrogeological system, and how aspects of the geosphere, or the engineered system, can influence it. This approach was taken during the 1980s in the UK when a number of 'types' of site were identified by the British Geological Survey starting with the identification of general principles (Gray 1976). The reasoning behind each type of generic site is summarized elsewhere (Robins 1980; Chapman & McEwen 1986; NIREX 1989). All these generic sites share the characteristic of inferred slow groundwater flow directed away from the biosphere. Very general hydrogeological characteristics were described for each such type of site, along with comments on other selection factors, such as the rock types, the infrastructure, the local economy and political matters. Of the site types which have been described, none have a specific, 'real' example location identified, and so these sites remain, in the UK, merely hypothetical.

Using a range of research results (Garven 1995; Person *et al.* 1996; Toth & Sheng 1996), it is now possible to predict the general nature of a hydrogeological system based on its geological setting. The 'setting' includes the geometries of the land surface, and the configuration of the sub-surface rocks, along with information of fluid-type distribution, heat sources, or other factors which may influence the hydraulic energy. Such hydrogeological predictions are made using numerical simulations, and this approach requires the specification of material-property distributions, and other parameters, which are not 'known' for a hypothetical site. Nevertheless, geoscientists can make reasonable assumptions about such hypothetical cases and thereby provide a means to assess the general characteristics of each such system.

A principal benefit from such a numerical simulation approach is the identification of those aspects of each system which are most important in controlling the critical behaviour (e.g. the ground-water movement, geochemistry or rock mechanics). Once these key geosphere behaviours are understood, the final stage of exploration is for the geoscientist to seek real-world settings which meet the primary geosphere criteria established through the analyses noted above. If, for example, it is concluded that the 'ideal' site is a granitic upland area which has not suffered extension, then geoscientists can set about the task of identifying potential sites, or discovering that this type of site does not exist in a given country.

It is at the stage of site selection that the different pressures acting upon geoscientists start to become apparent. The costs of site evaluation are large; consequently there is a commercial pressure to investigate those sites which are known to be politically feasible – should they prove to be geologically suitable. It is now known (NIREX 1989; RWMAC 1995) that the present site at Sellafield was not in the original short list of geologically suitable sites, but was added later as a 'variant' of the original geological concept. The choice was then made to evaluate only two of the twelve sites: Sellafield, Cumbria, and Dounreay in northern Scotland (the investigation of Dounreay was later suspended). Both of these sites were later admitted by NIREX to be less than ideal geologically, and seem to have been chosen primarily for political considerations. As stated by the Planning Inquiry Inspector: 'It seems that the process was affected by a strong desire to locate the repository close to Sellafield' (McDonald *et al.* 1996). This strategic choice later proved to greatly complicate and limit the success of the evaluation phase and led ultimately to the site's rejection, although this has not (yet) led to a release of the locations of the other ten sites.

Evaluation

The issue of nuclear waste disposal is one demanding an independent scientific assessment by independent geoscientists. By 'independent' we mean those (usually in the academic world) who are not normally obliged to uphold the views of commercial firms as a consequence of financial support or other commitments. This notion is similar to the restrictions against material interest enjoined against prospective jurors, and the other 'conflict of interest' restrictions with which we are all familiar in our professional codes of practice.

The idea here is one which reaches to the very heart of the scientific endeavour. According to Popper (1963) the scientific method is a process by which new understanding emerges: (1) discovery of a problem (usually as a rebuff to existing theory); (2) bold solution (new theory); (3) deduction of testable propositions; (4) tests (i.e. attempted refutations); (5) preference between competing theories; and (6) back to (1) discovery of a problem. Scientific 'truth' emerges when claims withstand challenges and, especially, when others independently support those claims (Mermin 1996). The process requires the mutual availability of information, and it further requires an established mechanism for discussion and debate. The notion of 'independence' is inherent to the activities of scientists. Although the end result of science is not guaranteed to be universal agreement on interpretations, science does – following debate – produce a consensus on what is agreed, and what is still unknown. This consensus of agreement, and agreement to disagree, extends to interpretations of earth processes as well as of data.

We suggest that the scientific approach is appropriate for the evaluation of a potential nuclear waste disposal site. In terms of the evaluation stage, the 'claim' to be tested is that the site is suitable for nuclear waste disposal. If the claim, that the site is suitable, is shown to be weak, or wrong, and the challenges to the claim are upheld by further independent scrutiny, then the viability of the site falls into serious question. Alternatively, if academic geoscientists are able to independently converge on an interpretation of the geosphere at the site such that these independent interpretations support the site's viability, then the proposal gains credibility.

As is true for most of science, most geoscientists who are interested in the nuclear problem must choose the option of challenging a claim (i.e. that some site is suitable for waste disposal). An attempt to support the claim would require facilities and capabilities which are beyond the scope of academic institutions. Instead, academics most often choose to seek weaknesses in claims – perhaps this is because the scientific method is so intimately linked with challenges. The following section is an example of this approach as taken by academic environmental geoscientists concerning the (now abandoned) site at Sellafield.

Examples from Sellafield

UK NIREX Ltd proposed the construction of a nuclear waste disposal site at a location near the Sellafield reprocessing plant in Cumbria, UK. NIREX scientists evaluated the suitability of this site for many years, and later used this knowledge to apply for planning permission to construct an underground laboratory. A substantial amount of replicate work was carried out by contractors funded by Her Majesty's Inspectorate of Pollution. Along with brief overviews in annual RWMAC (Radioactive Waste Management Advisory Committee) reports (RWMAC 1994), peer reviews have been undertaken the Royal Society (Royal Society, 1994) and other groups commissioned by NIREX. RWMAC advises government on strategy and progress but states that a full peer review of NIREX work is beyond its scope.

Fluid/rock/waste geochemistry

The geochemical containment of radioactive waste in the UK is difficult for several reasons, including its complexity: (1) it contains long-lived radionuclides, such as uranium; (2) it contains radionuclides with contrasting geochemistries; and (3) it comprises a physical mix of many different elements produced by different 'waste streams', where information on the mineralogical forms, redox states and valences is not readily available. The radionuclide uranium forms an important part of the waste streams which are awaiting disposal. In terms of inventoried mass in the waste, uranium amounts to around 2.5×10^6 kg (NIREX 1992*a,b,c*). The longevity of its radioactivity (Weigel 1986) and the potential risk from the repository (Nuclear Energy Agency 1989) dictate that the retention of uranium forms a major component of the safety assessment. Although radionuclides such as plutonium, caesium and iodine are important contributors to the radioactive waste inventory, uranium is by far the most abundant and better understood, and we emphasize its behaviour in this chapter.

The task of the geoscientist here is to predict the possible concentration of radionuclides throughout the nearby geosphere during the required 10^8 years containment period (NIREX 1992*c*; Royal Society 1994). This formidable task is associated with a requirement to construct industrial-scale systems for containing the waste within the disposal facility. The inevitable slow leakage of waste from a disposal site through time produces a source term to input into models of dispersion in the geosphere. A robust approach to waste isolation also needs to consider the containment of radionuclides by the surrounding geosphere, which requires information on the geochemistry of the natural groundwaters, and on the geochemistry of the rock in contact with (and potentially reactive with) those waters. The success of such retention is measured by a residence time in the surrounding rock. Source term and residence time are seriously degraded by a large flux of natural groundwater through the disposal facility, thus geochemical containment cannot be considered in isolation from hydrogeology.

Repositories must include a number of 'near field' barriers which are designed to prevent or retard the release of radionuclides from the waste into the geosphere, and then to the biosphere (NIREX 1993*b*). In common with ILW repository concepts elsewhere (Miller *et al.* 1994), wastes were to be immobilized in the Sellafield repository, usually in a cementitious material, packaged in concrete or stainless steel/mild steel containers, emplaced in repository vaults and back-filled with hundreds of thousands of cubic metres of cement-based material (known as NIREX Reference Vault Backfill, or NRVB; Atkinson *et al.* 1988*a,b*; Atkinson & Guppy 1988; Atkinson 1995). NIREX Reference Vault Backfill is mainly composed of hydrated, blended Portland cements (Bennet *et al.* 1992). The detailed constitution of the phases in these cements is complicated but dominantly they are poorly crystalline and represented in modelling studies by the mineral portlandite (Atkinson *et al.* 1993).

The main rationale behind the ‘cement and steel’ approach is to provide physical containment in the short term (hundreds of years) and chemical containment for radionuclides over longer timescales (hundreds of thousands of years; Hooper 1995). The reducing ambient conditions produced by the corrosion of the steel and the high pH produced by the dissolution of cement were intended to promote precipitation of low-solubility phases and to promote sorption of ionic species onto mineral surfaces. Many questions still remain concerning the roles of: sorption of radionuclides to mineral surfaces; bacterial control; and the formation of organic compounds which can both aid and hinder radionuclide migration.

At the expected 80°C repository temperatures, the ambient fluid pH would be around 10 (Atkinson *et al.* 1991). During construction and operation of the repository, oxygen would be introduced, and when finally backfilled, the cement grout would contain both gaseous and dissolved oxygen in its pores (NIREX 1995*d*). This oxygen would cause the aerobic corrosion of the 10^9 moles of iron present in the repository (from mild and stainless steel packaging containers, construction materials and waste inventory; Naish *et al.* 1990). This mechanism would remove oxygen from the system and it would be expected that repository conditions would become anaerobic within 100 years of closure (Atkinson *et al.* 1993). It is claimed that corrosion of the steel would then proceed anaerobically and produce hydrogen gas and a very reducing Eh (theoretical Eh is as low as -780 mV; Haworth & Sharland 1995), although the long-term duration of this Eh is questionable (McKeown & Haszeldine 1996). However such hydrogen gas along with carbon dioxide and methane exsolving from the wastes would create a pressure inside the repository vault; this energy could force radioactive water to move away from the site, possibly to the surface (Horseman 1996).

The problems of designing a suitable geochemical containment scheme can be illustrated by two examples: iodine and uranium. The UK waste mix will produce substantial quantities of ^{129}I, which is both highly metabolizable and radioactive; hence, it is potentially carcinogenic, and needs to be excluded from the biosphere for geologically long times. Iodine gas and iodine compounds are very soluble in water, unless combined into compounds such as CuI_2.

Attempting to control the pH and Eh, or to otherwise buffer the near-field geochemistry, are unlikely to be effective at retaining iodine in the vicinity of a disposal site since its solubility is not strongly affected by variations in these characteristics (Bishop *et al.* 1989). The only factor which might have contributed to the local retention of iodine would be the slow flow of groundwater through the alkaline concrete backfill. Once groundwaters containing iodine moved away from the facility, this element would be transported in solution wherever the groundwaters flow. If iodine gas exsolved from the groundwaters, this would be extremely mobile. NIREX (1994*b*) simulations show that iodine can form one of the major components of radionuclide release in the early life of the repository. Some NIREX simulations, which consider the excavation damage related to vertical access shafts (NIREX 1994*b*), showed that radioactive iodine could return to the surface within forty years, in an adverse case.

A full review of the large body of work associated with the geochemistry of uranium is beyond the scope of this chapter. There are many documents giving a broad account of its properties in mineral, fuel and aqueous form (Weigel 1986; Finch & Ewing 1992; Janeczek & Ewing 1992; Harvey 1995). The complexities of uranium thermodynamics are also dealt with in great detail in a wide variety of papers and reports (Langmuir 1978; Morss 1986; Lemire 1988; Bruno *et al.* 1987, 1993; Cross & Ewart 1990; Pearson & Berner 1991; Fuger 1992; Grenthe *et al.* 1992). The most important and fundamental property of the actinide element uranium is that it can exist in more than one oxidation state. This has a direct effect on its solubility in aqueous solution (Weigel 1986). Uranium in aqueous solution is known to exist in oxidation states from U^{2+} to U^{6+} However, only U^{4+} and U^{6+} are significant in nature (Basham & Kemp 1993). In oxidizing fluids (relatively high Eh) uranium in the hexavalent form (generally agreed to be the uranyl ion (UO^{2+}; Grenthe *et al.* 1992) can complex with a very large number of ligands such as hydroxide, carbonate and sulphate. Such complexes become more soluble in aqueous solution than UO^{2+} (Langmuir 1978; Weigel 1986). These complexes can be mobilized and transported in groundwater. Conversely, reducing conditions (low Eh) encourage the U^{4+} oxidation state to dominate, uranium precipitates from solution and results in very low concentrations in mobile solutions (Nash *et al.* 1981).

Attempting to force a reduced uranium oxidation state within the repository, by means of the 'cement and steel' approach is a sensible option to lower the solubility of uranium-complex ions. However there is a problem – although the presence of iron is likely to produce an initial reduced Eh within the repository, the durability of this condition is open to serious question. The natural groundwater will pervade the repository. NIREX claim this groundwater to be chemically reducing, based on the assumption that the mineral pyrite (FeS_2) is at present coating the walls of fractures in the Borrowdale Volcanic Group. This appears to a be a strange assumption, for the Sellafield site is in the centre of Britain's largest iron oxide (hematite) ore deposit – with no record of pyrite in the ores (Rose & Dunham 1977). When we attempted to recreate these reducing conditions (using rock data from NIREX reports), we discovered that NIREX's own data did not demonstrate any pyrite at the depth of the repository (NIREX 1995*b*; Haszeldine 1996). We found that the Eh of the deep BVG groundwaters is not easily derived, a view shared by the Planning Inquiry Assessor (McDonald *et al.* 1996).

Using a geochemical modelling code, we have shown that the BVG waters – in common with many natural waters (Lindberg & Runnels 1984) – display strong redox disequilibrium, and are most probably oxidizing (McKeown 1997). We have also shown it is highly unrealistic that BVG groundwater Eh would be controlled by equilibrium with pyrite (Haszeldine 1996). Other studies which make the same assumptions as NIREX have also predicted very reducing Eh conditions in the BVG (Metcalfe & Crawford 1994). If such assumed conditions were carried through to safety case assessments, this would result in gross underestimates of risk.

If the BVG waters are oxidizing – as we maintain – the native iron buffer in the repository (e.g. the steel) would be consumed much more rapidly than NIREX have predicted. Instead of lasting 10 000 years (NIREX 1994*c*), the buffer may only last for hundreds of years. Consequently, there would be no effective long-term engineered barrier against the mobilisation of uranium, nor any natural chemical barrier to the release of uranium from this site (McKeown & Haszeldine 1995). In summary, it can be appreciated that creating a geochemical barrier to prevent the release of radionuclides is not a simple task; this effort needs to reconcile the various reactive chemical behaviours of different radionuclides, and it still needs to consider the fundamental geological setting in which the disposal site is located. The concept of generating suitable geochemical conditions seems to be intended to make virtually any sub-surface site suitable for radioactive waste disposal. We have shown that this does not work at Sellafield, which is geochemically poorly suited for a disposal site, with no inherent natural retention capacity for some of the most problematic radionuclides.

Hydrogeology

A hydrogeological system can be segmented into three component parts – the fluid, the rock, and the energy (Fig. 12.1). Each of these aspects relates to one or more of the parameters in Darcy's porous medium equation:

$$Q=(k \bullet \nabla H)/v,$$

where Q is the flux, k is the permeability, ∇H is the gradient of the hydraulic head, and v is the fluid viscosity. Fluid energy is directly related to the ∇H term. Fluid characteristics can impact the fluid energy (buoyancy variations) as well as the viscosity.

The rock framework is directly related to the permeability distribution, but it additionally affects the fluid energy in several ways:

(1) an irregular water table will occur when the ground surface

Fluid energies
Rock framework
Fluid properties

Darcy Flow

$$Q = \frac{k \bullet \nabla H}{v}$$

Fig. 12.1. Three of the components of a hydrogeological system in relation to Darcy's flow law.

(top boundary) is both saturated with pore water and topographically uneven;

(2) permeability heterogeneities (faults and fractures, or modern changes to pore geometries due to mineral precipitation) alter the transmission of fluid energy, by acting as flow baffles (pressure drop), or conduits;

(3) any mechanical change in the rock framework (such as neotectonics, or thermal expansion, or volumetric changes associated with construction) will alter the pore volume, and hence will affect the pressures of the contained pore fluids.

Other geosphere factors which may affect the fluid energy include: water-table variations resulting from climate change; brine intrusion; and fluctuations in heat energy. Although design temperatures at Sellafield are stated to be 'low' (80°C), even small amounts of extra heat can be sufficient to initiate convective circulation of groundwaters; such circulations would seriously affect all safety aspects related to water flux.

The simulation of fluid flow through rocks is a subject of great complexity (Huyakorn & Pinder 1983; de Marsily 1986; Cathles 1990; Deming 1994). Flow between the pore space of rocks enables groundwater flow to be described using a porous-medium continuum approach (Bear & Yehuda 1990). The presence of fluid flow in fractures can add further complications. In fractured rocks such as those of the BVG at Sellafield, the interconnected fractures are considered to be the main passages for fluid flow, with the solid rock matrix considered to be almost impermeable (Domenico & Schwartz 1990). This situation results in wide differences in hydraulic conductivity, with fracture conductivity many orders of magnitude greater than that of the matrix (Brace 1980; Neumann 1990; Clauser 1992). Modelling flow in such a fractured system is obviously problematic (Freeze & Cherry 1979; de Marsily 1986). However, it can be achieved by: (1) modelling the fluid transport through each fracture or network of discrete fractures (Moreno & Neretnieks 1993); and (2) modelling the transport regime in the fractured mass as an equivalent porous and anisotropic continuum (Follin & Thunkin 1994).

The problem with the first approach is that the fractures/fracture networks have to be extensively mapped, with information regarding the hydraulic conductivity, connectivity and geometry exhaustively recorded (Garven 1994). For large-scale regional modelling this is impossibly arduous, although some numerical models use stochastic methods to generate statistical fracture networks (Long *et al.* 1991). In the second approach the fractured medium is viewed as an equivalent porous medium. The obvious advantage is that models of this type are well understood by a wide array of geoscientists and there are numerous implementations available (Domenico & Schwartz 1990). The high density of fracturing in the crystalline rocks of the Sellafield area precludes the need to consider individual fractures within such a regional system. Beyond such practical matters, if the spacing of fractures or fracture zones is less than the scale of numerical discretization (or grid size) in the model, then the equivalent porous-medium approach is justifiable (Garven 1995). The porous-medium approach is widely favoured for regional 2D modelling of fluid movements (e.g. Garven 1995), and it has been applied by others to the Sellafield site (Nicholls 1995; NIREX 1995*c*; Heathcote *et al.* 1996).

To investigate the NIREX proposal for a disposal site at Sellafield site, we initiated a programme of hydrogeological research. We took a very different approach to that used by NIREX in their investigations: we chose to focus on the sensitivities of the whole fluid flow system, and we sought to understand the main processes operating and to identify the key variables affecting the behaviour of the system. From NIREX reports and public, mainly British Geological Survey, information we constructed a geological cross-section through the site. This was converted into a finite element grid which was used in a 2D computer model which simulated steady state fluid flow (Haszeldine & McKeown 1995). Each element was assigned numerical values representing the equivalent petrophysical properties (porosity and permeability). Simulations of fluid flow were run on this model; these established the likely pathways of fluid flow, and the rates of flow, and they allowed us to investigate the sensitivity of the model to changes in rock properties – both within and outwith the measured ranges of hydraulic conductivity (permeability) of the various rock units.

All steady state simulations showed that topographically-driven groundwaters would descend from the BVG uplands and then pass upwards through the repository site, above which there is less than 200 m of BVG rocks. This 'barrier' of BVG rocks is all that exists to retard the transport of radionuclides carried by the flow issuing from the repository before these materials enter the permeable clastic rocks overlying the BVG. Key parameters in the model include the regional hydraulic conductivities of the BVG, the Permian breccias and conglomerates, and the faults. These regional values are poorly known from borehole tests, with local measurements in the BVG ranging over eight orders of magnitude (NIREX 1993*a*). As noted below, this range of values, or even the lower mode of this range used by NIREX in their modelling, did not in itself assist in choosing the most realistic or 'best' values with which to represent the BVG.

In order that we could define the most appropriate rock parameters to represent the fluid flow system, we attempted to constrain our models to 'reality' by calibrating the sub-surface pressures calculated by the model, against the sub-surface pressures measured by NIREX in their boreholes (expressed as hydraulic heads). We found that the best fit to the borehole data was obtained by using a BVG regional hydraulic conductivity some 500-1000 times greater than that used by NIREX (McKeown 1997). This 'calibrated' BVG permeability was much greater than that which has been used by NIREX in establishing their safety case (NIREX 1995*c*). Consequently, the more rapid water flows we envisage to be occurring in the sub-surface (i.e. as predicted by this model) will permit leaked radionuclides to move more rapidly than has been suggested by NIREX, and therefore the site would fail to meet safety targets (Wallace 1996; Haszeldine & Smythe 1997).

This concern relating to rapid, upwards groundwater flow through the proposed site was a common thread in evidence given in support of Cumbria County Council at the Inquiry (Haszeldine & Smythe 1996) and, even after NIREX presented all of their hydrogeological evidence, the Inspector was not convinced that the best understanding of groundwater flow rates and paths had been reached (McDonald *et al.* 1996). The hydrogeological work we have undertaken (Haszeldine & McKeown 1995; Haszeldine 1996; McKeown 1997) indicated that the Sellafield site is fundamentally a poor location, being at the outflow end of a regional hydrogeological system, rather than the recharge end (Toth & Sheng 1996). The site is also complex, with flow through rock fractures resulting in wide ranges of measured hydraulic conductivity. We consider that our approach to determining the regional, bulk, 'effective' permeability of the rock units at this site is good practice (i.e. by calibrating the model against measured data) and that BVG hydraulic conductivity values we derived represent 'upscaled' permeabilities incorporating highly-permeable fracture zones. The predictions arising from the model effectively falsify the claim that the site is suitable for nuclear waste disposal.

Although regulatory bodies such as Her Majesty's Inspectorate of Pollution (HMIP) have for many years undertaken parallel research on the suitability of the Sellafield site, it is very interesting to note that two key reports were not obtainable until the Inquiry began. Marked 'commercial-in-confidence' they detailed results from research commissioned by NIREX to look at both fluid flow (Nicholls 1995) and geochemistry (Tyrer *et al.* 1995) at the site. After NIREX ceased funding these projects they remained unpublished until it was discovered by Friends of the Earth that the information was of sufficient public interest to merit free availability. Such problems in obtaining important information highlight the possibility that, without the approach taken by independent earth scientists, there might possibly not have been a freely available alternative view on the site's hydrogeological and geochemical safety to that proposed by NIREX.

Faults and rock structure
NIREX also predicted the structure of the rocks in the area around the proposed disposal site (NIREX 1994*a*, 1995*a*). This entailed mapping the depth and orientation of the bounding surfaces of the different rock units, for example, the top of the BVG; it also required mapping the faults which cut these rocks, including their orientations and their throws. These structural interpretations were essential to the site evaluation because the safety of any disposal site rests on the ability of the geosphere to contain any released radionuclides. As we emphasized above, a critical component of the containment process is the rate of flow of groundwater past the waste, and a crucial element of hydrogeological models is the location and permeability of faults and their associated fractures. Therefore, it is absolutely essential to be able to confidently identify faults within boreholes, and to correlate these faults between individual boreholes. It is also necessary to be able to make similar interpretations of faults from other data (e.g. seismic, structure contour maps).

The first surveys and interpretations of faulting at this site were undertaken by the British Geological Survey; they identified two faults with opposing dips. A further five different interpretations of fault configurations have been published by NIREX in 1990, 1991, 1993, 1993 and 1995. Each interpretation is different, and each shows a varying number of faults, and a range of fault orientations (Smythe 1996; Haszeldine & Smythe 1997). The geological interpretation of the site is apparently subject to substantial revision each year. In a scientific context, such changes of interpretation resulting from new data or further analysis are common; but in the context of assessing a potential nuclear waste repository, these changes are worrying. Consequently, there is no reason to expect that the version of the structural interpretation which NIREX submitted (in 1995) to the Planning Inquiry is anything like the true picture.

As part of the effort to determine fault positions, NIREX funded the acquisition of nine 2D seismic reflection lines. However, NIREX reports do not contain any seismic data, but merely interpretations of those data. This means that it is impossible to make any judgement about the quality of those interpretations, or to attempt independent analysis. Another approach which is commonly used to evaluate faulting is to correlate faults between boreholes. In the area around the proposed disposal site at Sellafield, approximately twenty boreholes have been drilled. Both from wireline logs, and from cores recovered from these boreholes, it is possible to identify faults. In some cases a fault surface can be identifiable in the core, and often the actual fault plane is surrounded by a large number of brittle fractures associated with the fault 'damage zone' (Knipe *et al.* 1997). However, the identification of a fault in a borehole does not give direct information on inter-borehole correlation of that fault. This is important for assessing the true geometry of the fault plane, and for making use of any information concerning the flow characteristics of the fault.

In order to improve the inter-borehole interpretations, NIREX also acquired seismic tomograms between pairs of boreholes. These surveys image the velocity structure of the rocks between boreholes which, in principle, would enable the changed velocity associated with a fault plane and its damage zone to be mapped, from borehole to borehole. NIREX made such interpretations of these data (NIREX 1994*a*), but there seem to be major problems. An estimate of the reliability of the existing tomograms can be made by the simple method of abutting pairs of tomograms which share a common borehole. This test reveals severe mismatches in the interpretation: that is the same fault is interpreted to cut a single borehole in different places when imaged by different tomograms. An additional problem is that, in some cases, a fault interpreted from the tomogram does not cut the borehole at the depth where core or wireline information suggests the existence of a fault. In addition, the tomogram surveys cannot be matched satisfactorily with the 2D seismic sections. Therefore, the current tomogram surveys are inconsistent and must be considered inconclusive.

In the last decade, the most important and cost-effective breakthrough in sub-surface imaging technology has been the widespread introduction of 3D seismic methods, mainly in oil field settings. This technique provides a means of imaging complex geology, resulting in a depth model of the rocks at resolutions much better than can be obtained with 2D seismic. NIREX did not carry out a 3D seismic investigation until 1994, when Glasgow University under contract to NIREX undertook the densest such survey in the world over the proposed location of the RCF construction shaft. To date, these seismic data have not been used by NIREX to update their structural interpretation, nor have the data been released to others.

Preliminary interpretation of the 3D seismic data (Smythe *et al.* 1995) indicates that the dips on the top of the BVG (e.g. the unconformity between the BVG and the overlying cover) are consistent with the earlier 2D seismic interpretations, but dips of the lithological units within the BVG differ radically from those suggested by NIREX tomograms or NIREX structural maps. These comparisons suggest that NIREX's present maps are invalid, and that the geological interpretation of the entire site area needs to be revised. Of particular concern is the fact that the positions, orientations, and even very existence of major faults, as depicted in the current NIREX interpretation, may be unreliable. Since faults have a major impact on hydrogeological simulations, the uncertainties concerning the structural interpretation of this site mean that a safety case cannot be reliably supported.

Rock mechanics concerns arising from construction

We have focused in this chapter on hydrogeological aspects and we will not address topics concerning the design and construction of underground cavities. Instead, we address two points which are a consequence of such construction: the first is the increase of permeability in the volume around the disposal facility; and the second concerns perturbations of the water system caused by pumping of inflowing waters away from the works area during the construction phase.

Underground excavation inevitably leads to dilatancy of the rock mass surrounding the works. This is true whether the excavations are achieved through blasting or by tunnelling machine. The increased dilation of the rock mass will result in an increase in the bulk permeability. Simulations of a simple 3D hydrogeological system which is similar to that which applies to the Sellafield site reveal that such a local permeability anomaly leads to locally increased flow rates (Garven & Toptygina 1993). This increase in groundwater flow will alter the assumptions about the total water in contact with waste containers, as a consequence of the larger number of pore-volume exchanges. More worrying is the possibility that the volume of rock with increased permeability will significantly alter the rock framework in ways that affect the larger hydrogeological system. The specific configuration of the Sellafield site is such that the volume of rock with increased dilatancy may be sufficiently large to affect the 200 m of BVG between the repository and the base of the cover sequence above. This breach could lead to a much enhanced rate of exchange of groundwaters between the 'basement' domain, the proposed disposal site, and that of the layered rocks above, which are hydraulically connected to the surface.

A further hydrogeological concern associated with the construction, and related to rock mechanics matters, is the necessity to draw down the water table during construction. This would certainly occur in the shallow rock sequence, and it might have happened in the BVG if 'open' fractures were encountered since these might have large-volume inflows. This pressure draw down would have some effect on the hydrogeological system – an effect which could be simulated for a range of assumptions about its magnitude. Less easy to simulate are the possible mechanical effects of changes in pore pressure associated with such a draw down. These simulations would require assumptions regarding the detailed state of stress in the entire rock mass – before, and during, the construction. They would also require assumptions concerning the non-linear processes of stress relief. Given that tunnelling experts are still working to understand how to improve such predictions in their work, it goes without saying that there are remaining uncertainties. All of these uncertainties multiply if we also ask whether there will be any changes in the permeability distribution as a consequence of minor structural movements associated with the pore-pressure excursions. Based on the simple situation described by others (Garven & Toptygina 1993), it becomes apparent that this matter of linked processes (e.g. alteration of the groundwater system, and consequent alteration of the rock mechanics leading to changes in the groundwater system) requires considerably more attention.

Discussion

Academic geoscientists are, by and large, supported by the public purse. Their daily task is to engage in the 'science' process which we described above. Until recently, the academic's choice of a topic for research has been largely unconstrained, subject only to the availability of time, and the necessary resources. Political and economic changes have altered this situation, such that funding, and indeed the assessment of research quality, now give strong emphasis to the benefits to society of that research, generally perceived in terms of 'wealth creation'. These circumstances are forcing a major change in the directions taken by academic geologists, and in the way that they approach their vocation.

Many academic geoscientists are adopting the mantle of 'environmental geoscientists'. We applaud this change, and we feel confident that well-trained geoscientists can readily make this transition. However, 'environmental geoscience' demands a proactive approach. Scientists can adopt either of two strategies when assessing a proposition: to seek to support the claim by means of separate studies; or to seek to falsify that claim. Both of these avenues are notionally open to us, but both require access to relevant information, and to the necessary tools. The decision to do something about an environmental issue is what distinguishes an environmental geoscientist from other geoscientists. Some academic geoscientists must continue to be in a position to serve as public 'watchdogs' on issues of importance to the public at large.

In this chapter we have focused on the 'environmental' subject of nuclear waste disposal. We have sought to make the case that there is a role for independent geoscientific analysis of proposals which are made by commercial firms, and that this independent expertise largely resides in the academic community. What is also apparent is that the 'political' issues associated with nuclear waste disposal are difficult to separate from the scientific ones. The challenges brought by independent, academic environmental geoscientists (Haszeldine & Smythe 1996), in conjunction with research by such bodies as HMIP, have highlighted potentially dangerous shortcomings of the Sellafield plans. These challenges have necessarily acknowledged political aspects of the issue while seeking to focus on the scientific concerns.

It is now known that further site work at Sellafield has been abandoned. In part, this change of plan is related to the scientific arguments made by several parties at the Planning Inquiry, and especially to the challenges raised against the claims made by NIREX. In cost terms, the academic studies we have undertaken, and the other challenges, have consumed less than £200 000 of funding. This total compares favourably with the NIREX budget of some £400 million expended to date on the Sellafield investigation. If one takes into account the large sums which were expected to be spent on the RCF (another £400 million?), but which now will not be wasted on this unsuitable site, it would seem that an argument could be made that environmental geoscientists have proven to be of considerable benefit to society. The benefits arising from health and safety issues further emphasize the importance of having an independent point of view involved in the evaluation process.

What emerges from this argument, in terms of the Sellafield disposal scheme or others which may follow, is that society should demand the establishment of a proper scientific forum in which issues concerning nuclear waste disposal are openly debated. This process of debate has served science extremely well over many centuries, and we submit that it would serve society well in this matter of major environmental concern. The forum should be open to all interested parties. Both the availability of information, and funding, are issues which need to be resolved in order to make this a viable approach.

The discussions in the forum would almost certainly be vigorous, and the issues might not be quickly resolved. Indeed, the debate would be likely to reveal new research questions which are critical to the issue of finding a suitable site. Nevertheless, we all now recognize that something must be done with the existing waste, and the entire community can seek a solution which is the best possible one in the circumstances which exist. As a society we owe our descendants no less than this.

The plea we make concerning a forum to debate nuclear waste disposal will necessitate a supply of qualified academic, and commercial, environmental geoscientists. Other waste and pollution matters – both existing and yet to be uncovered – will likewise require both knowledge, and the skills to apply that knowledge to solve real problems. We see a solid future for well-trained geoscientists who understand the way the Earth works, its processes, and particularly, how processes interact, and how to do something about environmental issues.

Conclusions

1. Decisions made at the exploration stage when seeking a disposal site can be subject to political considerations which may ultimately jeopardize the technical success of the entire exploration, appraisal and licensing effort.
2. The scientific approach (i.e. refutation or verification) offers the best method of peer review. This process is achieved by means of distinct groups of workers undertaking parallel investigations to evaluate the interpretations and claims made by the companies which wish to develop a waste disposal site. In the early stages, there is likely to be an emphasis on refutation but, as the proposals are modified in light of the debate, it is hoped that there will be a convergence in assessments.
3. A judicial approach in deciding whether a safe site has been selected needs well-informed 'defence' and 'prosecution' cases to be formulated.
4. The waste disposal site at Sellafield was poorly chosen at the exploration stage. Insufficient consideration was given to geological factors, particularly the hydrogeology. Subsequent investigations during the evaluation of the site have proven expensive, and complex. Simulated groundwater flow at this site is upwards, with the potential for carrying radionuclides towards the surface.
5. The geochemistry of the site was not adequately considered at the exploration stage. The natural chemical environment is possibly adverse to the retention of uranium, even in the engineered repository. The potential for retention of other radioactive elements is also unclear (e.g. iodine).
6. Sellafield was a very poor site which, if efforts had continued there, would have been unnecessarily expensive to character-

ize, and would still not have met the regulatory target for safety, due to fundamental flaws in the geological characteristics of this location.

7. Academic earth scientists have had a key role in evaluating this site and in synthesizing and interpreting information from the evaluation investigation. They have been important in upholding the public interest on topics which are beyond the knowledge of much of society.

Acknowledgements

CMcK was funded by the Greenpeace Environmental Trust. Grant Garven kindly allowed us the use of his computer simulation code OILGEN, and he contributed many useful discussions concerning modelling approaches. NIREX provided data from the Sellafield site. We thank Mike Russell, Helen Lewis, Chris Burton, Iain Allison and two anonymous reviewers for helpful comments which have improved the manuscript.

References

Atkinson, A. 1995. *Buffering of pH in an inhomogenous repository*. UK NIREX Ltd., Harwell, UK. Report No. NSS/R287.

——— & Guppy, R. M. 1988. *Evolution of pH in a radwaste repository: leaching of modified cements and reactions in groundwater*. Department of Environment, London, UK. Report No. DoE/RW/89-025 Part 3.

———, Everitt, N. M. & Guppy, R. 1988*a*. *Evolution of pH in a radwaste repository: experimental simulations of cement leaching*. Department of Environment, London, UK. Report No. DoE/RW/89-025 Part 1.

———, ——— & ——— 1988*b*. *Evolution of pH in a radwaste repository: internal reactions between concrete constituents*. Department of Environment, London, UK. Report No. DoE/RW/89-025 Part 2.

———, Hearne, J. A. & Knights, C. F. 1991. *Aqueous geochemistry and thermodynamic modelling of* $CaO–SiO_2–H_2O$ *gels*. AEA Technology, Harwell. Report No. AEA-D&R0153.

———, Williams, S. J. & Wisbey, S. J. 1993. *NSARP reference document : the near field : January 1992.* UK NIREX Ltd., Harwell, UK. Report No. NSS/G117.

Basham, I. R. & Kemp, S. J. 1993. *A review of natural uranium and thorium minerals.* Her Majesty's Inspectorate of Pollution, London, UK. Report No. DOE/HMIP/RR/94/007.

Bath, A. H., McCartney, R. A., Richards, H. G., Metcalfe, R. & Crawford, M. B. 1996. Groundwater chemistry of the Sellafield area: a preliminary investigation. *Quarterly Journal of Engineering Geology*, **29**, Suppl. 1, S39–S57.

Bear, J. & Yehuda, B. 1990. *Introduction to Modelling of Transport Phenomena in Porous Media. Theory and applications of transport in porous media*, Kluwer Academic Publishers, Dordecht.

Bennet, D. G., Read, D., Atkins, M. & Glasser, F. P. 1992. A thermodynamic model for blended cements. II: Cement hydrate phases; thermodynamic models and modelling studies. *Journal of Nuclear Materials*, **190**, 315–325.

Billington, D. E., Lever, D. A. & Wisbey, S. J. 1989. *Safety assessment of radioactive waste repositories.* Organisation for Economic Co-operation and Development, Paris, France.

Bishop, G. P., Beetham, C. J. & Cuff, Y. S. 1989. *Review of literature for chlorine, technetium, iodine and neptunium.* UK NIREX Ltd., Harwell, UK. Report No. NSS/R193.

Black, J. H. & Brightman, M. A. 1996. Conceptual model of hydrogeology of Sellafield. *Quarterly Journal of Engineering Geology*, **29**, Suppl. 1, S83–S93.

Brace, W. F., 1980. Permeability of crystalline and argillaceous rocks. *International Journal of Rock Mechanics, Mineral Science & Geomechanics Abstracts*, **17**, 241–251.

Bredehoeft, J. D. & Maini, T. 1981. Strategy for radioactive waste disposal in crystalline rocks. *Science*, **213**, 293–296.

Bruno, J., Casas, I., Lagerman, B. & Munoz, M. 1987. The determination of the solubility of amorphous $UO_{2(s)}$ and the mononuclear hydrolysis constants of uranium (IV) at 25°C. *In:* Materials Research Society (eds) *Scientific basis for nuclear waste management X.* 84, Materials Research Society, Boston, Massachusetts, USA, 153–160.

———, Crawford, M., Fabriol, R., Jamet, P., Lang, H., Read, D.,

Tweed, C. & Warwick, P. 1993. *Status review of CHEMVAL2 technical areas, June 1992*. Her Majesty's Inspectorate of Pollution, London, UK. Report No. DoE/HMIP/RR/93.014.

Cathles, L. M. 1990. Scales and effects of fluid flow in the upper crust. *Science*, **248**, 323–329.

Chapman, N. A. 1994. The geologist's dilemma: predicting the future behaviour of buried radioactive wastes. *Terra Nova*, **6**, 5–19.

——— & McEwen, T. J. 1986. Geological environments for deep disposal of intermediate level wastes in the United Kingdom. *In:* IAEA (eds) *Siting, design and construction of underground repositories for radioactive wastes*. IAEA-SM-289/37, International Atomic Energy Authority, Vienna, Austria, 311–328.

——— & McKinley, I. G. 1987. *The geological disposal of nuclear waste*. John Wiley & Sons, Chichester.

Clauser, C. 1992. Permeability of crystalline rocks. *Transactions of the American Geophysical Union*, **73**(21), 233–238.

Cross, J. E. & Ewart, F. T. 1990. *HATCHES – a thermodynamic database management system*. UK NIREX Ltd., Harwell, UK. Report No. NSS/R212.

De Marsily, G. 1986. *Quantitative hydrogeology. Groundwater hydrology for engineers*. Academic Press Inc. (London) Ltd., London.

——— 1987. An overview of coupled processes with emphasis on geohydrology. *In: Coupled processes associated with nuclear waste repositories*, Academic Press Inc., London, 27–37.

Deming, D. 1994. Fluid flow and heat transport in the upper continental crust. *In:* Parnell, J. (eds) *Geofluids: Origin, Migration and Evolution of Fluids in Sedimentary Basins*. Geological Society, London, Special Publications, **78**, 27–42.

Domenico, P. A. & Schwartz, F. W. 1990. *Physical and Chemical Hydrogeology*. John Wiley & Sons, New York.

Finch, R. J. & Ewing, R. C. 1992. The corrosion of uraninite under oxidising conditions. *Journal of Nuclear Materials*, **190**, 133–116.

Follin, S. & Thunkin, R. 1994. On the use of continuum approximations for regional modelling of groundwater flow in crystalline rocks. *Advances in Water Resources*, **17**, 133–14.

Freeze, R. A. & Cherry, J. A., 1979. *Groundwater*. Prentice-Hall,

Englewood Cliffs, N.J.
Fuger, J. 1992. Thermodynamic properties of actinide aqueous species relevant to geochemical problems. *Radiochimica Acta*, **58/59**, 81–91.
Garven, G. 1994. Genesis of stratabound ore deposits in the midcontinent basins of North America. 1. The role of regional groundwater flow – a reply. *American Journal of Science*, **294**, 760–765.
——— 1995. Continental-scale groundwater flow and geologic processes. *Annual Review in Earth and Planetary Sciences*, **23**, 89–117.
Garven, G. & Toptygina, V. I. 1993. Numerical modelling of three dimensional variable density groundwater flow systems and heat transport in large sedimentary basins. *Geological Society of America Abstracts with Programs*, **25**, 182.
Gray, D. A. 1976. *Disposal of highly active, solid radioactive wastes into geological formations- relevant geological criteria for the United Kingdom*. Institute of Geological Sciences, report No. 76/12.
Grenthe, I., Fuger, J., Lemire, R.J., Muller, A.B., Nguyen-Trung, C. & Wanner, H. 1992. *Chemical Thermodynamics of Uranium*. Chemical thermodynamics, 1, Elsevier, Amsterdam.
Harvey, B. R. 1995. Speciation of radionuclides. *In:* Ure, A. M. & Davidson, C. M. (eds) *Chemical Speciation in the Environment*. Blackie Academic and Professional, Glasgow, 276–306.
Haszeldine, R. S. 1996. Subsurface geology, geochemistry and water flow at a rock characterisation facility (RCF) at Longlands Farm. Proof of evidence. *In:* Haszeldine, R. S. & Smythe, D. K. (eds) *Radioactive Waste Disposal at Sellafield, UK. Site Selection, geological and engineering problems*. University of Glasgow, Glasgow, UK, 121–174.
——— & McKeown, C. 1995. A model approach to radioactive waste disposal at Sellafield. *Terra Nova*, **7**(1), 87–96.
——— & Smythe, D. K. 1996. *Radioactive Waste Disposal at Sellafield, UK. Site selection, geological and engineering problems*. University of Glasgow, Glasgow, UK.
——— & ——— 1997. Why was Sellafield rejected as a disposal site for radioactive waste? *Geoscientist*, **7**, 18–20.

Haworth, A. & Sharland, S. M. 1995. *The evolution of the Eh in the pore water of a radioactive waste repository*. UK NIREX Ltd., Harwell, UK. Report No. NSS/R308.

Heathcote, J. A., Jones, M. A. & Herbert, A. W. 1996. Modelling groundwater flow in the Sellafield area. *Quarterly Journal of Engineering Geology*, **29**, Suppl. 1, S59–S81.

Holmes, J. 1995. *Proof of Evidence. Science overview*. RCF Planning Inquiry, PE/NRX/13, UK NIREX Ltd., Harwell, UK.

Hooper, A. 1995. The NIREX repository concept. In: *The geological disposal of radioactive waste*, Royal Lancaster Hotel, London, IBC Technical Services Ltd, Gilmoora House, 57–61 Mortimer St., London.

Horseman, S. 1996. Generation and migration of repository gases: some key considerations. In: *International Conference on Waste Disposal*, 21–22 Nov 1996, London, IBC Technical Services Ltd, Gilmoora House, 57–61 Mortimer St., London.

Huyakorn, P. S. & Pinder, G. F. 1983. *Computational methods in Subsurface Flow*. Academic Press Inc., London.

Janeczek, J. & Ewing, R. C. 1992. Structural formula of uraninite. *Journal of Nuclear Materials*, **190**, 128–132.

Karlsson, F. 1995. The Swedish approach to near-field issues. In: *The geological disposal of radioactive waste*, Royal Lancaster Hotel, London, IBC Technical Services Ltd, Gilmoora House, 57–61 Mortimer St., London.

Knipe, R. J., Fisher, Q. J., Jones, G., Clennell, M. R., Farmer, A. B. *et al.* 1997. Fault seal analysis: successful methodologies, applications and future directions. In: Moller-Pedersen, P. & Koestler, A. G. (eds) *Hydrocarbon Seals – Importance for Exploration and Production*. Norwegian Petroleum Society Special Publication, **7**, Elsevier, Singapore, 15–40.

Langmuir, D. 1978. Uranium-solution mineral equilibria at low temperatures with applications to sedimentary ore deposits. *Geochimica et Cosmochimica Acta*, **42**, 547–569.

——— 1995. Nuclear waste management in the United States. In: *The geological disposal of radioactive waste*, Royal Lancaster Hotel, London, IBC Technical Services Ltd, Gilmoora House, 57–61 Mortimer St., London.

Lemire, R. J. 1988. *Effects of high ionic strength groundwaters on calculated equilibrium concentrations in the uranium-water system.* Atomic Energy of Canada Ltd. Report No. AECL-9549.

Lindberg, R. D. & Runnels, D. D. 1984. Groundwater redox reactions: an analysis of equlibrium state applied to Eh measurements and geochemical modelling. *Science*, **225**, 925–927.

Long, J. C. S., Karasaki, K., Davey, A., Peterson, J., Landsfeld, M, Kemeny, J. & Martel, S. 1991. An inverse approach to the construction of fracture hydrology models conditioned by geophysical data. An example from the validation exercises at the Stripa Mine. *International Journal of Rock Mechanics, Mineral Science & Geomechanics Abstracts*, **28**(2/3), 121–142.

McDonald, C. S., Jarvis, C. & Knipe, C. V. 1996. *RCF planning appeal by UK NIREX Ltd.* DOE. Report No. APP/HO900/A/94/247019.

McKeown, C. 1997. *A model approach to radioactive waste disposal at Sellafield.* PhD thesis, University of Glasgow.

——— & Haszeldine, R. S. 1995. Modelling groundwater flow and chemistry in the proposed repository zone. In: *The geological disposal of radioactive waste*, Royal Lancaster Hotel, London, IBC Technical Services Ltd, Gilmoora House, 57–61 Mortimer St., London.

McKeown, C. & Haszeldine, R. S. 1996. *A model approach to waste disposal. Progress report for Greenpeace Environmental Trust.* University of Glasgow. Report No. 3.

Mermin, N. D. 1996. Reference Frame: What's wrong with this sustaining myth? *Physics Today*, **49**(3), 11–13.

Metcalfe, R. & Crawford, M. B. 1994. *Models of water/rock interactions in the Borrowdale Volcanic Group within the potential repository zone at Sellafield.* British Geological Survey, Keyworth, UK. Report No. WE/94/26C.

Michie, U. 1996. The geological framework of the Sellafield area and its relationship to hydrogeology. *Quarterly Journal of Engineering Geology*, **29**, Suppl. 1, S13–S27.

Miller, W., Alexander, R., Chapman, N., McKinley, I. & Smellie, J. 1994. *Natural analogue studies in the geological disposal of radioactive wastes.* Nagra, Wettingen, Switzerland. Report No. NTB 93–03.

Moreno, L. & Neretnieks, I. 1993. Fluid flow and solute transport in a network of channels. *Journal of Contaminant Hydrology*, **14**, 163–192.

Morss, L. R. 1986. Thermodynamic Properties. *In:* Katz, J. J., Seaborg, G. T. & Morss, L. R. (eds) *The Chemistry of the Actinide Elements*. 2, Chapman & Hall, London, 1278–1360.

Naish, C. C., Balkwill, P. H., O'Brien, T. M., Taylor, K. J. & Marsh, G. P. 1990. *The anaerobic corrosion of carbon steel in concrete*. UK NIREX Ltd., Harwell, UK. Report No. NSS/R273.

Nash, J. T., Grainger, H. C. & Adams, S. S. 1981. Geology and concepts of genesis of important types of uranium deposit. *Economic Geology 75th Anniversary Volume*, 63–116.

Neumann, S. P. 1990. Universal scaling of hydraulic conductivities and dispersivities in geologic media. *Water Resources Research*, **26**, 1749–1758.

Neuzil, C. E. 1995. Abnormal pressures as hydrodynamic phenomena. *American Journal of Science*, **295**, 742–786.

Nicholls, D. B. 1995. *Hydrogeological modelling of the Sellafield site, Vols I & II – Text and Appendices*. Her Majesty's Inspectorate of Pollution, London, UK. Report No. TR-Z2-7.

NIREX 1989. *Deep repository project, preliminary environmental and radiological assessment and preliminary safety report*. UK NIREX Ltd., Harwell, UK. Report No. 71.

——— 1992*a*. *The 1991 UK radioactive waste inventory*. UK NIREX Ltd., Harwell, UK. Report No. 284.

——— 1992*b*. *The physical and chemical characteristics of UK radioactive wastes*. UK NIREX Ltd., Harwell, UK. Report No. 286.

——— 1992*c*. *The radionuclide content of UK radioactive wastes*. UK NIREX Ltd., Harwell, UK. Report No. 285.

——— 1993*a*. *The geology and hydrogeology of the Sellafield area: Interim assessment*. UK NIREX Ltd., Harwell, UK. Report No. 524 (4 vols).

——— 1993*b*. *Scientific update 1993 : NIREX deep waste repository project*. UK NIREX Ltd., Harwell, UK. Report No. 525.

——— 1994*a*. *The 2-D interpretation of tomogram data from the Rock Characterisation facility area, Sellafield. UK*. UK NIREX Ltd., Harwell, UK. Report No. S/94/007.

——— 1994*b*. *An assessment of the impact of the rock characterisation facility on groundwater flow and on risk from the groundwater pathway*. UK NIREX Ltd., Harwell. Report No. 560.
——— 1994*c*. *Post-closure performance assessment, gas generation and migration*. UK NIREX Ltd., Harwell, UK. Report No. S/94/003.
——— 1995*a*. *The 3D geological structure of the PRZ: summary report*. UK NIREX Ltd., Harwell, UK. Report No. S/95/005.
——— 1995*b*. *The hydrochemistry of Sellafield, 1995 update*. UK NIREX Ltd., Harwell, UK. Report No. S/95/008.
——— 1995*c*. *NIREX 95: A Preliminary analysis of the groundwater pathway for a deep repository at Sellafield*. UK NIREX Ltd., Harwell, UK. Report No. S/95/012 (3 vols).
——— 1995*d*. *Post-closure performance assessment, near-field evolution*. UK NIREX Ltd., Harwell, UK. Report No. S/95/009.
Nuclear Energy Agency 1989. *Safety assessment of radioactive waste repositories*. Organisation for Economic Co-operation and Development, Paris, France.
Pearson, F. J. & Berner, U. 1991. *Nagra thermochemical database I: Core data*. Nagra, Wettingen, Switzerland. Report No. NTB 91–17.
Person, M., Raffensberger, J. P., Ge, S. & Garven, G. 1996. Basin-scale hydrogeologic modelling. *Reviews of Geophysics*, **34**, 61–87.
Popper, K. 1963. *Conjectures and Refutations: the Growth of Scientific Knowledge*. Routledge and Kegan Paul, London.
Robins, N. S. 1980. *The geology of some United Kingdom nuclear sites related to the disposal of low and medium level radioactive wastes: Part 1*. Institute of Geological Sciences, Report No. ENPU 80-5.
Rose, W. C. C. & Dunham, K. C. 1977. *Geology and hematite deposits of South Cumbria*. Institute of Geological Sciences, HMSO, London.
Royal Society 1994. *Disposal of radioactive waste in deep repositories*. Report of a Royal Society study group, Royal Society, London.
RWMAC 1994. *Fourteenth annual report of the radioactive waste management advisory committee*. Report No. 14. HMSO, London.

——— 1995. *Site selection for radioactive waste disposal facilities and the protection of human health. Report of RWMAC and ACSNI Study Group, March 1995*. Department of the Environment, London.

Smythe, D. K. 1996. The 3-D structural geology of the PRZ. Proof of Evidence. *In:* Haszeldine, R. S. & Smythe, D. K. (eds) *Radioactive Waste Disposal at Sellafield, UK. Site selection, geological and engineering problems*. University of Glasgow, Glasgow, UK, 237–278.

———, Holt, J. M., Elstob, M. & Robson, C. 1995. *A high-resolution 3D vibroseis survey of a potential nuclear waste repository*. 57th Conference and Technical Exhibition, Glasgow, UK, European Association of Geoscientists and Engineers.

Toth, J. 1995. Hydraulic continuity in large sedimentary basins. *Hydrogeology Journal*, **3**, 4–16.

——— & Sheng, G. 1996. Enhancing safety of nuclear waste disposal by exploiting regional groundwater flow: the recharge area concept. *Hydrogeology Journal*, **4**, 4–25.

Tyrer, M., Bennet, D. G., Read, D. & Yunus, I. 1995. *Near field and chemical transport modelling*. Her Majesty's Inspectorate of Pollution, London, UK. Report No. TR-Z2-9.

Wallace, H. 1996. Model validations and the role of the proposed Rock Characterisation Facility at Sellafield. *In*: Haszeldine, R. S. & Smythe, D. K. (eds) *Radioactive Waste Disposal at Sellafield, UK. Site selection, geological and engineering problems*. University of Glasgow, Glasgow, UK, 189–196.

Weigel, F. 1986. Uranium. *In:* Katz, J. J., Seaborg, G. T. & Morss, L. R. (eds) *The Chemistry of the Actinide Elements*. 1, Chapman & Hall, London, 169–442.

Wood, W. W. 1997. Fluxes: a new paradigm for geologic education? *Ground Water*, **35**, 1.

13 Airborne particulate characterization for environmental regulation

John Merefield, Ian Stone, Jo Roberts, Jeff Jones & Jan Barron

Summary

- World-wide concerns over air quality have accelerated research in the growing field of airborne particulate characterization.
- The Earth Resources Centre, University of Exeter has developed novel dust characterization procedures during a major research programme with British Coal Opencast and has subsequently implemented the methodology whilst working with local authorities in England and Wales.
- This chapter describes the implementation of these airborne dust studies for effective regulation and illustrates how the science of environmental geology in air quality is disseminated to both the specialist decision-makers and to the non-specialist.

Airborne particulates reach the Earth's ambient atmosphere from a wide range of sources. A relatively small contribution is derived from the destruction of meteorites whilst terrestrial sources include aerosols from oceanic sea sprays and breaking waves, minerals derived from agriculture and exposed soils, fires and volcanic activity, as well as anthropogenic sources, such as civil engineering, industrial and traffic emissions (Pye 1987).

The size of airborne particulates can range from 0.05 to 1000 μm (microns). Although particles from a few microns to around 100 μm are difficult to resolve by the naked eye, they can cause discoloration of surfaces such as cars, washing and window ledges and are, therefore, categorised as nuisance dust and as such are of ongoing concern to the regulatory authorities (Anon. 1996).

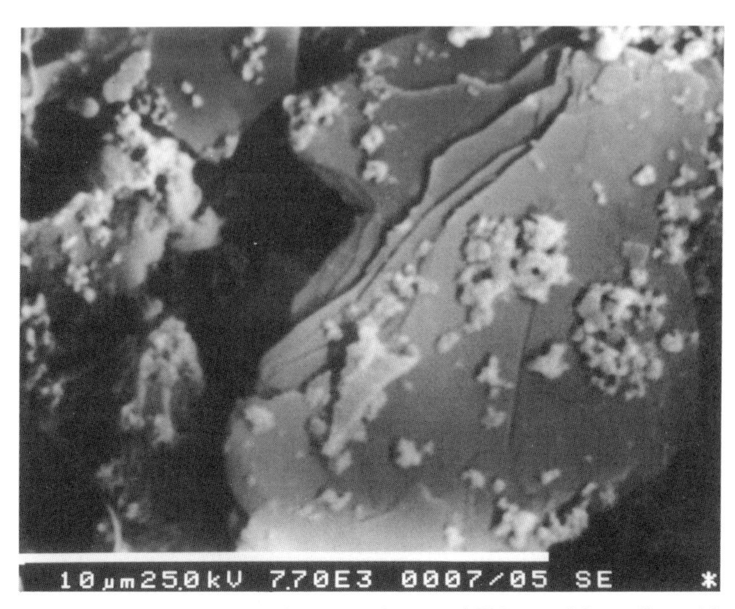

Fig. 13.1. Scanning electron microscope image of PM_{10} particles collected from a TEOM sampler at an opencast coal mine, showing shale lithoclasts hosting diesel deposits. White scale bar is 10 μm.

The finest particulates have mean aerodynamic diameters of less than 0.1 μm and arise largely from combustion processes or from gas to particulate conversions as the nuclei mode. When these combine to form particles in the size range from 0.1–2 μm they achieve accumulation status. The coarse category with diameters from 2–100 μm largely consists of particulates derived from the dispersion of aeolian dusts, soil erosion and industrial activity. Airborne particulates are classified according to their origin into the following categories.

1. *Primary airborne particulates:* those directly emitted from a source. These include rock and mineral particles from mining, quarrying and construction, as well as flyash from coal combustion and soot particles from other carbon-based combustion processes, such as from diesel fuel and wood burning. They can be readily resolved using the scanning electron microscope (SEM) (Fig. 13.1).

Fig. 13.2. Scanning electron microscope image of PM_{10} particles collected from a TEOM sampler showing the influence of secondary calcium-, ammonium- and sodium sulphate particles. Energy dispersive chemical analysis supports identification of compounds. White scale bar is 10 μm.

2. *Secondary airborne particulates:* produced as a result of nucleated condensation from within the gas phase. These may include the oxidation of sulphur dioxide to produce sulphates such as ammonium sulphate and gypsum. Other secondary particulates include chloride and nitrate salts derived respectively from sea spray and the oxidation of nitrogen dioxide. Energy dispersive x-ray chemical analysis (EDS) with SEM can aid identification of such compounds (Fig. 13.2).

Nuisance and health

Although recent studies have demonstrated that particles from submicron to 100 μm can enter the body and deposit in various parts of the respiratory tract, current medical opinion considers that particles with an aerodynamic diameter of less than 10 μm size (PM_{10}S) and even smaller at less than 2.5 μm ($PM_{2.5}$S) represent the greatest risk to human health. Ultimately, PM_1 could prove the key particle size factor in assessment of health risk. Bronchitis was proven as the health effect in the London smogs of the 1950s whilst the recent increase in asthmatic morbidity conditions may be associated with present-day air quality.

The larger particles generally associated with nuisance result from anthropogenic activities such as civil engineering, quarrying and opencast mining. They generally contain a dominant proportion of geological minerals and are appropriate for investigation by environmental geologists. The methods of sampling and analysis for air quality regulation, therefore, does depend upon the type of problem being investigated and the manner in which the results will be used (Tomas & Stone 1996).

Fugitive dusts

Sampling

Dust deposit sampling is normally associated with nuisance complaints whilst sampling of the total suspended particulates (TSP) or dust flux, is necessary if the source of the emission is to be determined. In general, dust sampling will either be passive, that is relying on the wind to bring the particulates to the dust gauge, or active, utilizing a pump to introduce airborne particles onto a filter system (Mark 1994).

Passive

Deposit sampling systems used by regulatory authorities traditionally include the British Standard Deposit Gauge (Anon. 1969) which simply consists of a collection bowl. Unfortunately, this bowl is not aerodynamically sound and has generally been replaced by use of the upturned frisbee (Hall & Upton 1988) with a foam insert and has proved to be an efficient dust deposit collector, providing samples for weighing and analysis (Chiu *et al.* 1996). When a series of frisbees are used in a traverse, this allows the rate of dust drop-off, with distance from a source, to be established. The British Standard Directional Gauge (Anon. 1972) consists of slotted pipe collectors facing the four directions of the compass and collects coarse dust suitable for provenance investigations. Further techniques include using glass slides to collect dust. This is measured by assessing the decrease in their reflectivity as soiling increases. Passive collectors are relatively inexpensive but rely on long periods of exposure for reliable sampling, typically of up to one month in duration.

Active

Forced samplers use a pump with a capacity from a few litres per minute (low volume) to those with a few hundred litres per minute (high volume). With a known flow-rate and known sampling time, the mass of particles collected on the filter can be used to determine ambient dust levels. Examples of active samplers range from personal samplers usually employed in studies of occupational (indoor) health to the Graesby Andersen High Volume Sampler (Wight 1994). Total sampling heads are used in nuisance investigations whilst some pumped samplers can be modified to collect the finest fractions, such as PM_{10} and $PM_{2.5}$ categories, for health studies. The Tapered Element Oscillating Microbalance (TEOM) provides an added dimension of real-time monitoring. The frequency of vibration of the glass element supporting the dust collecting filter directly reflects the mass deposited on it which is displayed instantly and temporally.

Methods of monitoring

Dust fingerprinting has been established for the monitoring of fugitive particulate emissions in nuisance complaints (Merefield *et al.* 1994*a*,*b*, 1995*a*,*b*). Where information is required for impact zonation around a dust source clusters, and/or traverses, of frisbee deposit gauges are used to establish dust levels, the rate of drop-off from a source and to characterize the dust itself. British Standard four-way directional gauges allow the particles to be characterized and can also provide information on provenance. Window ledges are often used to sample dust deposits in communities adjacent to the source of a dust problem in the first instance to determine the sampling strategy.

Sample preparation

Samples collected in 5 litre frisbee sampling bottles, normally over a one-month sampling period, are prepared using vacuum filtration. Usually 0.1 μm cellulose nitrate filters and a multi-port vacuum filtration system are employed to separate the material from rainwater before air-drying and subsequent analysis.

Subsamples are prepared for X-ray diffraction (XRD) analysis by crushing and smearing with propan-2-ol onto a 1.5 cm^2 set area of a glass slide. This provides maximum preferred crystallite orientation, especially for clay minerals along their 001 series of basal reflections. Due to the small sample size involved and to allow for resulting variations in peak intensities for all phases, a set of 16 shale standards (0.3–50 mg) have been produced. These were smeared over a 1.5 cm^2 set area of the glass slide and analysed 3 times using standardized excitation conditions. Reproducibility proved about ± 10% from the mean. The resulting calibration curve now permits correction of kaolinite and quartz peak heights which is important for reliable inter-mineral comparisons such as the K_{100}/Q_{101} ratio.

Subsamples for SEM analysis are routinely prepared by re-suspension in 5 ml of H_2O and re-filtering onto 0.1 μm polycarbonate membrane filters. These are mounted on 25 mm size stubs and coated with gold to make them conduct to prevent charging effects during analysis. This provides a dispersed sample for subsequent examinations.

Analysis

Analysis consists of XRD for qualitative analysis of mineral species and SEM to obtain an estimation of particle size, sphericity and angularity, with energy dispersive X-ray spectrometry (EDS) to determine the discrete particle chemistry (Merefield *et al.* 1995*b*).

Individual grains are examined by SEM/EDS to determine their discrete particle size, chemistry and morphology. About 200 particle examinations are considered the optimum. The data are logged on a spreadsheet (currently Novell QuattroPro for Windows) for statistical analysis and graphical presentation. A photographic record of sample characteristics is also captured by the SEM/EDS system.

Characterization

Dust samples are a reflection of their source material, their size and morphology and a function of distance from source. Physical and chemical examination of these characteristics provides an environmental fingerprint which is then available for use in regulation. This application uses output obtained from the analytical techniques employed in fugitive dust monitoring.

Mineralogy

X-ray diffraction of crystalline phases and their interrelationships is used to determine dust provenance. This is illustrated by the monitoring of airborne dust around opencast coal sites. The dominant type of particulate produced from this source is from the removal of the shale overburden, transport and storage. The shale contributes kaolinite to the dust mineral assemblage. The kaolinite basal diffractogram reflection from a highly orientated sample can be used to generate a ratio with a relatively constant ambient phase such as quartz. Since the shale also has a quartz component, the kaolinite–quartz ratio is not linear and will eventually saturate. However, as the relative quantity of shale decreases with distance from the source, the relative amount of ambient quartz increases. Consequently, the decrease in kaolinite–quartz (K/Q) ratio with distance from a site provides a reflection of the drop-off in dust from an opencast site. Figure 13.3 demonstrates this process. The K/Q ratio of 0.321 at a source of industrial dust on an opencast coal site decreases in value with distance from the source and at a distance of 500 m the ratio has fallen to 0.091.

Particle size, morphology and chemistry

Both SEM and EDS are used to examine discrete particle size, sphericity, form, angularity and chemistry. This information is used primarily to describe changes in particle size with distance from a

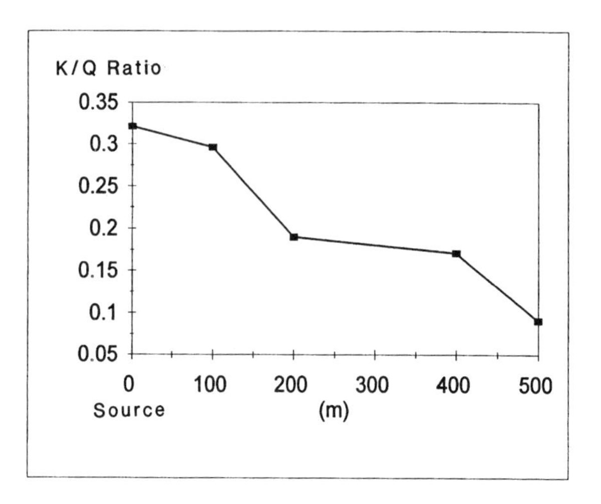

Fig. 13.3. The relationship of kaolinite to quartz in dust collected by frisbee deposit gauges from 0 to 500 m away from an opencast coal site illustrates the decreasing impact of fugitive dust emissions.

source. Frequency and grain-size of shale, quartz, coal and other dust particles collected by a transect of frisbee deposit gauges at an opencast coal site is used to illustrate this point (Fig. 13.4). The effect of decrease in particle size in relation to distance (source to 500 m) from the point of fugitive emissions is shown. At the source of industrial dust, ambient quartz (representing regional background particulates) is fine grained (2–10 μm) whilst shale particles are relatively coarse (3–30 μm). The kaolinite to quartz ratio is 0.321. Frequency and grain-size of shale, quartz, coal and other particles collected 100 m away from the opencast coal site show the ambient quartz to be fine in size again. The shale particles are similar to those collected at source (3–50 μm). The kaolinite to quartz ratio is 0.296. Two hundred metres away from the source, ambient quartz is fine (2–10 μm) but some coarse shale particles are present (3–30 μm). The kaolinite to quartz ratio is 0.190. Four hundred metres away from the site, the ambient quartz is again fine but shale particles are skewed towards finer fractions (2–5 μm). The kaolinite to quartz ratio is 0.171. Finally, the most distant sample collected 500 m away from the site shows ambient quartz fine (2–

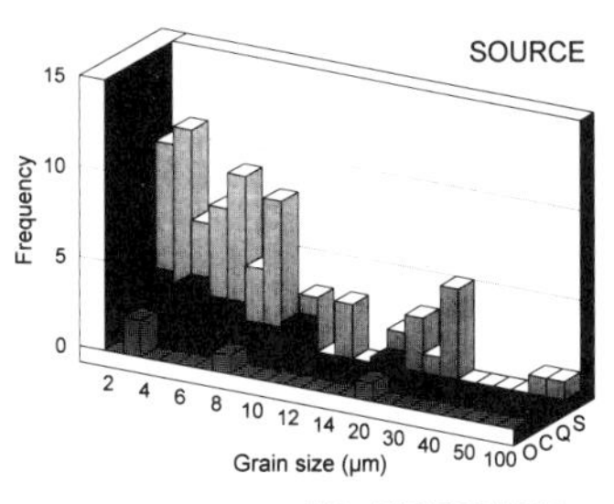
SOURCE
Frequency
0
5
10
15
2 4 6 8 10 12 14 20 30 40 50 100 O C Q S
Grain size (μm)

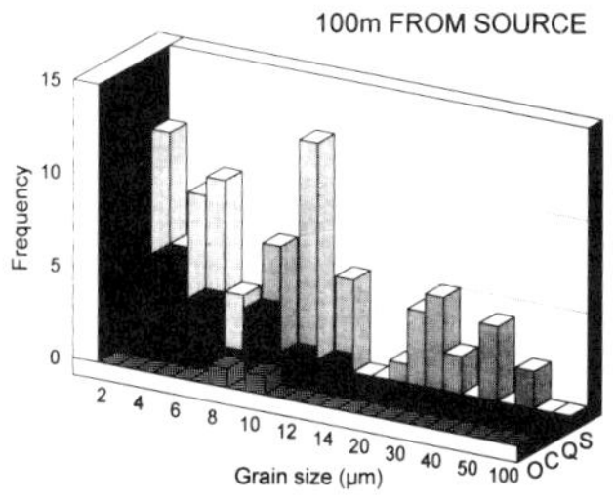
100m FROM SOURCE
Frequency
0
5
10
15
2 4 6 8 10 12 14 20 30 40 50 100 O C Q S
Grain size (μm)

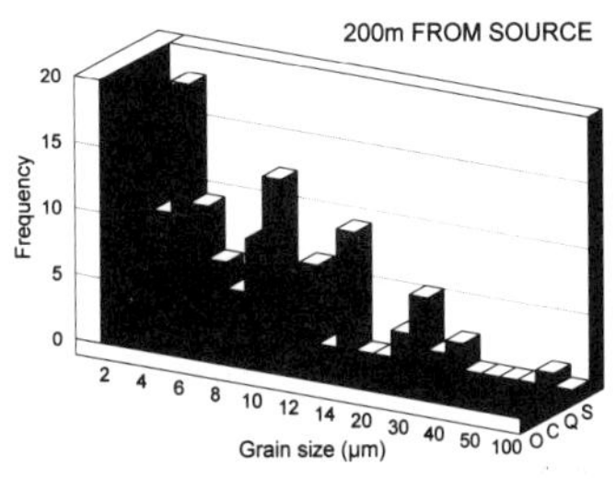
200m FROM SOURCE
Frequency
0
5
10
15
20
2 4 6 8 10 12 14 20 30 40 50 100 O C Q S
Grain size (μm)

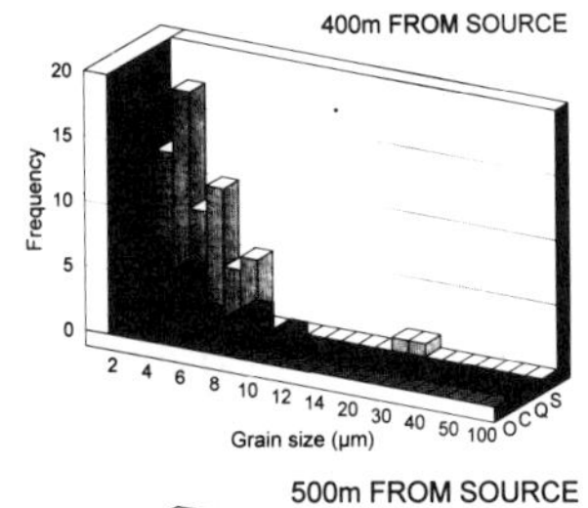
400m FROM SOURCE
Frequency
0
5
10
15
20
2 4 6 8 10 12 14 20 30 40 50 100 O C Q S
Grain size (μm)

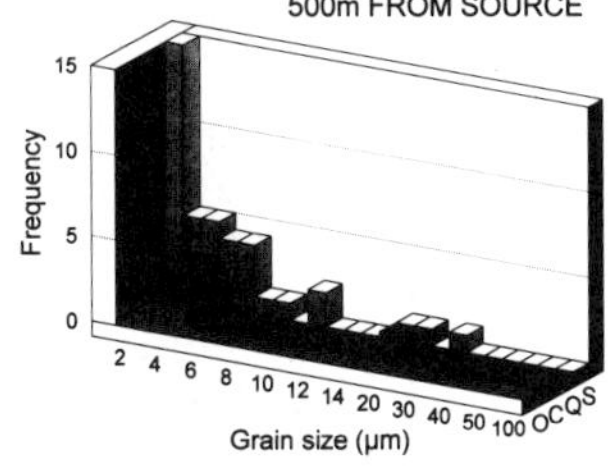
500m FROM SOURCE
Frequency
0
5
10
15
2 4 6 8 10 12 14 20 30 40 50 100 O C Q S
Grain size (μm)

6 μm) and shale particles also heavily skewed towards finest fractions (2–3 μm). The kaolinite to quartz ratio is 0.091. The discrete quartz, therefore, retains a relatively constant size in all samples, whilst the site-generated shale particles become skewed towards the fine size category with increased distance from site. This size related drop-off mirrors that of the inter-mineral ratios from 0.321 to 0.091. Whilst these samples are dominated by opencast-derived dust, additional particulates, and therefore additional sample weight are derived from such material as flakes of paint, plastic, carbonates and organics.

The characterization of dust samples therefore enables the impact zonation of a dust source. Increases in inter-mineral ratios combined with an increase in the particle size of specific particle phases provides reliable evidence to establish provenance and its environmental impact.

Problems in regulation

Currently, regulations associated with dust are limited in scope. A recommended standard of 50 $\mu g\,m^{-3}$ per rolling 24 hour mean has been suggested for PM_{10}. Whilst this may be subject to criticism for the relatively large size range of particles collected, geographical location of monitoring, and frequent lack of chemical and physical characterization, it is primarily aimed at measuring the fine particles within the urban environment which may have an adverse health impact (Anon. 1995). Legislation for nuisance particulates is more limited. An unofficial guideline for a weight of 200 $mg\,m^{-2}$

Fig. 13.4. Frequency and grain-size of shale (S), quartz (Q), coal and (C) and other particles (O) collected by frisbee deposit gauge on an opencast coal site at increasing distances from source. (**a**) At source: ambient quartz is fine grained, whilst shale particles are coarse (3–30 μm); the kaolinite to quartz ratio is 0.321. (**b**) 100 m away from the source: ambient quartz is again fine and shale particles are similar to those collected at source (3–50 μm); the kaolinite to quartz ratio is 0.296. (**c**) 200 m away from source: ambient quartz is again fine but some coarse shale particles are present (3–30 μm); the kaolinite to quartz ratio is 0.190. (**d**) 400 m away from source: ambient quartz is again fine but shale particles are skewed towards finer fractions (2–5 μm); the kaolinite to quartz ratio is 0.171. (**e**) 500 m away from source: ambient quartz is again fine but shale particles are heavily skewed towards finest fractions (2–3 μm); the kaolinite to quartz ratio is 0.091.

day^{-1} is suggested for deposited particles in the United Kingdom. However, the justification for using this figure is not adequately made.

Sample weight is not a thoroughly reliable indicator of nuisance as dusts can be derived from a variety of sources at any one location (Anon. 1996). A large proportion of sample weight can be composed of spores, fine organic matter in addition to a variety of mineral sources including roads and buildings. Consequently, the physical and chemical composition of the dust samples should be examined additionally to determine their characteristics and thereby identify potential sources and their impact.

Effective regulation

The method described here provides two key elements for the regulators: (1) dust fingerprinting and (2) provenance.

Environmental Health Officers investigating dust nuisance complaints, for example, are able to submit samples for characterization. Once this is achieved, follow-up strategic sampling and analysis enables the mineral components to be traced back to the dust's mineral origin. The performance of remedial action enforced on the polluter by the regulator can then be examined during subsequent monitoring. Graphical and visual presentation of results are an important part of this process. The photographic images obtained by SEM/EDS analysis greatly assist in dissemination of the results to industry and to the public.

Resources required for a local authority dust monitoring programme comprise a series of low-cost frisbee deposit gauges. These are installed along a transect(s) running from the source (or sources) of dust emissions, their transport path and into the community where dust emissions have been reported. Calendar monthly collection of samples during the dryer summer months (the main period of nuisance complaints) for particulate characterization ensures continuous monitoring. Window-ledge samples taken as a result of new complaints can then be compared with those established for this local dust database resource.

Airborne particle characterization is proving essential in dust and other types of airborne particulate monitoring and control. The weighing of samples without recourse to provenance is valueless.

Dust fingerprinting through environmental geology provides the means to establish the source of dusts from complex industrial and urban situations. It will be established increasingly as air particulate limits are set and responsibility is apportioned through litigation proceedings.

Acknowledgements

The authors are particularly grateful to Torfaen, Carmarthenshire and Powys County environmental health authorities for funding the research projects that have enabled development of airborne particulate characterization for regulation in environmental health.

References

Anon. 1969. *Methods for the Measurements of Air Pollution*. British Standard 1747, Part 1. Depositional Gauges. HMSO, London

——— 1972. *Directional Dust Gauge*. British Standard 1747, Part 5. HMSO, London.

——— 1995. *Non-biological particles and health*. The Committee on the Medical Effects of Air Pollution, Department of Health. HMSO, London.

——— 1996. *Airborne particulate matter in the United Kingdom*. Third Report of the Quality of Urban Air Review Group (QUARG), Department of Environment. HMSO, London.

Chiu, T. W., Grainger, P., Merefield, J. R. & Stone, I. M. 1996. The aerodynamics of the Frisbee dust collectors. *Proceedings of International Symposium on Air Pollution by Particulates, Prague, Czech Republic, October 1995*. Czech Geological Survey, Prague; 95–115.

Hall, D. J. & Upton, S. L. 1988. A wind tunnel study of the particle collection efficiency of an inverted frisbee used as a dust deposit gauge. *Atmospheric Environment*, **22**, 1383–1394.

Mark, D. 1994. *The sampling of aerosols in the ambient atmosphere*. Valid Analytical Measurement Initiative: Project 14, Annex D: the development of sampling guidelines, AEA-TPD-353.

Merefield, J. R., Stone, I., Jarman, P. J., Roberts, J., Jones J. & Dean, A. 1994*a*. Fugitive dust characterisation in opencast mining areas. *In*: Paithankar, A. G. (ed.) *The Impact of Mining on the Environment: problems and solutions*. Proceedings of an International Symposium, Nagpur, India, January, 1994. A.A. Balkema, New Delhi, 3–10.

———, ———, Rees, G., Roberts, J., Dean, A. & Jones, J. 1994*b*. Mineralogy and provenance of airborne dust in opencast coal mining areas of South Wales. *Proceedings of the Ussher Society*, **8**, 13–316.

———, ———, Roberts, J., Dean, A. & Jones, J. 1995*a*. Monitoring airborne dust from quarrying and surface mining operations. *Transactions of the Institute of Mining & Metallurgy*, **104**, A76–78.

———, ———, Jarman, P., Rees, G., Roberts, J., Jones, J. & Dean, A. 1995*b*. Environmental dust analysis in opencast mining areas. *In*: Whately, M. K. G. & Spears, D. A. (eds) *European Coal Geology*. Geological Society, London, Special Publications, **82**, 181–188.

Pye, K. 1987. *Aeolian Dust and Dust Deposition*. Academic Press, London.

Tomas, J. & Stone, I. 1996. *Proceedings of the International Symposium on Air Pollution by Particulates*. Czech Geological Survey, Prague.

Wight, G. D. 1994. *Fundamentals of Air Sampling*. CRC Press Inc., Boca Raton.

14 Anthropogenic platinum group elements in the environment

Stephen J. Edwards & Fergal Quinn

Summary

- The environment preserves a dynamic legacy of human activity in the form of metal contamination.
- In the present study, the platinum group elements were chosen for investigation for two reasons: (1) they are understudied and information on their global redistribution by humans is needed to enable future calculation of loading rates necessary for the development of global mass balance models; and (2) their very low background abundances in the environment make them very sensitive tracers of human activity and subsequent contamination.
- With significant and increasing quantities of platinum group elements being added to the environment from smelters, autocatalysts, waste water and sewage sludge, it is proposed that in the future these elements could be recovered from such sludge and contaminated soil and sediment.

Increases in human population place ever greater demands on the resources of the Earth. Not only are resources being used, but their very use is often responsible for global reduction in the quality of water, air, soil and sediment. Consequently, in modern society we have a situation where humans have arguably become one of the most important components in the global and regional cycling of trace metals and much of the surface and near-surface environment is contaminated with at least one metal (Nriagu & Pacyna 1988; Nriagu 1990). The extent to which anthropogenic metals – metals introduced into the environment by human activity – are distributed throughout the Earth's surface system depends on many factors, including: (1) date of the first use of the metal; (2) demand for the metal; (3) the extent of its use; and (4) the type of use.

The aim of the present study is to compile and review the limited data available for anthropogenic platinum group elements (ruthenium (Ru), rhodium (Rh), palladium (Pd), osmium (Os), iridium (Ir) and platinum (Pt)) in order to determine if these newly discovered metals, which have not been used extensively by humans until recently, are already accumulating in the environment. The study also forms the first attempt to gain an overview of human activities impacting on the global abundance and distribution of platinum group elements and paves the way for future calculation of loading rates and the development of global mass balance models for these elements.

'Old' versus 'new' metals

Metals may be classified as either 'old' or 'new' depending on their date of discovery. A classic example of an 'old' element is lead. Over the past 5500 years the world production of lead has increased by almost seven orders of magnitude (Settle & Patterson 1980). For this reason, anthropogenic lead has spread over the surface of the Earth and has entered all environmental systems (e.g. Nriagu 1990). The aim of this chapter is to ask whether the 'new' platinum group elements have started to follow the same trend.

Although platinum was worked by South American Indians at the time they were conquered by the Spanish conquistadors, the platinum group elements were officially discovered between 1763 and 1844 (Craig *et al.* 1996). The use of these metals has been extensive over the past few decades with recent global demand increasing every year (Johnson Matthey 1996). Today the platinum group elements are utilized in autocatalyst, electrical, chemical, dental, medical/biomedical, petroleum, glass and jewellery applications, and there is substantial investment demand for platinum (Fogg & Cornellisson 1993; Johnson Matthey 1996). Owing to their recent discovery, their extensive use in modern society, and their very low background concentrations in most components of the environment, the platinum group elements have great potential to define where and by how much the environment is becoming contaminated by them.

Non-anthropogenic platinum group elements

In all studies attempting to define the impact human activity has on metal concentrations in the environment, it is essential to define a non-anthropogenic background for each metal. In an attempt to do this for the platinum group elements, abundances of these elements in the crust and in sediments have been compiled (Table 14.1). These values indicate that normal background concentrations of these elements are very low, of the order of parts per billion (ppb) or less. Concentrations of platinum group elements in the Bushveld and Stillwater complexes (Table 14.1) indicate that background concentrations are generally of the order of a thousand times below those normally required for a platinum group element deposit (MacDonald 1987).

Table 14.1. Non-anthropogenic concentrations of platinum group elements (ppb)

	Pt	Pd	Os	Ir	Ru	Rh
Average crust	5[a]	1.05[b]	0.05[c]	0.025[b]	<0.5[d]	<0.5[d]
Deep marine sediment[e]	0.3–21.9	0.7–11.27	<0.05–0.81	0.02–1.2	<0.2–7.0	
Organic-rich marine sediment[f]			0.095–0.693			
Bushveld Complex[g]	3740	1530	63	74	430	240
Stillwater Complex[h]	5000	17300	22	19	50	150

[a] Naldrett (1989); [b] Crocket & Kuo (1979); [c] Esser (1991); [d] Koide *et al.* (1986); [e] Crocket & Kuo (1979), Hodge *et al.* (1985, 1986), Koide *et al.* (1991), Colodner *et al.* (1992); [f] Ravizza *et al.* (1991), Ravizza & Turekian (1992); [g] Hiemstra (1979), composite sample of the Merensky Reef; [h] Barnes (1983), J-M Reef.

Anthropogenic platinum group elements

The data available on anthropogenic platinum group element concentrations are limited. Enough does, however, exist to begin to define how and where anthropogenic platinum group element contamination is occurring.

Smelting
No data have been obtained for the processing of primary platinum group element ore, but two notable studies have been undertaken on smelting operations involving nickel–copper sulphide ores containing traces of platinum group elements. An investigation by Crocket & Teruta (1976) of the International Nickel Company's Copper Cliff smoke stack in the Sudbury area of Ontario, Canada, provided convincing evidence that platinum group elements may be discharged to the environment during smelting. Particulate matter recovered from the stack contained 11 000 ppb platinum, 2500 ppb palladium and 250 ppb iridium. In Kelley Lake, 4 km south-southwest of the stack, contaminated sediments yielded 1000–2800 ppb platinum, 110–230 ppb palladium and 14–56 ppb iridium. Edwards (1996) demonstrates that similar levels of contamination are present in topsoil in the Kola Peninsula of Russia. The pollution of the Kola Peninsula by mining and mineral processing operations is amongst the worst in the world. The area around the worst polluter – the smelter at Monchegorsk – is so highly polluted that it is now classified as a 'technogenic desert' by Russian scientists; this area covered 760 km^2 in 1988. It is estimated that the Monchegorsk smelter emits several hundred kilogrammes of platinum and palladium each year, and 5 km south of the smelter topsoils on average contain 50 ppb platinum (maximum 460 ppb) and 188 ppb palladium (maximum 1760 ppb).

Coal combustion
It is well known that the combustion of fossil fuels results in release of metals to the atmosphere (e.g. Nriagu & Pacyna 1988). The study of Chyi (1982) implies that significant amounts of platinum group elements may enter the atmosphere as a result of coal combustion. Chyi (1982) measured <1.0–210 ppb platinum in whole coal samples from western Kentucky coal, USA, and ≤ 2000 ppb platinum in low-temperature ash derived from this coal. These concentrations suggest that significant quantities of platinum may be discharged to the environment via fly ash in many, if not all, coal-burning nations.

Autocatalyst emissions
Large quantities of platinum and palladium are being discharged

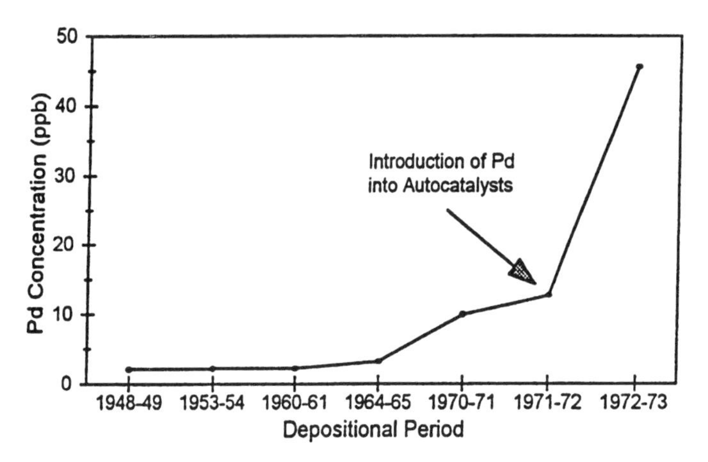

Fig. 14.1. Graph using the data of Lee (1983) to show the rise in palladium concentrations in dated sediment in Palace Moat, Tokyo, Japan. The sudden increase in palladium concentration is attributed to the introduction of autocatalysts containing palladium.

from autocatalysts and König *et al.* (1992) have shown that the discharge is predominantly in solid form as a consequence of mechanical abrasion of the catalyst. Studies have demonstrated that this discharge is accumulating in dust and other sediment along roads and in urban environments. For example, road dust samples from San Diego, USA, contain 37–680 ppb platinum and 15–280 ppb palladium (Hodge & Stallard 1986), and those from London, England, $\leq$33 ppb platinum (Anon. 1995). Figure 14.1 demonstrates that this accumulation appears to be progressive over time. This observation is corroborated by Wei & Morrison (1994) who demonstrated that the average concentration of platinum in dust collected from a kerbside and a parking area in Göteborg, Sweden, increased between 1984 and 1991 due to the increased use of autocatalysts. The platinum group element-rich dust may remain where it was deposited, especially if it can be fixed in roadside soil, or it may be washed by rain into storm drains, urban rivers and coastal waters (Hodge & Stallard 1986; Wei & Morrison 1994).

Sewage sludge

Sewage sludges are anaerobically digested or aerated by-products of waste water treatment and purification. They accumulate, therefore, as waste products in sewage treatment plants. Sludges in Germany and the USA are highly enriched in platinum group elements as the following data indicate (Lottermoser & Morteani 1993): <10–1070 ppb platinum, 38–16200 ppb palladium, <3–3180 ppb osmium, 0.6–460 ppb iridium, <2–7050 ppb ruthenium and <2–2700 ppb rhodium. The main sources of these elements in sludge include industrial and domestic waste and autocatalysts (Lottermoser & Morteani 1993). Hence, we have an interesting situation whereby platinum group elements are mined, processed and used in society, and then they become concentrated, perhaps to ore grade, in sewage sludge (cf. Table 14.1). This sludge may be considered an anthropogenic metal deposit (Lottermoser & Morteani 1993). Currently there is no acceptable process for the large scale extraction of metals from sewage sludge and the sludge is simply treated as a waste product which is disposed of by either: (1) ocean dumping; (2) burial in landfill; (3) incineration or pyrolysis and subsequent burial of slags and ashes in suitable waste sites; or (4) composting and use as a fertiliser on agricultural land. Through treating the sludge purely as a waste product, the four disposal techniques effectively spread vast quantities of platinum group elements throughout into the environment. To put this in context, Lottermoser & Morteani (1993) estimated that world-wide 8 tonnes of platinum and 80 tonnes of palladium are accumulating in sewage sludge each year. These amounts are put back into the environment via the disposal of sludge.

Municipal wastewater

Several studies have demonstrated that the discharge of municipal waste water, containing sewage, in estuarine and coastal areas is resulting in the accumulation of platinum group elements with the sediments of these environments. For example, the Whites Point deposit located near a large municipal waste disposal outflow in Los Angeles, USA, contains 2000 ppb platinum, 13.7 ppb palladium and 47 ppb osmium (Koide *et al.* 1991). The concentration of osmium is very high when compared with results from contaminated sediments in Long Island Sound and New Haven Harbour, USA (0.040–0.774 ppb, Esser & Turekian 1993), and in

Massachusetts and Cape Cod bays, USA (0.022-0.286 ppb, Ravizza & Bothner 1996). Although these osmium concentrations are much lower than that measured at Whites Point, much of the osmium is still attributed to sewage contamination of sediment on the basis of the isotopic composition of osmium in the sediment (Esser & Turekian 1993; Ravizza & Bothner 1996). This provides evidence that disposal of municipal waste is putting platinum group elements back into the environment.

Summary

There is unequivocal evidence that anthropogenic platinum group elements are accumulating in the environment. Hotspots of these elements will typically be found: (1) around and downwind of smelters processing platinum group element-bearing ore and downwind of coal burning installations; (2) along and adjacent to roads; (3) in car parks; (4) in waste water and waste water treatment systems; (5) in sewage sludge; and (6) in rivers, estuaries and coastal areas receiving municipal waste and road run-off. These hotspots are most pronounced in the developed world, but will grow in number as developing nations become more industrialised. With the legacy of lead and other 'old' metals in mind, it is predicted that the whole environment will eventually exhibit an anthropogenic platinum group element signature.

The future

With ever-increasing demand for platinum group elements in the developed and developing nations (Johnson Matthey 1996), the rate and amount of movement of platinum group elements by humans will increase. Autocatalysts are predicted to increasingly contribute to anthropogenic platinum group element concentrations in the environment as more vehicles fitted with autocatalysts are sold world-wide. For example, in Britain alone, 2.6 million cars were fitted with catalytic converters in 1993 and this number is expected to reach 13 million by the year 2000 (Anon. 1995). Until now platinum and palladium have been the dominant platinum group elements released from autocatalysts, but rhodium is expected to figure more prominently in the future as tighter vehicle emission standards require higher rhodium loadings on autocatalysts. There are two positive aspects, however, of this environmental contamination: (1) the potential mineral reserves provided by anthropogenic platinum group element deposits; and (2) the use of platinum group elements to pinpoint sources of pollution.

Anthropogenic platinum group element deposits are forming in sewage sludges (Lottermoser & Morteani 1993) and, possibly, in estuarine and coastal sediments accumulating in areas linked to highly populated and industrialized centres (Quinn & Edwards 1996). The rate of formation of these deposits is incredibly high because of the massive scale at which humans mine and process platinum group elements and the speed at which they generate waste bearing these elements. It is estimated that 8 tonnes of platinum and 80 tonnes of palladium are accumulating annually world-wide in sludge (Lottermoser & Morteani 1993). Based on 1995 data for the demand of platinum group elements (by the Western World) and for the average London metal price fixing (Johnson Matthey 1996), these tonnages equate to 5.4% of the platinum demand, which is worth \$109 million, and 42% of the palladium demand, which is worth \$389 million. From these figures it is quite obvious that if environmentally sound and cost-effective methods can be developed for the extraction of platinum group elements and other metals from sludge, this medium will become an extremely valuable and readily available metal resource.

In the Kola Peninsula it is estimated that nickel smelters emit \$50 million of metals per year, of which half of this value may be attributed to emissions of platinum, palladium and gold (Edwards 1996). As a large proportion of these metals enter and accumulate in topsoil, it is feasible that these soils could be mined in the future (C. Reimann, reported by Edwards 1996). Alternatively, the metals could be filtered out at source, making the filtrate a highly valuable commodity.

The addition of anthropogenic platinum group elements to the environment may be used to pinpoint sources of pollution because only certain industries emit platinum group elements and normal background concentrations of these elements are very low. This possibility is enhanced by the fact that other metals generally have relatively high normal background concentrations and their pollution is far reaching (e.g. lead). Osmium isotopes demonstrably exhibit the greatest potential for tying down sources of particular pollution (Esser & Turekian 1993; Ravizza & Bothner 1996), but the future should see more research in developing the other platinum group elements as tracers of anthropogenic discharge to the environment.

In conclusion, it is worth remembering that many scientists now believe there is an intimate link between elevated concentrations of platinum group elements and catastrophic episodes in Earth history, involving global environmental change and mass extinction (e.g. Alvarez *et al.* 1980). With this possible linkage, one is left to ponder the possibility that history is repeating itself and as anthropogenic platinum group elements spread throughout the environment they are signalling catastrophic environmental change. It is easy to make this connection as human activity increasingly redistributes huge volumes of the Earth's materials, alters the biogeochemical balance of the Earth, destroys ecosystems and species, and causes extinction. Whether or not the linkage is real or coincidental remains to be seen.

Acknowledgements

We thank B. Lottermoser for helpful discussions and information to support an earlier part of this study.

References

Alvarez, L. W., Alvarez, W., Asaro, F. & Michel, H. V. 1980. Extraterrestrial cause for the Cretaceous–Tertiary extinction. *Science*, **208**, 1095–1108.

Anon. 1995. Plenty of platinum. *New Scientist*, **12 August**, 11.

Barnes, S. J. 1983. *Petrology and geochemistry of a portion of the Howland (J-M) reef of the Stillwater complex, Montana.* PhD thesis, University of Toronto.

Chyi, L. L. 1982. The distribution of gold and platinum in bituminous coal. *Economic Geology*, **77**, 1592–1597.

Colodner, D. C., Boyle, E. A., Edmond, J. M. & Thomson, J. 1992. Post-depositional mobility of platinum, iridium and rhenium in marine sediments. *Nature*, **358**, 402–404.

Craig, J. R., Vaughan, D. J. & Skinner, B. J. 1996. *Resources of the Earth*. Prentice Hall, New York.
Crocket, J. H. & Kuo, H. Y. 1979. Sources for gold, palladium and iridium in deep-sea sediments. *Geochimica et Cosmochimica Acta*, **43**, 831–842.
——— & Teruta, Y. 1976. Pt, Pd, Au and Ir content of Kelley Lake bottom sediments. *Canadian Mineralogist*, **14**, 58–61.
Edwards, R. 1996. Rich pickings from Russia's polluted soils. *New Scientist*, 28 September, 5.
Esser, B. K. 1991. *Osmium isotope geochemistry of terrigenous and marine sediment*. PhD thesis, Yale University.
——— & Turekian, K. K. 1993. Anthropogenic osmium in coastal deposits. *Environmental Science and Technology*, **27**, 2719–2724.
Fogg, C. T. & Cornellisson, J. L. 1993. *Availability of platinum and platinum-group metals*. Bureau of Mines Information Circular IC 9338, United States Department of the Interior.
Hiemstra, S. A. 1979. The role of collectors in the formation of platinum deposits in the Bushveld Complex. *Canadian Mineralogist*, **17**, 469–482.
Hodge, V. F. & Stallard, M. O. 1986. Platinum and palladium in roadside dust. *Environmental Science and Technology*, **20**, 1058–1060.
———, ———, Koide, M. & Goldberg, E. D. 1985. Platinum and the platinum anomaly in the marine environment. *Earth and Planetary Science Letters*, **72**, 158–162.
———, ———, ——— & ——— 1986. Determination of platinum and iridium in marine waters, sediments and organisms. *Analytical Chemistry*, **58**, 616–620.
Johnson Matthey. 1996. *Platinum 1996*. Johnson Matthey, London.
Koide, M., Goldberg, E. D., Niemeyer, S., Gerlach, D., Hodge, V., Bertine, K. K. & Padova, A. 1991. Osmium in marine sediments. *Geochimica et Cosmochimica Acta*, **55**, 1641–1648.
———, Hodge, V. F., Yang, J. S., Stallard, M., Goldberg, E. D., Calhoun, J. & Bertine, K. K. 1986. Some comparative marine chemistries of rhenium, gold, silver and molybdenum. *Applied Geochemistry*, **1**, 705–714.
König, H. P., Hertel, R. F., Koch, W. & Rosner, G. 1992. Determination of platinum emissions from a three-way catalyst-

equipped gasoline engine. *Atmospheric Environment*, **26*A***, 741–745.

Lee, D. S. 1983. Palladium and nickel in north-east Pacific waters. *Nature*, **305**, 47–48.

Lottermoser, B. G. & Morteani, G. 1993. Sewage sludges: Toxic substances, fertilisers, or secondary metal resources? *Episodes*, **16**, 329–333.

MacDonald, A. J. 1987. Ore deposit models #12. The platinum group element deposits: Classification and genesis. *Geoscience Canada*, **14**, p155–166.

Naldrett, A. J. 1989. *Magmatic sulfide deposits*. Oxford University Press, Oxford.

Nriagu, J. O. 1990. Global metal pollution. *Environment*, **32**, 7–33.

——— & Pacyna, J. M. 1988. Quantitative assessment of world-wide contamination of air, water and soils by trace metals. *Nature*, **333**, 134–139.

Quinn, F. & Edwards, S. J. 1996. Increased flux of anthropogenic platinum and palladium to estuarine and coastal environments. *In*: Bottrell, S.H. (ed.), *Proceedings of the 4th International Symposium on the Geochemistry of the Earth's Surface*, Ilkley, England, 397–401.

Ravizza, G. E. & Bothner, M. H. 1996. Osmium isotopes and silver as tracers of anthropogenic metals in sediments from Massachusetts and Cape Cod bays. *Geochimica et Cosmochimica Acta*, **60**, 2753–2763.

——— & Turekian, K. K. 1992. The osmium isotopic composition of organic-rich marine sediments. *Earth and Planetary Science Letters*, **110**, 1-6.

———, ——— & Hay, B. J. 1991. The geochemistry of rhenium and osmium in recent sediments from the Black Sea. *Geochimica et Cosmochimica Acta*, **55**, 3741–3752.

Settle, D. M. & Patterson, C. C. 1980. Lead in Albacore: guide to lead pollution in Americans. *Science*, **207**, 1167–1176.

Wei, C. & Morrison, G. M. 1994. Platinum in road dusts and urban river sediments. *The Science of the Total Environment*, **146/147**, 169–174.

Part Three

Natural hazards in the coastal zone

The management of natural hazards is a major issue in many developed and developing countries of the world, particularly in active tectonic and volcanic zones. In Britain, flooding and mass movements in the coastal zone pose a major hazard to coastal infrastructure. Part Three examines current British perspectives of coastal hazard management. In reality, only a small percentage of the coastline of Britain is eroding at an appreciable rate; management of this part of the coastline requires a full understanding of coastal processes.

15 The rate and distribution of coastal cliff erosion in England: a cause for concern?

Anthony R. P. Cosgrove, Matthew R. Bennett & Peter Doyle

Summary

- A national survey of coastal cliff top erosion along England's coast was carried out by sampling 757 km of cliffed coast, yielding 12 112 individual estimates of erosion during the last 66 to 99 years.
- This dataset suggests that only 10% of the coastline sampled is experiencing rates of erosion in excess of 0.5 m a^{-1}, while only 5% has erosion rates in excess of 1.0 m a^{-1}.
- The pattern of erosion reflects the outcrop of engineering soils – Quaternary sediments and pre-Quaternary clays – and a strong link between the rate of erosion, the size of cliff failure and the frequency of failure is established.
- The highest erosion rates are associated with low soft-rock cliffs, the next highest with tall, soft-rock cliffs and the lowest erosion rates appear to be associated with hard-rock cliffs.
- A predictive model is established by which process and erosion rate can be deduced from the changing pattern of cliff top morphology through time.

In recent years there has been a heightened awareness that 'Island Britain' is shrinking. This concern has been fuelled by at least three issues. First, by spectacular cliff failure and its human impacts, such as the loss of Holbeck Hall in Scarborough, North Yorkshire (West 1994). Second, by an increased awareness that certain types of coastal defences may have a detrimental impact on coastal erosion (e.g. Granja & de Carvalho 1995; Shuskiy & Schwartz 1988; House of Commons Environment Select Committee 1992; Bray *et al.* 1995;

Brunn 1995). Third, due to concern about the acceleration of erosion by sea-level rise, induced by the current phase of global warming (Bird 1985; Carter & Bartlett 1990; Clayton 1990; Nichols & Leatherman 1994; Wray *et al.* 1995), and finally by the continued urbanization of the coastal zone (Lumsden 1992). Despite this, few studies have attempted to quantify the nature and extent of erosion in a national context. Clearly, a study of this kind is important in order to determine the veracity of these environmental concerns.

Most coastal erosion studies have simply focused on small sections or regions of coast in order elucidate the role, on the rate of erosion, of such variables as lithology, structure, cliff height and variation in process (e.g. Agar 1960; May 1966, 1971; Brunsden & Jones 1972; Hutchinson 1973; Brunsden 1974; Cambers 1976; Wood 1978; McGreal 1979; Robinson 1979; May & Heeps 1985; Coad *et al.* 1987; Gray 1988; Barton 1990; Jones & Williams 1991; Weller 1991). Although there are clearly a great number of papers in this vein, there is a need for a national view of erosion pattern and process, particularly since in recent years there has been a shift from piecemeal coastal defence, based on local government boundaries, to a more regional approach based on natural geomorphological boundaries or coastal cells (Bray *et al.* 1995; McCue 1998). This shift in emphasis has been primarily fuelled by the production of Shoreline Management Plans – developed as an aid to fostering regional coastal planning – and their construction demands detailed base-line data on such things as the rate and process of coastal erosion (MAFF 1995; McCue 1998).

This chapter is intended to provide some of the basic data required in such base-line surveys and to attempt to place the problem of coastal erosion in a national context. Our aims are to: (1) determine the importance of coastal cliff recession along the whole of England's coast; (2) identify the rate and distribution of erosion; and (3) to examine the link between the rate of erosion and the magnitude and frequency of the processes involved.

Methods

A wide range of methods have previously been employed to monitor cliff erosion, including: (1) the use of direct field observations, normally repetitive surveys (e.g. Valentin 1954; May 1971; Williams & Davis 1987; Dias & Neal 1992); (2) the use of oblique and vertical photographs, and, increasingly, satellite images (e.g. Horikawa & Sunamura 1967; Griggs & Johnson 1979; Shepard & Kuhn 1983; Dolan *et al.* 1991; Jones *et al.* 1993; Komar & Shih 1993; Moon & Healy 1994; Wray *et al.* 1995); and (3) cartographic investigations using sequential historical maps (e.g. Valentin 1954; So 1967; McGreal 1979; Sunamura 1982; May & Heeps 1985; Gray 1988). Direct data collection is problematical, particularly when ensuring that monitoring occurs over a sufficiently long enough period, in order to sample the full range of magnitude and frequencies experienced on a given length of coast (e.g. Cambers 1976; Hooke & Kain 1982). It is also clearly impractical for large national investigations such as this. The use of sequential aerial photographs provides a good method of obtaining recession data, and also allows process observations to be made as well. However, vertical aerial photographs have only been widely available in Britain since the late 1940s, and it is difficult to obtain a regular sequence of air photographs of a similar scale and quality for the whole of Britain. In addition, it is often difficult to obtain access to the photographs, since most modern surveys are piecemeal and have been flown by commercial companies, in contrast to earlier, systematic, RAF sorties.

Historical maps provide perhaps the most comprehensive data source. Since the Ordnance Survey was established in 1791, Britain has been regularly resurveyed and a suite of maps at consistent scales has been published at regular intervals (Hooke & Kain 1982). Most urban and populated rural areas, which includes most of the coast, have been surveyed at 1:2500 approximately four times, in the 1870s, 1890s, 1920s, and 1970s, although precise survey dates vary regionally. However, remote and unpopulated areas have only been surveyed to a scale of 1:10560 or 1:10000. It is, therefore, possible to obtain erosion rates for the last 100 years at a scale of 1:2500 for most, although not all, of the coastline of England. Ordnance Survey maps provide a consistent suite of maps which were constructed to similar standards, using similar techniques, and more recently (1938 onwards) to a common national projection (Hooke & Kain 1982). In addition, these maps are accessible since

Fig. 15.1 (a). The extent of cliffed coast sampled in this study. In areas of cliffed coast not shown data were unavailable either due to poor map coverage, or associated with problems of defining the cliff top. **(b)** Areas of cliffed coast with an erosion rate in excess of 0.5 m a^{-1}. The sample period varies from 64 to 99 years depending on the region involved (Table 15.3).

they are held centrally at the National Map library in the British Museum. A wide range of pre-Ordnance Survey maps also exist for earlier periods, and typical examples include: tithe maps, enclosure maps, county maps and estate plans. These vary in availability and accuracy and were not usually constructed to a common national standard, and as such are of limited value in a national study such as this (Harley 1968; Hooke & Perry 1976; Hooke & Kain 1982). Therefore, for the purpose of this study, the 1:2500 Ordnance Survey maps have been used.

Data were sampled for cliffed coast only, as identified from Ordnance Survey maps. For the purpose of this study, cliffed coast is defined as any coastal section elevated above High Water Mark, which is not obviously part of a dune, salt marsh, shingle ridge, or other low lying coastal feature. The coast was examined in kilometre long sections, and classed accordingly as 'cliffed' or 'non-cliffed'. A kilometre section of coast was classed as cliffed, even if it contained low lying areas, provided there were some elevated sections and that measurements of cliff recession were consequently possible.

Measurements were not possible in Northumbria, Cumbria and Lancashire due to absence of a continuous cover of the four sequential 1:2500 Ordnance Survey maps. A second problem was encountered in those areas where there is a wide landward zone of coastal failures of varying age. Here it was impossible to effectively define a cliff top using the maps, and therefore no estimates of recession were made. Examples of this include Folkestone Warren in Kent, the Black Ven area of Dorset and the Ventnor region of the Isle of Wight.

The length of coast sampled is shown in Figure 15.1a. For each section of cliffed coast the following observations were made from a series of original and sequential Ordnance Survey maps at a scale of 1:2500.

1. The line of the cliff top was traced from the oldest available map and a series of fixed points, such as building and field boundaries, were established. These fixed points were used to register the tracing on subsequent maps. Prominent grid lines were drawn on the map using later editions, as these employ the current national projection.
2. For each kilometre of cliffed coast 16 equally spaced measurements were made from the map tracing of cliff recession. Coastal recession was measured between each successive map and between the first and last map within a sequence. Readings were only made where four maps were available. These data were stored in a database using four figure grid references. A total of 757 km of coast were sampled, involving approximately 12 112 measurements of cliff recession.
3. For each kilometre of cliffed coast the style of coastal retreat was recorded using the classification shown in Fig. 15.2. In addition, the dominant lithologies were recorded using 1:50 000 geological maps, where available.

The cliff top was measured in preference to the cliff base since: (1) it is more easily defined on the maps; (2) its position is not affected by

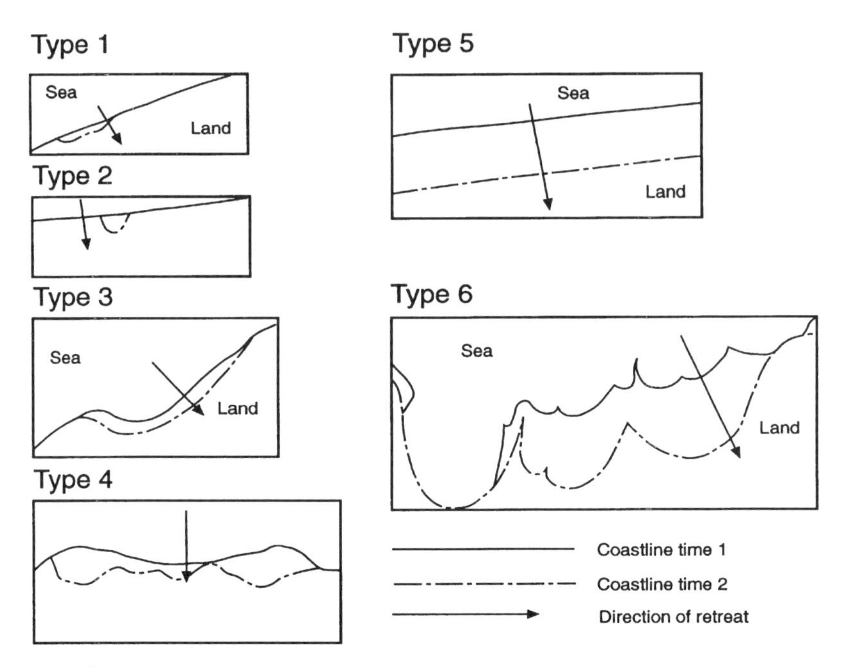

Fig. 15.2. The classification of changing cliff top morphology.

fallen debris (Komar & Shih 1993); (3) it is of greater practical importance in land management, such as hazard zonation or in controlled retreat; and (4) the use of the cliff top is consistent with previous work (e.g. Steers 1951; Valentin 1954; Cambers 1976; May 1966, 1971; May & Heeps 1985; Gray 1988; Coad *et al.* 1987; Barton 1990).

There are two main areas of error associated with measurement of cliff tops. First, there may be errors in the Ordnance Survey maps themselves, as several authors have recognized surveying errors of up to 1 m (e.g. Hooke & Kain 1982; Gray 1988; Nicholls & Webber 1987). A particularly important example is Chapel Point in Cornwall, were a surveying error in 1880 skewed the headland to the north with respect to its location on the 1970 edition of the same map (Fig. 15.3a). Cliff advances between sequential Ordnance Survey map editions at Walton-on-the-Naze in Essex have been attributed to problem in cliff edge definition during different

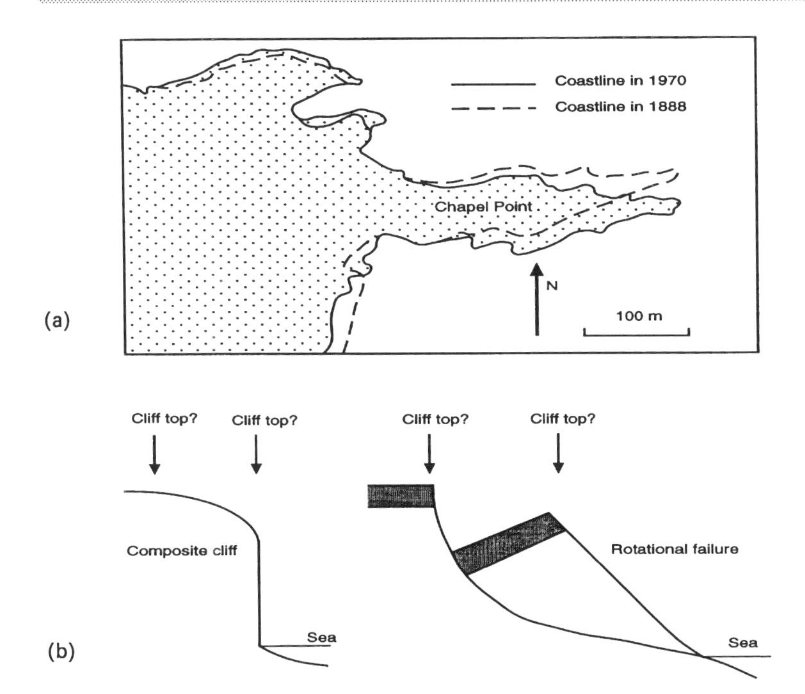

Fig. 15.3. Error associated with 1:2500 Ordnance Survey maps: (**a**) surveying errors at Chapel Point; (**b**) problems associated with defining the cliff edge.

surveys (Fig. 15.3b; Gray 1988). Second, measurement errors by the operator may result in some anomalies within the dataset. For example, a measurement error of 1 mm on a 1:2500 map corresponds to a 2.5 m error on the ground. Despite these problems the dataset gives an approximation of the rate of recession and more importantly it is internally consistent, allowing the relative importance of coastal erosion and its distribution to be established. Similar survey methods were used for each map edition, the surveyors were trained to common standard, and each used a similar protocol. In addition a single operator made all the measurements from the Ordnance Survey maps for this study (Cosgrove 1997). A comparison of published recession rates (Table 15.1), using a variety of different methods, for specific localities was compared to those obtained in this study and good correlation, explaining 66.3% of the variance, was obtained.

Table 15.1. Comparison of cliff erosion rates obtained in previous studies relative to those obtained here.

Author	Data source	Location	Rate ($m\,a^{-1}$)	Rate: this study ($m\,a^{-1}$)
Agar 1960	Maps	Teesside	0.09	
West 1994	Field work	Scarborough	80m/3 days	
Valentin 1954	OS Maps	Holderness	1.2	1.11
Cambers 1976	OS Maps	Norfolk	0.9	0.61
		Suffolk	0.8	1.6
Gray 1988	OS Maps	Essex	0.78	0.8
May 1966	OS Maps	Isle of Sheppey	0.95	0.89
		Reculver	0.68	0.3
		N. Isle of Thanet	0.25	0.16
		E. Isle of Thanet	0.25	
		Kingsdown	0.8	0.17
		St. Margarets Bay	0.1	0.15
		Dover	0.11	0.1
		Folkestone	0.28	0.1
		Fairlight	0.65	0.2
		Covehurst	0.2	0.35
		Seven Sisters	0.46	0.33
		Newhaven	0.46	0.32
		Middleton	0.23	
		S. Isle of Wight	0.0	
		Brightstone Bay	0.5	
		Newtown River	0.3	
Hutchinson 1973	OS Maps	Sheppey	2.2	
So 1967	OS Maps	Studd Hill, Kent	1.2	
Brunsden & Jones 1980	Field work	Fairy Dell, Devon	0.5	
May 1971	Field work	Birling Gap	0.68	0.67

Coastal cliff erosion in England

Rates of erosion: a national perspective

Concern about coastal erosion is usually generated by high magnitude, but low frequency events with a high media profile such as the loss of Holbeck Hall, Scarborough, in 1993 (West 1994). However, in common with most geomorphological systems (cf.

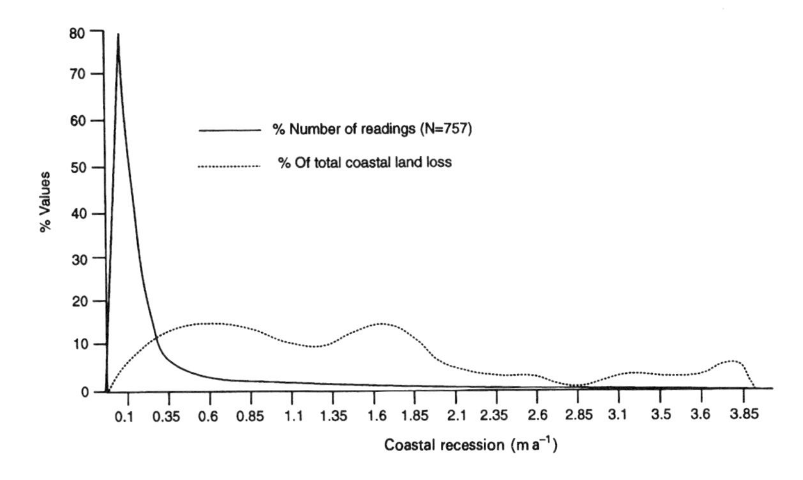

Fig. 15.4. Magnitude and frequency plot for cliff erosion along the 757 km of the English cliffed coast sampled.

Wolman & Miller 1960), the data collected here suggest that nationally most coastal erosion is actually associated with relatively small, but high frequency, events. As Figure 15.4 shows, most land loss is associated with erosion rates of between 0.4 and 1.6 m a^{-1}. Higher erosion rates (max 4.0 m a^{-1}) do occur, but rarely with a sufficient frequency to make a significant contribution to the total amount of land lost nationally.

Table 15.2. Proportion of cliffed coast sampled in relation to five arbitrary erosion classes.

Class range (m a^{-1})	Descriptor	% of cliffed coast sampled	% of total cliff loss
0.0.1	Negligible	77	2
0.1–0.5	Moderate	13	22
0.5–1.0	Intense	5	23
1.0–1.5	Severe	3	27
>1.5	Very severe	2	26

Table 15.3. Regional variation in the rate of erosion.

Region	Number of readings (%)	% of total cliff loss	Average $m\,a^{-1}$	Minimum $m\,a^{-1}$	Maximum $m\,a^{-1}$	Standard deviation	Coefficient of variance (%)	Length of record (years)	Coastal cliff top morphology %					
									Type 1	Type 2	Type 3	Type 4	Type 5	Type 6
Yorkshire	10.6	4.32	0.05	0.0	0.3	0.08	160	79	4	16	39	33	6	2
Holderness	4.5	38.61	1.11	0.3	2.18	0.45	40.5	79	0	0	0	0	100	0
Norfolk	3.5	15.5	0.61	0.12	1.69	0.44	72.1	84	0	14.2	39.7	3.4	25	17.8
Suffolk	1.3	17.45	1.6	0.47	3.51	0.83	51.8	87	0	0	0	0	100	0
Kent	4.0	7.98	0.25	0.0	1.2	0.31	124.0	86	0	12	22	44	3	19
Sussex	2.6	6.63	0.28	0.0	0.74	0.21	77.3	87	15	0	0	55	20	0
Isle of Wight	3.6	3.5	0.2	0.0	0.44	0.16	79	64	0	0	17	31	34	17
Hampshire	1.7	3.92	0.3	0.0	0.85	0.27	88	900	0	0	50	50	0	
Dorset	6.8	2.46	0.08	0.0	0.5	0.1	124	82	16	21	37	26	0	0
South Devon	15.2	1.2	0.01	0.0	0.37	0.05	422	91	33	64	0	0	0	0
North Devon	9.0	0.17	0.01	0.0	0.08	0.01	423	91	75	0	12.5	12.5	0	0
Cornwall	37.3	0.16	0.01	0.0	0.11	0.01	100	99	50	12	0	38	0	0
England	100	100	0.148	0.0	3.51			64–99	7	8	20	30	30	5

The regions are defined geographically on the basis of county boundaries with the exception of: Yorkshire – Staithes Estuary to Flamborough Head; Holderness – Flamborough Head to Spurn Point; Hampshire – Hurst Castle Spit to Hengistbury Head; Dorset – Hengistbury Head to the Dorset/Devon border; N. Devon – from the North Devon/Cornwall boundary to Avonmouth.

Table 15.4. Average erosion rates and characteristics for the principal lithologies recorded along the sampled coast.

Lithology	Number of reading on the coast (%)	% of total erosion recorded	Average $m\ a^{-1}$	Minimum $m\ a^{-1}$	Maximum $m\ a^{-1}$	Coastal cliff top morphology (%)					
						Type 1	Type 2	Type 3	Type 4	Type 5	Type 6
Chalk	9.3	8.9	0.14	0	0.74	15	18.8	13.2	47.2	5.6	0
Slate	20.30	0.06	0.01	0	0.04	37.5	25	0	37.5	0	0
Sandstone	18.7	6.8	0.06	0	0.8	50	7.4	11.1	81.5	0	0
Crystalline rocks	23	0.11	0.01	0	0.01	70	30	0	0	0	0
Limestone	2.8	0.03	0.01	0	0.02	20	0	80	0	0	0
Shale	4.6	0.55	0.01	0	0.27	22.2	0	77.8	0	0	0
Clay	4.5	8.7	0.27	0	1.69	0	0	18	22.9	40.9	18
Quaternary sediments	14.1	71.3	0.73	0	3.51	2.2	0	15.3	20	52.9	10.6

The rate of cliff erosion for 757 km of the English cliffed coastline sampled is shown in Table 15.2 in relation to five erosion classes. The figure for total cliff loss, during the time period sampled, was obtained by summing all the land lost to erosion at each sample point along the coast during the study period (64 to 99 years). The most striking point within these data is that 77% of the coast sampled experienced little or no erosion ($<0.1\,\mathrm{m\,a^{-1}}$), accounting for only 1% of all the land loss during the study period (Table 15.2). On only 10% of the sampled coastline was erosion greater than $0.5\,\mathrm{m\,a^{-1}}$, a rate which would give a land loss of 50 m in 100 years, and which would therefore be of significance to cliff top infrastructure. At higher rates the proportion of coastline affected falls even further, as for example, only 5% has an erosion rate in excess of $1.0\,\mathrm{m\,a^{-1}}$.

The regional distribution of coastal cliff erosion is presented in Table 15.3 and areas with an erosion rate above $0.5\,\mathrm{m\,a^{-1}}$ are shown in Fig. 15.1b. Areas of rapid coastal cliff recession are confined to just three regions, Holderness, Suffolk/Essex and Norfolk (Fig. 15.1b; Table 15.3), which together account for 71.56% of all the cliff loss recorded in this study. In contrast other regions, particularly Cornwall and Devon, have very low erosion rates, although the levels of variance are high, which indicates a degree of along-coast variability with high erosion rates occurring locally (Table 15.3). This distribution is controlled, at a national scale, by lithology. As the data in Table 15.4 indicate, high erosion rates usually occur where soft rocks, such as pre-Quaternary clays and Quaternary sediments, form the cliffline. Conversely, erosion is very low on more competent rocks such as limestones, shales, slates and crystalline rocks. This is supported by the observation that over 71% of all the cliff loss within the study occurred in cliffs composed of Quaternary sediments. Clearly, the outcrop pattern of these erosion susceptible lithologies determines, to a large extent, the distribution of rapid coastal cliff erosion in England.

In addition to the above observations the dataset does not reveal any systematic increase or decrease in erosion over the timescale sampled (Fig. 15.5). The regions of Yorkshire, Holderness and Norfolk all show a slight increase in erosion at the start of the current century, although this trend is not picked out elsewhere.

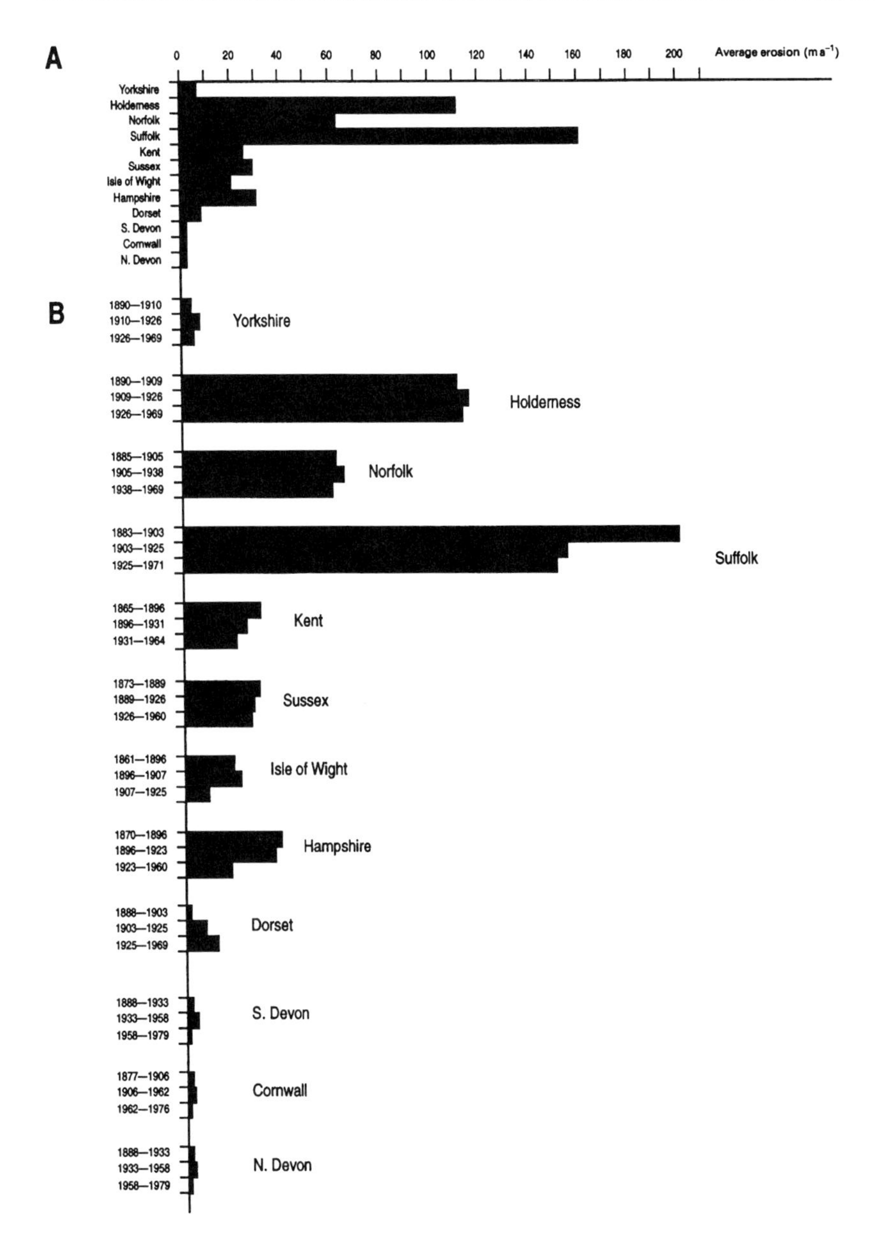

Fig. 15.5. Variation in erosion rate by region (**A**) and through time (**B**) for each of the regions identified in Table 15.3.

Several areas, Suffolk, Kent, Sussex and Hampshire, show a slight decrease, while Dorset is the only region to show a marked increase during the last 100 years (Fig. 15.5). In general, however, these data are too coarse (*c.* three 30 year time intervals) to pick out any subtle changes in erosion with time, which may be associated with increased coastal urbanization, the construction of coastal defences or resulting from slight sea-level rise.

These figures suggest that coastal cliff erosion is not as significant a problem as is sometimes implied in Britain. They demonstrate that only 75 km of English coastline would experience land loss greater or equal to 50 m in a 100 year period, and that just 15 km would lose more than 150 m. Clearly, at locations with land loss of 50 m or greater the problem is serious, both in terms of the impact on cliff top infrastructure and loss of agricultural land, although it is clear that erosion is not, according to this dataset, an endemic problem on the coastline of England. In practice, however, local factors are critical in determining the significance of even the smallest erosion rate. If essential infrastructure exists close to the cliff edge, then a rate of a few centimetres a year is just as significant as one of several metres over the same timescale. Consequently, any evaluation of the importance of land loss to coastal erosion in England must be combined with a knowledge of human vulnerability to even the most minor erosion rate. In this way it is significant that the highest rates of erosion in England occur in Holderness and Suffolk/Essex, which are not heavily urbanized, although agricultural land and associated infrastructure is often threatened.

Rates of erosion: a regional perspective

So far in this discussion we have only considered national variation in coastal erosion. However, the data collected reveal strong local variability in erosion along a given length of coast. This is illustrated in Fig. 15.6 for sections of the Holderness and Norfolk coast, and is true of most of the coastal sections sampled. The cause of such variability has been assigned, on hard-rock coasts, to variation in the susceptibility of cliffs to erosion and failure, and is often associated with joints, faults or other fracture zones (e.g. Middlemiss 1983; Carter & Bartlet 1990). Along soft-rock coasts

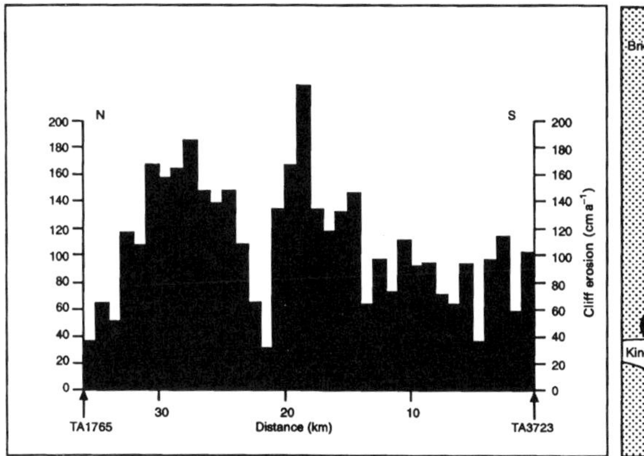

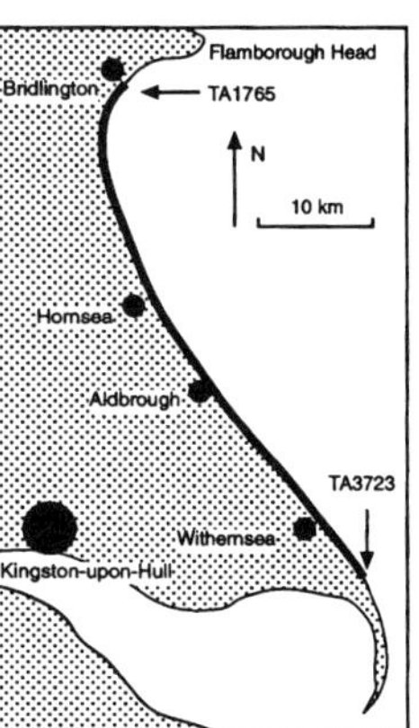

(a)

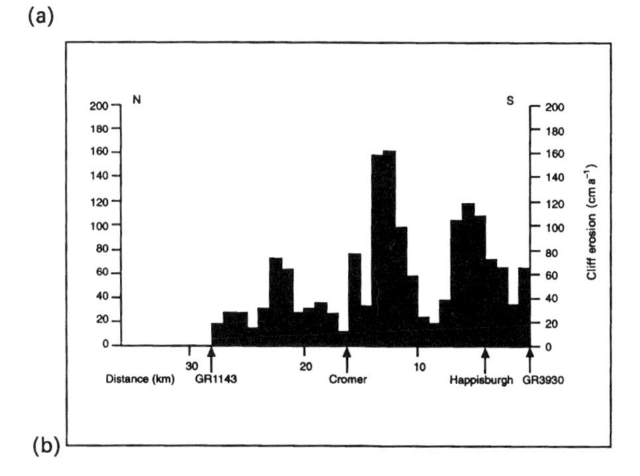

(b)

Fig. 15.6. Variation in the rate of coastal cliff erosion along the (**a**) Holderness and (**b**) Norfolk coasts. Each bar represents a one kilometre average based on 16 sample points.

variability in lithology and cliff height may also be important. In order to examine this last point, data on lithological sections and cliff height are required for a continuous section of coast along which cliff recession has been monitored.

Continuous sedimentological sections are available from the literature for some parts of the British coast, and most notably in Norfolk were the Quaternary sediments have been subject to detailed study (e.g. Eyles *et al.* 1989; Hart 1990; Lunka 1994). For example, the paper by Ehlers *et al.* (1991) contains a continuous scale section over 4 km long of the coastal stratigraphy from Sheringham in the east to Weybourne in the west (Figs 15.1 & 15.7). Here, the cliffs reach a maximum height of 38 m OD at Skelding Hill, just west of Sheringham. This has been simplified and is reproduced in Fig. 15.6, along with the erosion rates obtained in this study. The cliffs contain a range of fine-grained diamictons (glacial tills) as well as sand and silt units and has been heavily folded and fractured. These sediments overlie a chalk platform, which lies just beneath the beach crest in the east but is several metres above it in the west (Fig. 15.7). Several points emerge from this exercise.

1. Erosion rates decline east to west with increasing thickness of the chalk unit at the cliff base, which is located close to mean High Water Mark. The average for the non-chalk areas is 0.22 m a^{-1} compared to 0.18 m a^{-1}. In the section just east of Weybourne the undulating chalk rockhead causes variation in the erosion rate; rates are low where the chalk is high in the cliff and fast where it is low. Significantly, if sea-level was to rise by as little as 0.5 m the basal chalk would be submerged and erosion rates would increase as the softer sediments come under direct wave attack.
2. The fastest erosion rate on this section of coast (0.4 m a^{-1}, to the west of Gap A; Fig. 15.7) occurs were the cliff is low and composed of unlithified sediments.
3. Variation in the stratigraphy of the cliff sediments appears to have little or no systematic effect on the rate of erosion. Unconsolidated sands appear to erode as quickly as consolidated diamictons. For example, just west of Sheringham the occurrence of a sand unit at the cliff base, instead of consolidated diamicton, is not associated with an increase in erosion. However, in general, the variability in erosion rates is greater where the chalk base is absent, with a range of 0.36 m a^{-1}, which may be compared to 0.19 m a^{-1}, where the chalk is present.

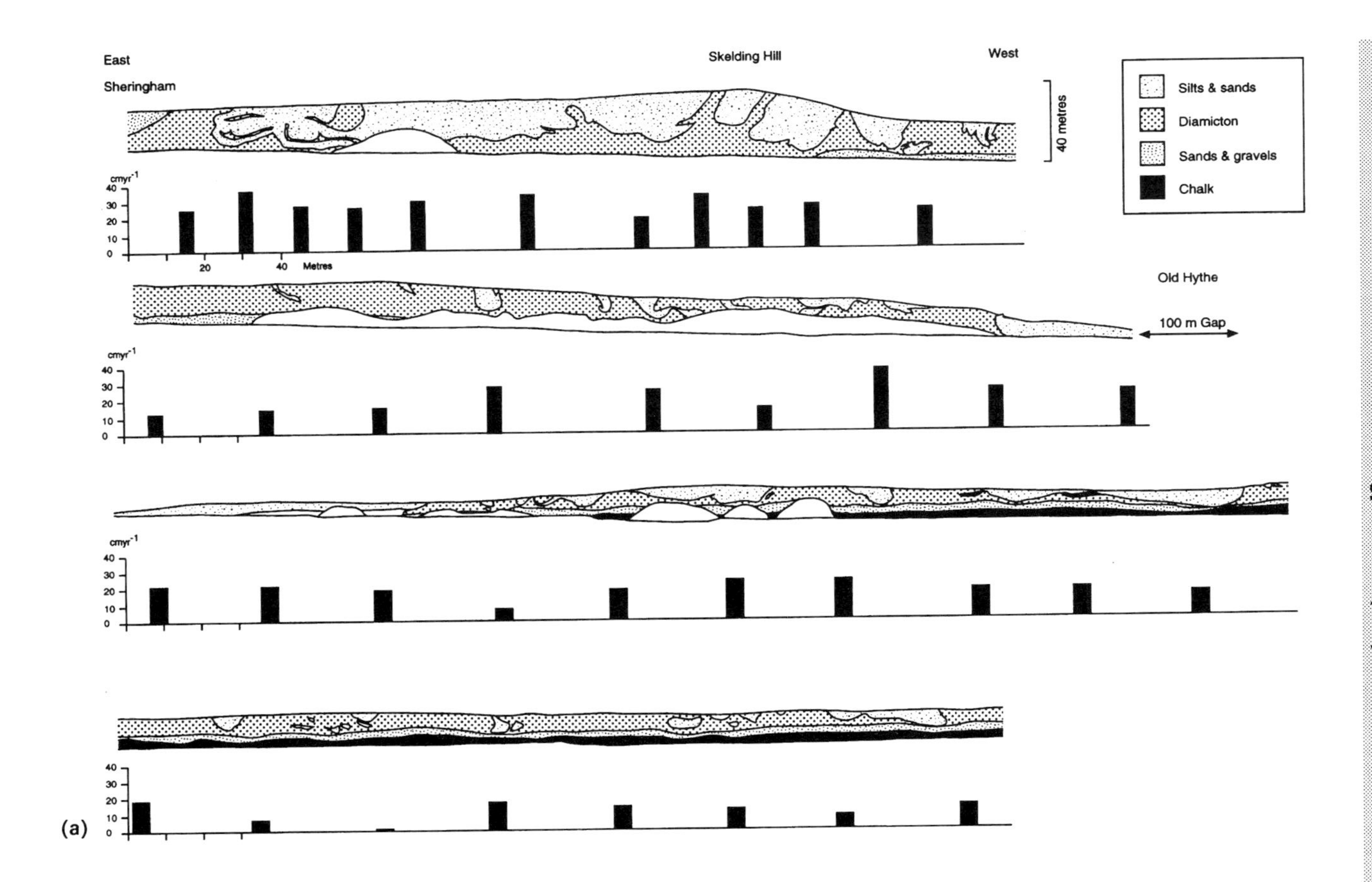

East
Sheringham
Skelding Hill
West
40 metres
Silts & sands
Diamicton
Sands & gravels
Chalk
Old Hythe
100 m Gap
Metres
cmyr-1
40 30 20 10 0
(a)

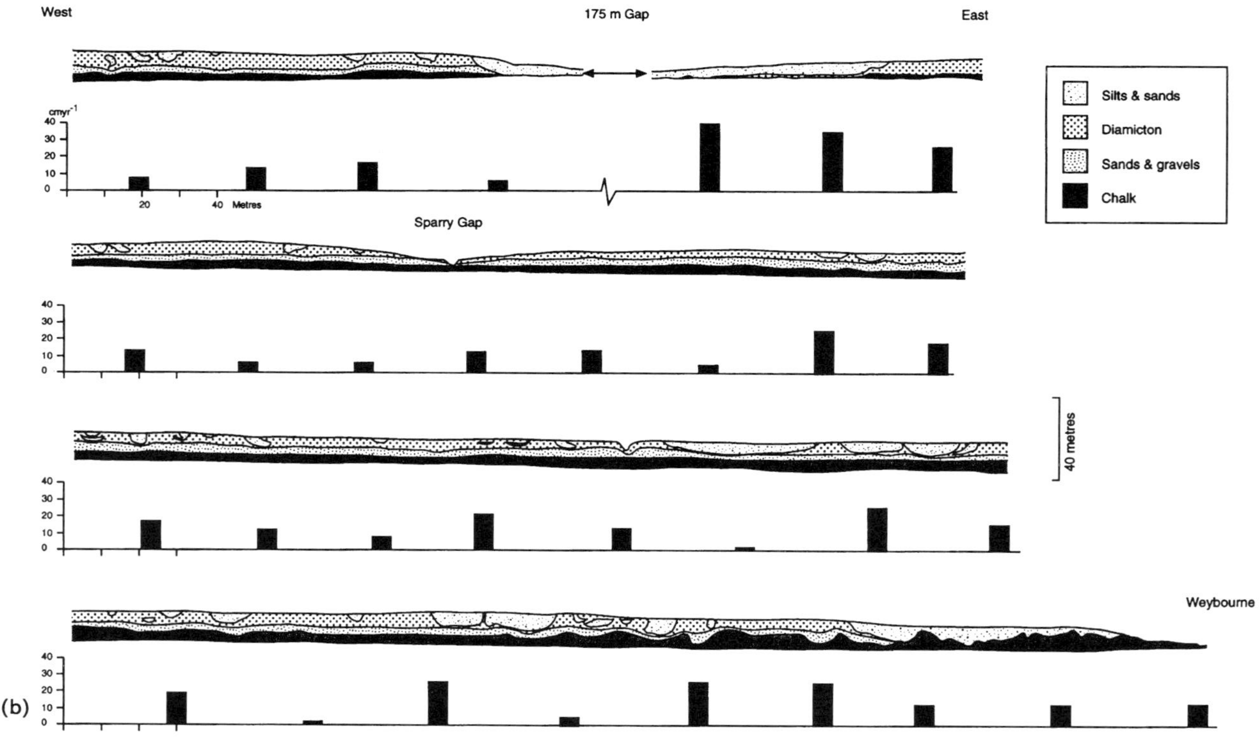

Fig. 15.7. Relationship between the rate of cliff recession and detailed cliff stratigraphy for a section of the Norfolk coast between Sheringham and Weybourne. The cliff stratigraphy is modified from Ehlers *et al.* (1991).

Variation in the detail of cliffline stratigraphy and changes in cliff height can therefore explain some of the along coast variability in otherwise seemingly homogeneous cliff sections. This exercise demonstrates the importance of the exposure of a relatively resistant rock at the cliff base in controlling the overall rate of cliff recession. This may, however, be highly sensitive to changing sea-level; a slight rise should accelerate the rate of erosion in Norfolk. The detailed stratigraphy of unlithified sediments above this resistant rock has little apparent effect on the rate of erosion. This supports the conclusion of Griggs & Johnson (1979) at Santa Cruz in California which also emphasized the importance of the basal lithological unit in determining cliff erosion.

Table 15.5. Cliff top morphology and erosion rates

Cliff top morphology	Average erosion ($m\,a^{-1}$)	% of cliffed coast sampled	% of total cliff loss
Type 1	0.08	7	1
Type 2	0.11	8	2
Type 3	0.19	20	9
Type 4	0.16	30	12
Type 5	0.88	30	66
Type 6	0.76	5	10

Cliffline morphology and process

Gray (1988) first suggested the idea that changes in the geometry of a cliff top line, through time, may provide an indication not only of process, but also of the magnitude and frequency of erosion. As a consequence in this study the character of the cliff top, through time, was classified into six broad types on the basis of visual distinction (Fig. 15.2). The link between the changing geometry of a cliff top and the rate of cliff recession is shown in Table 15.5. The highest rates of erosion are associated with cliff Types 5 and 6, and most of the national cliff loss reported in this study (66%) occurs on Type 5 cliffs. In general, Type 5 cliffs are located on all the most severely eroding cliff sections in England (Table 15.3), representing rapid, parallel, retreat of the cliff top. Type 6 is also associated with a high average erosion rate (Table 15.5), but is less widely distributed (5% of observations) and only accounts for 10% of the total amount of cliff recession observed. In this case cliff recession occurs via deep coastal crenulations and is not continuous along a stretch of coast (Fig. 15.2).

These observations can be explained in relation to cliff height and process which control the magnitude and frequency of erosive episodes along a section of coast. Types 1, 2 and 3 cliff top morphology all occur on relatively tall (> 20 m) cliffs composed of resistant rocks, such as chalk, sandstone and a range of crystalline lithologies (Table 15.4). These types of cliff top morphology are common along parts of the Yorkshire, Cornwall, Dorset and Devon coasts (Table 15.3), where cliff retreat is predominantly by rock fall of varying size (Agar 1960; May 1971; Robinson 1979; Middlemiss 1983; Barton 1990; Carter & Bartlet 1990). Variation in the type of cliff top morphology reflects the size and concentration of the rock fall activity. If it is localized and widely spaced a Type 1 cliff line is obtained (Fig. 15.2). If, however, rock fall is concentrated by faults, joints or fracture zones, as is common on such cliffs (May 1971; Middlemiss 1983; Barton 1990; Corbett 1990), then a Type 2 cliff top results. Finally, Type 3 cliff tops tend to occur in areas were large vertical slab failures and topples have been reported (Steers 1964).

Type 4 cliff tops also occur on relatively competent rock types (Table 15.4), such as the chalk cliff of the Seven Sisters in Sussex. May (1971) investigated cliff erosion in this area and found that it was a function of large rock falls and block failures along zones of weakness in the cliff where wave energy was concentrated. These zones of weakness were due to both lithological variation and jointing which allowed preferential weathering. Other areas of Type 4 cliff tops are also associated with large rock mass failure, often concentrated along pre-existing lines of weakness (Wilson 1952).

In contrast Type 5 and 6 cliff tops occur on 'soft-rock' cliffs composed of Quaternary sediments, and Tertiary clays (Table 15.4). Type 5 cliff tops tend to occur on relatively low cliffs (< 5 m), particularly on unconsolidated sediments, while Type 6 occurs on higher cliffs composed of over-consolidated clay units. The

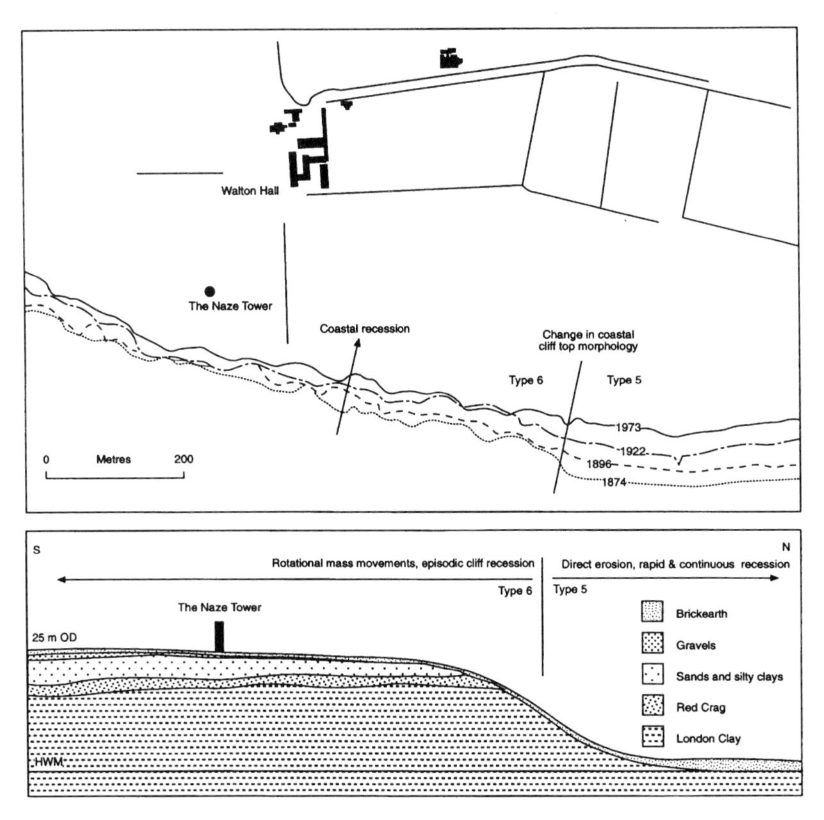

Fig. 15.8. A map of the cliffed coast at Walton-on-the-Naze showing the relationship between the type of cliff top morphology and rate of erosion. (Modified from Gray 1988.)

transition between these two cliff top morphologies is well illustrated at Walton-on-the-Naze in Essex (Fig. 15.8; Gray 1988). Here there is a south to north transition from a Type 6 cliff top to a Type 5, associated with: (1) a decrease in cliff height from 20 m to 2 m; (2) a change in composition from a sequence of London Clay, Red Crag and brickearth, to one composed completely of brickearth; (3) a transition from periodic rotational mass failure in the south, to direct coastal erosion and small block topples in the north; and (4) an increase in the average rate of cliff recession from $0.55\,m\,a^{-1}$ to $0.88\,m\,a^{-1}$. This type of contrast between Type 6 and

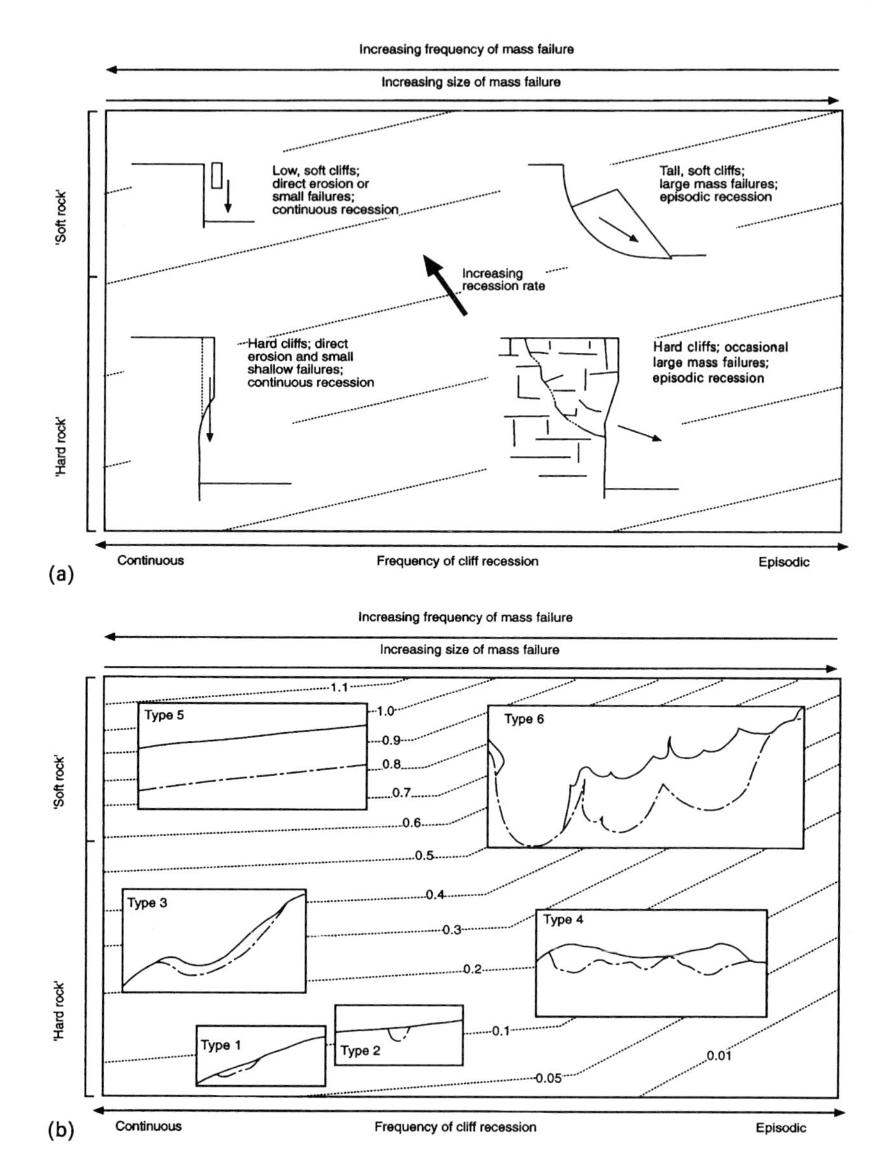

Fig. 15.9. Model of the relationship between the rate of erosion and the size, frequency and type of cliff failure. (**a**) Shows the hypothetical position of different types of cliff failure within the erosion matrix. (**b**) Shows the position of the different types of cliff top morphology within the erosion matrix. Recession contours are based on the average rate for each cliff top morphology.

Type 5 cliff tops is true of most areas. A Type 5 cliff tops occurs were the cliffs are low and eroded directly by marine action and minor cliff failure (Steers 1951; Robinson 1979; Pringle 1985), while Type 6 cliff tops occur where cliff recession occurs periodically either by rotational failure or some other type of mass movement (So 1967; Hutchinson 1973; Brunsden & Jones 1980).

Although somewhat generalized, this model of cliff top morphology and its relationship to process and lithology holds in the majority of situations examined. The relationship between cliff top type and the rate of coastal recession evident in Table 15.5 is a function of the process of cliff recession and size of mass movement involved. Large magnitude cliff failures cause dramatic cliff top loss, but occur only episodically since they produce a large amount of cliff foot debris which must be eroded before direct cliff recession can occur again. The time taken to remove this debris is greater in the case of hard, competent, rocks, since soft clays and sands are easily eroded by secondary failures (e.g. debris flows and earth avalanches) and by direct wave action. In contrast, small magnitude cliff failure causes small amounts of cliff top loss, but occurs with a greater frequency since there is little cliff foot debris to remove. This is summarized in the conceptual model in Fig. 15.9. The first part of this model places the four main types of cliff erosion process or failure within a matrix of: (1) frequency of cliff failure or recession; (2) size of mass movement or cliff recession event; and (3) lithology, soft rock (engineering soils) or hard rocks. Cliff recession is fastest in the top left-hand corner of the matrix, associated with low, soft rock, cliffs which fail continuously by small mass movements and is slowest at the bottom left associated with tall, hard rock, cliffs which fail via large but episodic mass movements (Fig. 15.9a). The second part of the model places the six cliff top morphologies recognized in this study (Fig. 15.2) within this matrix (Fig. 15.9b), linking them to the process of cliff recession. Erosion rates, within this second matrix, are based on the average rate obtained for each cliff top morphology and provide an indication of the link between process, magnitude, frequency and the rate of coastal cliff top loss. Although somewhat generalized, it provides an indication of the importance of cliff height and lithology in determining erosion rates. Low, soft-rock, cliffs erode more quickly than tall, soft-rock, cliffs which, in turn, erode faster in general than hard-rock cliffs.

Cliff top morphology (Fig. 15.2) emphaizes, therefore, the importance of magnitude and frequency in determining the rate of coastal recession. Large magnitude events, such as the loss of Holbeck Hall, in Scarborough (West 1994), may be dramatic but are of low frequency, and most land loss is the result of small continuous cliff failure. Significantly, it is possible to predict, at least in a general way, the rate and process of coastal erosion from the changing pattern of cliff top morphology through time (Fig. 15.9).

Conclusions

The principal conclusions obtained from this study of erosion along the cliffed coastline of England are listed below.

1. Severe coastal erosion affects only a small proportion of the cliffed coastline of England. Approximately 77% of the 757 km of coast sampled experiences little or no coastal erosion ($\leq 0.1\,\mathrm{m\,a^{-1}}$). Only 10% of the coast has erosion rates in excess of $0.5\,\mathrm{m\,a^{-1}}$. Areas of rapid erosion correspond to cliffs composed predominantly of Quaternary sediments and pre-Quaternary clays. At a national scale the erosion pattern reflects the distribution of soft, erodible, lithologies.
2. No systematic increase or decrease in erosion was recorded over the timescale sampled.
3. Local variation in the rate of cliff erosion is high and can be explained, for the Norfolk coast at least, by variation in cliff height and detailed cliff stratigraphy. Of critical importance is the presence of competent lithologies at the cliff base, in this case chalk. This has implications for long term erosion rates, since a rise in sea-level may elevate the wave base above the chalk leading to accelerated erosion in the future. Other examples of this type exist at several locations.
4. The character of the cliff top line through time can be classified into six categories, and linked to the rate of erosion. Cliff top morphology provides an indication of the processes and their magnitude and frequency along the coast. Erosion

rates are controlled by the size of the cliff failure and the frequency with which it occurs. Most erosion occurs on low cliffs, often composed of unlithified Quaternary sediments. These cliffs fail by small topples and block failures and the process is continuous. Higher soft-rock cliffs fail by large mass movements, particularly rotational failures; land loss is therefore large but only occurs episodically, since the fallen debris act as a barrier to continued coastal erosion. It is possible to categorize the cliffs on the basis of their cliff top morphology into an erosion matrix, which attempts to show the relationship between the process and frequency of cliff failure and the magnitude or size of the individual events (Fig. 15.9).

In summary, this exercise demonstrates that coastal cliff erosion is not necessarily as significant on England's coast as has been previously implied, although variation in the vulnerability of human infrastructure to erosion at even low levels has not been considered. It is possible, however, to predict the process, magnitude and frequency of erosion from changes in cliff top morphology with time. This demonstrates that rapid coastal erosion is associated with low cliffs composed of soft rocks rather than tall cliffs, even though the magnitude of each mass failure may be greater on such cliffs. Finally, as expected, erosion along hard-rock cliffs is of only minor importance.

References

Agar, R. 1960. Post-glacial erosion of the north Yorkshire coast from the Tees estuary to Ravenscar. *Proceedings of the Yorkshire Geological Society*, **32**, 409–428

Barton, M. E. 1990. Stability and recession of the chalk cliffs at Compton Down, Isle of Wight. In: *Proceedings of the International Chalk Symposium*. Thomas Telford, London, 541–544.

Bird, E. C. F. 1985. *Coastline changes: a global review*. Wiley, Chichester.

Bray, M. J., Carter, D. J. & Hooke, J. M. 1995. Littoral cell definition and budgets for central southern England. *Journal of*

Coastal Research, **11**, 381–400.
Brunn, P. 1995. The development of downdrift erosion. *Journal of Coastal Research*, **11**, 1242–1257.
Brunsden, D. 1974. The degradation of a coastal slope, Dorset, England. *In*: Brown, E. H. & Waters, R. S. (eds) *Progress in Geomorphology*. Institute of British Geographers Special Publication, **7**, London, 79–98.
——— & Jones, D. K. C. 1972. The morphology of degraded landslide slopes in South West Dorset. *Quarterly Journal of Engineering Geology*, **5**, 205–222.
——— & ——— 1980. Relative time scales and formative events in coastal landslide systems. *Zeitschrift für Geomorphologie*, **34**, 1–19.
Cambers, G. 1976. Temporal scales in coastal erosion systems. *Transactions of the Institute of British Geographers*, **1**, 246–256.
Carter, R. W. G. & Bartlet, D. J. 1990. Coastal erosion in Northern Ireland – Part 1: sand beaches, dunes and river mouths. *Irish Geographer*, **23**, 1–16.
Clayton, K. M. 1990. Sea-level rise and coastal defences in the UK. *Quarterly Journal of Engineering Geology*, **23**, 283–287.
Coad, M. A., Sims, P. C. & Ternan, J. L. 1987. Coastal erosion and slope instability at Downderry, south-east Cornwall: an outline of the problem and its implication for planning. *In*: Bell, M. G., Cripps, F. G., & O'Hara, M. (eds) *Planning and Engineering Geology*. Geological Society, London, Engineering Group Special Publications **4**, 529–532.
Corbett, B. O. 1990. Slif in the chalk cliffs at Brighton. In: *Proceedings of the International Chalk Symposium*, Thomas Telford, London, 527–531.
Cosgrove, A. R. P. 1997. *The rate and pattern of coastal erosion in England: a national appraisal*. MPhil thesis, University of Greenwich.
Dias, J. M. & Neal, W. J. 1992. Sea cliff retreat in southern Portugal: profiles, processes and problems. *Journal of Coastal Research*, **8**, 641–654.
Dolan, R., Fenster, M. S. & Holme, S. J. 1991. Temporal analysis of shoreline recession and accretion. *Journal of Coastal Research*, **7**, 723–744.

Ehlers, J., Gibbard, P. & Whitean, C. A. 1991. The glacial deposits of north-west Norfolk. *In*: Ehlers, J., Gibbard, P. & Rose, J. (eds) *Glacial deposits in Great Britain and Ireland*. AA Balkema, Rotterdam, 223–243.

Eyles, N., Eyles, C. H. & McCabe, A. M. 1989. Sedimentation in an ice-contact subaqueous setting: the Mid-Pleistocene 'North Sea Drifts' of Norfolk, UK. *Quaternary Science Reviews*, **8**, 57–74.

Granja, H.M. & De Carvalho, G. S. 1995. Is the coastline 'protection' of Portugal by hard engineering structures effective? *Journal of Coastal Research*, **11**, 1229–1241.

Gray, J. M. 1988. Coastal cliff retreat at the Naze, Essex since 1874: patterns, rates and processes. *Proceedings of the Geologists' Association*, **99**, 335–338.

Griggs, G. B. & Johnson, R. E. 1979. Coastline erosion, Santa Cruz County. *California Geology*, **April**, 67–76.

Harley, J. B. 1968. The evolution of early maps: towards a methodology. *Cartographic Journal*, **13**, 177–183.

Hart, J. K. 1990. Proglacial glaciotectonic deformation and the origin of the Cromer Ridge push moraine complex, North Norfolk, England. *Boreas*, **19**, 165–180.

Hooke, J. M. & Perry, R. A. 1976. The planimetric accuracy of tithe maps. *Cartographic Journal*, **13**, 177–183.

——— & Kain, R. J. P. 1982. *Historical Change in the Physical Environment: a guide to sources and techniques*. Butterworth, London.

Horikawa, K. & Ssunmura, T. 1967. An experimental study on coastal cliffs using aerial photographs. *Coastal Engineering Japan*, **10**, 67–83.

House of Commons Environment Select Committee, 1992. *Coastal zone protection and planning*. (Volumes I & II), House of Commons, HMSO, London.

Hutchinson, J.N. 1973. The response of London Clay cliffs to differing rates of toe erosion. *Geologia Applicata e Idrogeologia*, **8**, 221–239.

Jones, D.G. & Williams, A.T. 1991. Statistical analysis of factors influencing cliff erosion along a section of the west Wales coast, UK. *Earth Surface Processes and Landforms*, **16**, 95–111.

Jones, J. R., Cameron, B. & Fisher, J. J. 1993. Analysis of cliff retreat and shoreline erosion: Thomson Island, Mass., USA. *Journal of Coastal Research*, **9**, 87–96.

Komar, P. D. & Shih, S. M. 1993. Coastal erosion along the Oregon coast: a tectonic sea-level imprint plus local controls by beach processes. *Journal of Coastal Research*, **9**, 747–765.

Lumsden, G.I. 1992. *Geology and the environment in Western Europe*. Oxford University Press, Oxford.

Lunkka, J. P. 1994. Sedimentation and lithostratigraphy of the North Sea Drift and Lowestoft Till Formation in the coastal cliffs of north-east Norfolk, England. *Journal of Quaternary Science*, **9**, 209–233.

MAFF, 1995. *Guidelines for the construction of Shoreline Management Plans*. HMSO, London.

May, V. 1966. *A preliminary study of recent coastal changes and sea defences in south-east England*. Southampton Research Series in Geography, Southampton University, Southampton.

——— 1971. The retreat of chalk cliffs. *Geographical Journal*, **137**, 203–206.

——— & Heeps, C. 1985. The nature and rates of change on chalk coastlines. *Zeitschrift für Geomorphologie*, **57**, 81–97.

McCue, J. 1998. Sense and sustainability – achieving geological conservation objectives as part of the present shoreline management plan process. *This volume*.

McGreal, W. S. 1979. Marine erosion of glacial sediments from a low energy cliffline environment near Kilkeel, Northern Ireland. *Marine Geology*, **32**, 89–103.

Middlemiss, F. A. 1983. Instability of chalk cliffs between the South Foreland and Kingsdown, Kent, in relation to geological structure. *Proceedings of the Geologists' Association*, **94**, 115–122.

Moon, V.G. & Healy, T. 1994. Mechanisms of coastal cliff retreat and hazard zone delineation in soft flysch deposits. *Journal of Coastal Research*, **10,** 663–680.

Nicholls, R. J. & Webber, B. 1987. Coastal erosion in the eastern half of Christchurch Bay. *In*: Bell, M. G., Cripps, F. G., & O'Hara, M. (eds) *Planning and Engineering Geology*. Geological Society, London, Engineering Group Special Publications, **4**, 549–554.

——— & Leatherman, S. P. 1995. Sea-level rise and coastal

management. *In*: McGregor, D. F. M. & Thompson, D. A. (eds) *Geomorphology and Land Management in a Changing Environment*. Wiley, Chichester, 229–244.

Pringle, A. W. 1985. Holderness coast erosion and the significance of ords. *Earth Surface Processes and Landforms*, **10**, 107–124.

Robinson, A. H. W. 1979. Erosion and accretion along part of the Suffolk coast of East Anglia, England. *Marine Geology*, **37**, 133–146.

Shepard, F. P. & Kuhn, G. G. 1983. History of sea arches and remnant stacks of La Jolla, California, and their bearing on similar features elsewhere. *Marine Geology*, **51**, 139–161.

Shuskiy, Y. D. & Schwartz, M. L. 1988. Human impact and rates of shoreline retreat along the Black Sea coast. *Journal of Coastal Research*, **4**, 405–416.

So, C. L. 1967. Some coastal changes between Whitstable and Reculver, Kent. *Proceedings of the Geologists' Association*, **77**, 475–490.

Steers, J.A. 1951. Notes on erosion along the coast of Suffolk. *Geological Magazine*, **88**, 435–439.

——— 1964. *The Coastline of England and Wales*. Cambridge University Press, Cambridge.

Sunamura, T. 1982. A predictive model for wave-induced erosion, with application to Pacific coasts of Japan. *Journal of Geology*, **90**, 167–178.

Valentin, H. 1954. Der landverlust in Holderness, Ostengland von 1852 bis 1952. *Die Erde*, **6**, 296–315.

Weller, S. 1991. Coastal erosion along the north-west coast of the Isle of Man. *Swansea Geographer*, 80–88.

West, L. J. 1994. The Scarborough landslide. *Quarterly Journal of Engineering Geology*, **27**, 3–6.

Williams, A. T. & Davis, P. 1987. Rates and mechanics of coastal cliff erosion in Lower Lias rocks. *Proceedings of the conference on coastal sediments, 1987*. American Society of Civil Engineers, 1855–1870.

Wilson, G. 1952. The influence of rock structures on coastline and cliff development. *Proceedings of the Geologists' Association*, **63**, 20–48.

Wolman, M. G. & Miller, J. P. 1960. Magnitude and frequency of

forces in geomorphic processes. *Journal of Geology*, **68**, 54–74.
Wood, A. 1978. Coastal erosion at Aberystwyth: the geological and human factors involved. *Geological Journal*, **13**, 61–72.
Wray, R. D. Leatherman, S. P. & Nicholls R. J. 1995. Historic and future land loss for upland and marsh islands in the Chesapeake Bay, Maryland. *Journal of Coastal Research*, **11**, 1195–1203

16 Offshore sand banks and their role in coastal sedimentary processes: the Welsh coast of the outer Severn Estuary, SW Britain

Siegbert Otto

Summary

- Offshore sand bank and coastal morphology has an important role in determining the onshore distribution of wave energy within estuaries.
- The morphodynamics and sedimentary environments of the Severn Estuary in SW England are reviewed and the results of a wave modelling exercise are presented.

Sedimentary processes in the coastal zone are the result of complex interactions between tides and tidal currents, and waves and wave-induced currents (Hodgson 1966; Lamb 1987; Onishi 1987; Pattiaratchi & Collins 1988). When waves (or swells) propagate into shallow water they are modified by refraction, shoaling, and energy losses (Hamm *et al.* 1993; Dodd *et al.* 1995). Refraction of waves entering shallow coastal waters is often the overriding factor in governing sediment entrainment and deposition, sediment transport paths and ultimately landform evolution (King 1953; Ramanadham & Sastry 1959; Reddy & Varadachari 1972; Shrivastava 1979; Laitananda *et al.* 1981). The waxing and waning fortunes of beaches and spits can also often be explained by changing patterns of wave refraction and associated long shore currents (e.g. Hodgson 1966; Peterson *et al.* 1991). Natural and man-made obstructions result in gradients of incident wave energy along the coast, giving rise to areas of sediment deposition (Bascom 1954; Mii 1956; Hicks & Inman 1987) and/or erosion (Pierce *et al.* 1970; Lamb 1987; Brownlie & Calkin 1981; Otto 1996*a*). Long-term changes in relative sea-level and short-term variations, brought about by fluctuating tidal levels, not only determine the fetch over which wind can act on the water surface, and consequently potential maximum wave heights, but also the extent to which subaqueous topography interferes with the incident waves (Mii 1956; Al Mansi 1990). Offshore, subaqueous highs, such as sea mount ridges (Roberts & Chien 1965), dumped spoil heaps (Pye & Neal 1993) and sandbanks (Otto 1996*a,c*), are the most common cause for the generation of strong wave energy foci along the coastline. It is therefore by no means surprising that the most severe coastal erosion is occasionally found concentrated in the more sheltered parts of bays (Caston 1967) and estuaries (Otto 1996*c*).

The purpose of this chapter is: (1) to present the results of wave refraction modelling from two areas along the Welsh coast of the outer Severn Estuary in southwest Britain; (2) to illustrate the importance of offshore sand banks on controlling inshore wave climate; (3) to relate the modelled wave climate to coastal morphodynamics and sedimentary processes; (4) to consider the wider implications of this investigative approach; and (5) to recommend this type of approach for other coastal and estuarine environmental assessment studies. Prerequisite for the holistic appreciation of the results of the hydrodynamic modelling and their wider implications is a thorough understanding of the estuarine morphodynamics and sedimentary processes – past and present. Consequently, before the wave modelling is discussed a detailed review of the study area and two of the most important geomorphological/sedimentological environments, namely the sand environments and tidal wetlands, within the estuary are presented.

Study area

Geology, geomorphology and sedimentary environments

The present-day Severn Estuary lies in the broad Severn Vale which widens southwestwards before joining the west-facing inner Bristol Channel (Fig. 16.1). The estuary is underlain chiefly by soft, Triassic to Lower Jurassic mudrocks with the occasional more resistant Silurian, Devonian (Old Red Sandstone) and Carboniferous beds (Evans 1982; Hawkins 1990), with only the most recent Quaternary stages unequivocally identified in the study area (e.g. Worsley 1985). Lower sea-levels during the Wolstonian and

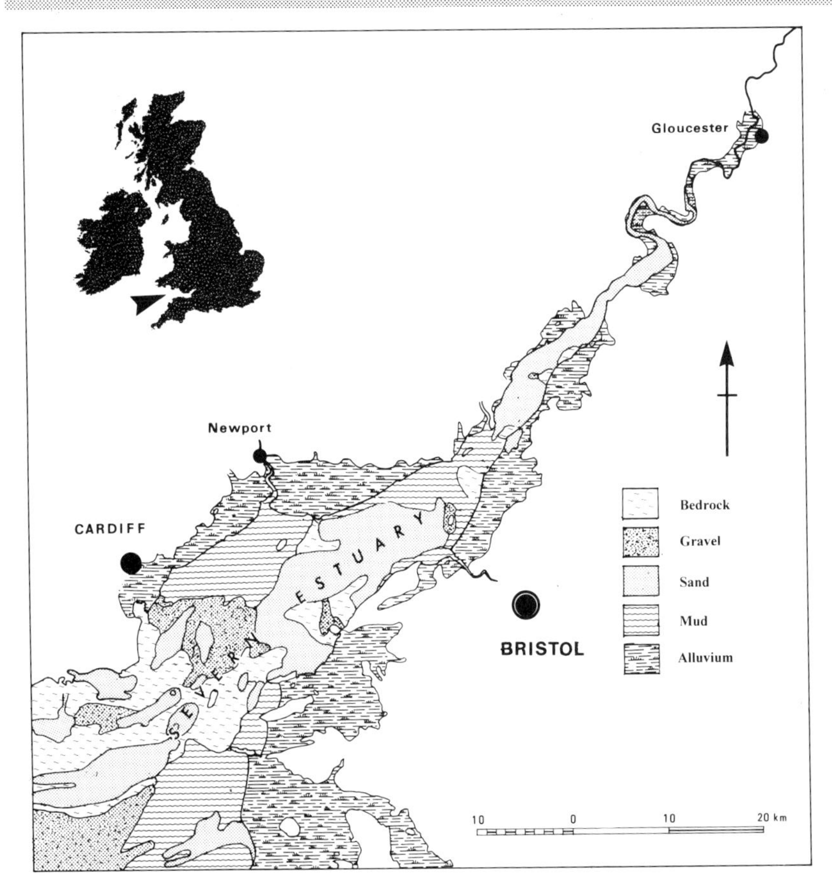

Fig. 16.1 (a). Location map of the Bristol Channel and Severn Estuary. **(b)** Generalized distribution of sedimentary environments in the inner Bristol Channel and Severn Estuary (modified from Allen 1990*a*).

Devensian glacial stages gave rise to steep and narrow gorges (e.g., 'The Shoots') cut into the Severn Valley by the Severn and its tributaries (e.g. Mitchel *et al.* 1973; Allen 1987; Hawkins 1990). Additionally, marine erosion created extensive wave-cut platforms in the outer Severn Valley (Andrews *et al.* 1984). The bedrock surface forms topographically a 'valley-within-a-valley'.

Five major sedimentary environments (i.e. bedrock, gravel, sand, mud and saltmarsh) are recognized in the inner Bristol Channel and the Severn Estuary (Fig. 16.1). The bedrock environment occurs where strong currents have swept the surficial sediments away and only patches of the most compact sand and gravel remain. The areal extent of the bedrock environment decreases up-estuary, and is absent in the inner Severn Estuary. Extensive areas of ice-emplaced Quaternary lag gravels can be found in the inner Bristol Channel, closely associated with the bedrock environment (Hawkins 1971; Culver & Banner 1979; Ferentinos & Collins 1985). The gravel distribution does not coincide with regions of maximum currents, and gravels, typically, do not occur in the channels. The sand environment is extensive and consists of a highly mobile sand population. The mud environment is also widespread and consists of settled mud, stationary and mobile suspensions (Kirby & Parker 1982*a,b*). The change in climate and subsequent rise in sea-level over the post-glacial Flandrian time is reflected in the widespread deposits of estuarine alluvium and buried peats, though only a small strip of tidally inundated saltmarsh remains today outside the sea defence, owing to extensive reclamation since Romano-British times (Williams 1970; Allen & Rae 1987, 1988).

Wind, waves and surges

The Bristol Channel and Severn Estuary face the prevailing westerly and southwesterly winds. In the Severn Estuary these tend to be funnelled together with northeasterly and northerly winds along its axis between the hills on either side. This localized steering of the coastal winds by topographic features (Oliver 1960) is less evident at Avonmouth (Allen 1992*a*), where all the wind data discussed and used in the wave climate modelling in this study were recorded. Figure 16.2 summarizes Meteorological Office data on the direction from which the wind blows, its speed and the respective yearly percentage frequency distribution over the period 1970 to 1988 inclusive. The direction from which the wind blows is bimodaly distributed at low wind speeds. One mode is southerly to northwesterly, the other is approximately northeasterly. With increasing wind speed the first mode grows in dominance, associated with a marked southerly skew, especially at higher speeds. The vector mean direction, however, does not vary much with speed and is 224° for near-gale and greater speeds. Autumn

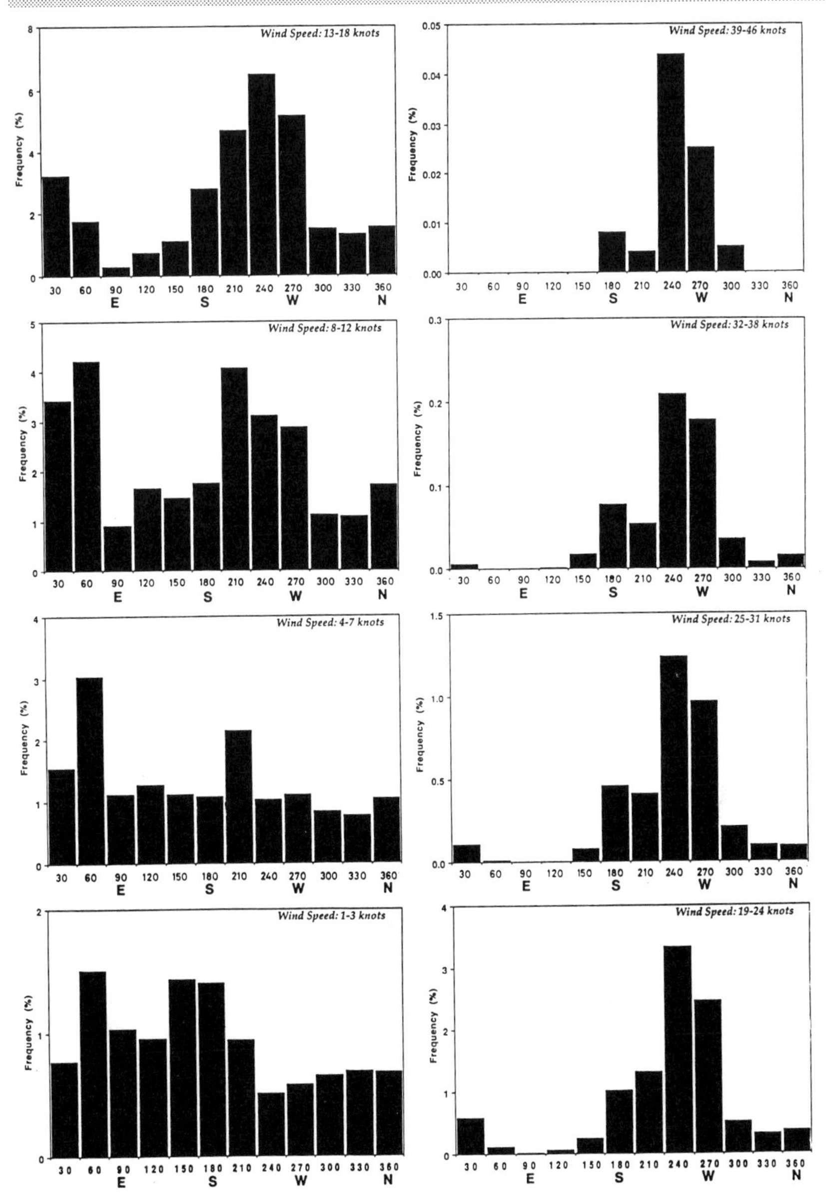

Fig. 16.2. Percentage frequency of average yearly wind speeds and directions at Avonmouth in the outer Severn Estuary (1970 to 1988 inclusive). Note the bimodal distribution of directions at low wind speeds. With increasing wind speed the westerly mode grows in dominance, associated with a marked southerly skew.

and winter tend to be the windiest seasons with frequent gales and occasional storms. Westerly storms are almost 10 times more frequent than storms from other directions (see also Shuttler 1982).

At Port Talbot, in the outer Bristol Channel, owing to an unrestricted fetch of 6000 km in the Atlantic Ocean, occasional wave heights of 7 m have been recorded and heights of up to 5–6 m are not uncommon in times of storms (Jackson & Norman 1979). Wave conditions in the Severn Estuary are determined by the directions and strengths of the above-mentioned 'local' winds. The longest fetches in the estuary are aligned SW–NE and W–E, the general direction along which the strongest winds blow. No field-based wave data are available for the Severn Estuary proper, and only limited information was gathered in the inner Bristol Channel, close to the mouth of the estuary by Shuttler (1982). Although the two-year recording period was of relatively low windiness, Shuttler (1982) recorded maximum significant wave heights of up to 3.5 m and 2.2 m measured in the inner channel and the outermost estuary (ENE of Flat Holm) respectively, associated with westerly and northeasterly storms. This illustrates a significant decrease in wave height from the outer to the inner Bristol Channel, and even more so (considering the short distance) from the inner channel into the outer estuary. Maximum predicted significant wave heights for the outer estuary do not exceed *c.* 1.7 m for Force 9 gales along the longest – generally southwesterly – fetches (HR Wallingford 1976).

Surge data at Cardiff, Newport and Avonmouth were only investigated for the period 1920 to 1960 (Lennon 1963*a,b*), an interval of relative quiescence and decreasing frequencies of westerly and southwesterly winds (Lamb 1982). These surges did not exceed the height of local HATs, and even the calculated 100- and 200-year maximum surge levels exceed the predicted HAT heights by only two decimetres (Otto 1996*a*). These storm surge data are certainly unrepresentatively low for conditions prevailing over the last quarter of the century, in respect of frequencies as well as surge heights.

Tides and tidal currents

Tides in the Bristol Channel/Severn Estuary are predominantly semi-diurnal. The strongly macrotidal regime of the Severn

Estuary, as predicted for the standard port of Avonmouth, is typified by a highest astronomical tide (HAT) of 14.8 m above tidal datum (i.e. lowest astronomical tide, LAT), which is not reached in every year; mean high water of spring and neap tides (MHWS, MHWN) of 13.2 and 10.0 m respectively, with associated spring and neap tide ranges of 12.3 and 6.5 m respectively (Hydrogapher of the Navy 1992). Because of convergence within the funnel-shaped and rapidly narrowing estuary, mean high and low water heights of spring and neap tides increase upstream.

Along with the extreme behaviour of tidal elevations along the estuary, tidal stream velocities are correspondingly vigorous. Recorded tidal current data are rare, particularly with respect to temporal and spatial variations (Griffiths 1974; Hamilton *et al.* 1975; Hamilton 1979; Crickmore 1982). Hamilton (1979) recorded maximum spring flows in the outer estuary of around 2.0 and $1.7\,m\,s^{-1}$ for a mid-channel and sandbank position respectively. Highest spring flow rates of over $5.5\,m\,s^{-1}$ occur in the narrow, deep channel of 'The Shoots', the first serious constriction in the upper part of the outer estuary.

Most of the tidal energy supplied to the estuary is dissipated in the rapidly converging outer estuary up to the constriction at Beachley (i.e. 6.89×10^{10} and 1.74×10^{10} watts for spring and neap tides respectively; Otto 1996*a*). The magnitude and pattern of tidal wave energy dissipation are governed foremost by the upstream decrease in amplitudes. Additionally, the area of intertidal flats (above LAT) relative to that part of the estuary covered permanently by water (i.e. channel) increases up-estuary. The channel to mudflat ratio diminishes most rapidly in the outer estuary (i.e. from *c.* 6 to 0.3) between the mouth of the estuary and Sudbrook, which aids in the dissipation of tidal wave energy (Otto 1996*a*).

Sea-level rise

Conventional estimates of the recent glacio-eustatic rates of sea-level rise in southern Britain range from $0.4–0.6\,mm\,a^{-1}$ (Pirazzoli 1989) to $1–2\,mm\,a^{-1}$ (Valentin 1953; Gornitz *et al.* 1982; Wigley & Raper 1987; Woodworth 1987, 1990; Trupin & Wahr 1990). The current crustal subsidence rate for the region is $0.2–0.5\,mm\,a^{-1}$

(Shennan 1989). All these rates are much smaller than the ones based on saltmarsh accretion rates (Allen & Rae 1988; French 1990; Allen 1991), which vary from 3.8–4.7 mm a^{-1}, with an apparent significant acceleration since the last century. Computed model projections by Clayton (1990) and Warrick & Oerlemans (1990) suggest rates in sea-level rise of up to 9.1 mm a^{-1} by the end of the twentyfirst century (Otto 1996b).

Sand environment

Distribution, morphology and provenance

Due to the severe hydrodynamic regime, the extensive and varied sand environment of the Severn Estuary and inner Bristol Channel is comprised of a highly mobile sand population. In the inner Bristol Channel the sands are chiefly bank-detached, subtidal shoals (e.g. Nash Sand, Culver Sand, Holm Sand: Fig. 16.1). In the outer Severn Estuary there are approximately equal areas of subtidal sands and mainly bank-detached, intertidal sandflats. The sandflats are covered with (large) sandwaves and (smaller) megaripples (Cornish 1901; Harris & Collins 1985). However, although the bedforms reverse their orientation during the tidal cycle, Hawkins & Sebbage (1972) could find no evidence of internal sedimentary structures and consequently suggested that the sandwaves were highly unstable. Further evidence of their instability is the generally quicksand nature of the sandshoals.

Little is known about the origin of sand-grade sediment, however a modern source is excluded. Based on the presence of large (i.e. $> 200\,\mu m$) stenohaline, marine foraminifera, Murray & Hawkins (1976) inferred an eastward transport of sand-grade material into the estuary. The gravel fraction at Holm Sand consists of Upper Cretaceous, Coal Measures and Lower Lias lithologies (Davies 1980), suggesting that the original deposit was derived from the Welsh hinterland and emplaced by the southward moving Welsh ice-sheet (Bowen 1970).

Heavy mineral assemblages of igneous and metamorphic derivation found in the sands of the Bristol Channel (Barrie 1980*a,b*, 1981) and Severn Estuary (Otto 1996*a*) move the provenance of the deposits even further afield. Possible source areas include Cornwall, with its plutonic intrusions and associated contact-metamorphism, and the high-grade regionally metamorphosed rocks of the Scottish highlands. Following the deposition of these sandy and gravelly sediments in the Irish Sea and outer Bristol Channel, extensive wave and tidal reworking – chiefly during Flandrian times – introduced the sands into the inner Bristol Channel and Severn Estuary. The material previously deposited was thereby also redistributed to the extent that today no *in situ* glacial deposits are left in the main part of the Bristol Channel, but only Wolstonian reworked material (Garrard & Dobson 1974; Culver & Banner 1979).

Sand transport pathways

The general trend of the sediment bodies parallels the strong, essentially rectilinear current streams which are directed along the main axis of the system (Fig. 16.3), with the fine sediment being distributed around the margins of the area and the axial regions being occupied by sandy, gravelly or rocky areas (cf. Fig. 16.1). The position of the sand banks coincide with narrow flood-dominated areas along the coastlines of the channel, which flank an ebb-dominated central zone. Consequently, flood and ebb streams are 'mutually evasive', resulting in an eastward directed sand transport along the northern channel margin into the estuary, and a westward transport in the centre (Harris 1982, 1988; Harris & Collins 1988, 1991). The Severn Estuary is flood-dominated throughout, and the 'mutually evasive' system of tidal streams is more pronounced and more clearly defined than in the Bristol Channel (Fig. 16.3).

Along the northern sand transport path the various banks form local clockwise (e.g. Culver and Holm Sands, and Middle Grounds) and anti-clockwise (e.g. Cardiff Grounds) recirculation cells. Based on McLaren *et al.* (1993) grain size trend analysis, Culver and Holm Sands and Cardiff Grounds are in a state of equilibrium, whereas the Middle Grounds show evidence of net accretion. Much of this sand appears indeed to be gathering in the Middle Ground–Denny Shoal area and then feeding further east into the middle estuary, although sand may be returning westward along the southern zone of the outer estuary to Culver Sand in the inner Bristol Channel (Parker & Kirby 1982). Grain size analysis has shown that the sands

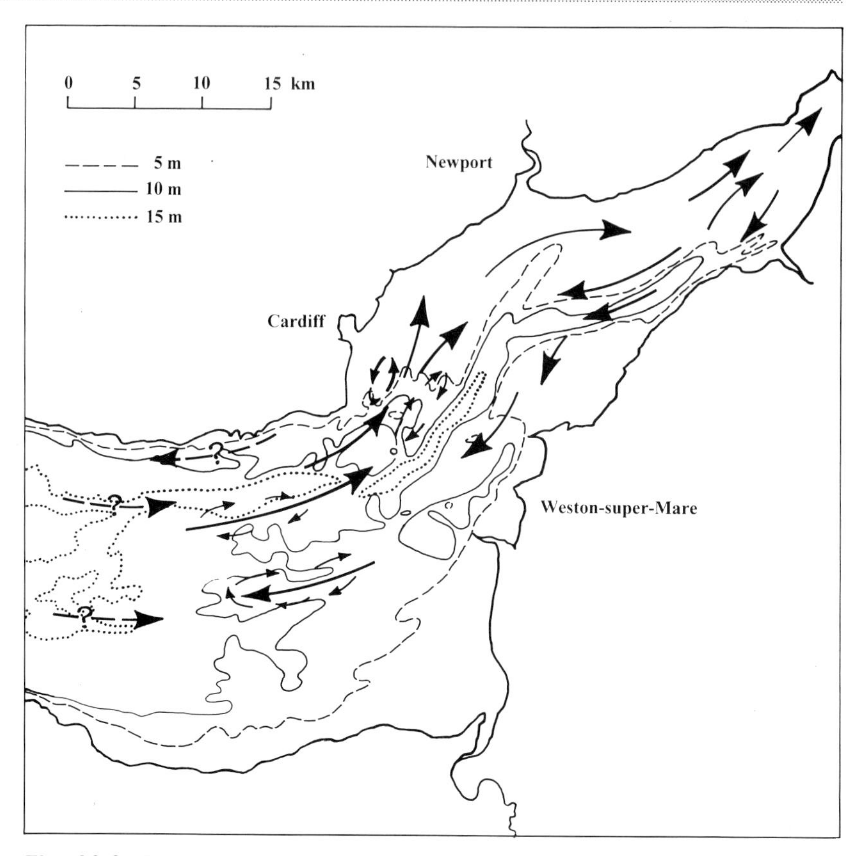

Fig. 16.3. Sand transport pathways in the inner Bristol Channel and outer Severn Estuary (modified from Parker & Kirby 1982).

are generally clean and very well sorted, and are coarsest in the flood dominated areas of Cardiff and Middle Grounds. Finer populations are present on English Grounds and Denny Shoal which are subject to both strong flood and ebb flows (Hamilton 1979; McLaren *et al.* 1993). There is an overall decrease up-estuary in the mean grainsize of the sand banks and sandflats from the inner Bristol Channel to the inner Severn Estuary (Hamilton 1979, 1982; McLaren *et al.* 1993).

In apparent contrast to the model suggested above, sand is being transported westwards from a curved and obliquely aligned

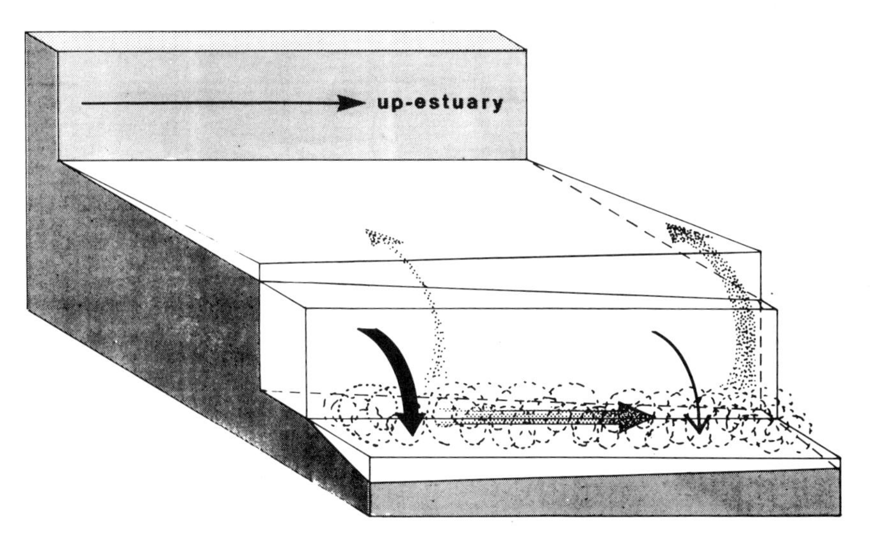

Fig. 16.4. Schematic presentation of the inland retreat of the Severn Estuary by means of a 'stratigraphic rollover' (modified from Allen 1990*a*).

'bedload parting' (Fig. 16.3) extending from Barry in a roughly southeastwards direction toward Bridgwater Bay, while sand is being transported eastwards from this zone into the estuary (Stride 1963; Kenyon & Stride 1970; Pingree & Griffiths 1979; Johnson *et al.* 1982). The continuing controversy between advocates of 'bedload parting' (Stride & Belderson 1990, 1991) and 'mutually evasive transport' models (Harris 1988; Harris & Collins 1988, 1991) serves to emphasize the inadequacy of our present state of knowledge for a full understanding of the sediment supply and transport within such a complex system. However, the two systems need not be mutually exclusive. The bedload partings exist mainly offshore from the mouth of estuaries and the mutually evasive paths mainly within estuaries (Stride & Belderson 1991; McLaren *et al.* 1993; Harris *et al.* 1995), suggesting that the former is a large scale mechanism whereas the latter operates in intermediate scale settings. Additionally, there can be little doubt that sediment is being transported into the middle and inner estuary (e.g. Hamilton 1979, 1982; Parker & Kirby 1982; McLaren *et al.* 1993), partly erosively derived from the outer estuary (Fig. 16.4), in order to account for the process of 'stratigraphic rollover' observed in the contemporary Severn Estuary (Allen 1990*a*; Otto 1996*d*).

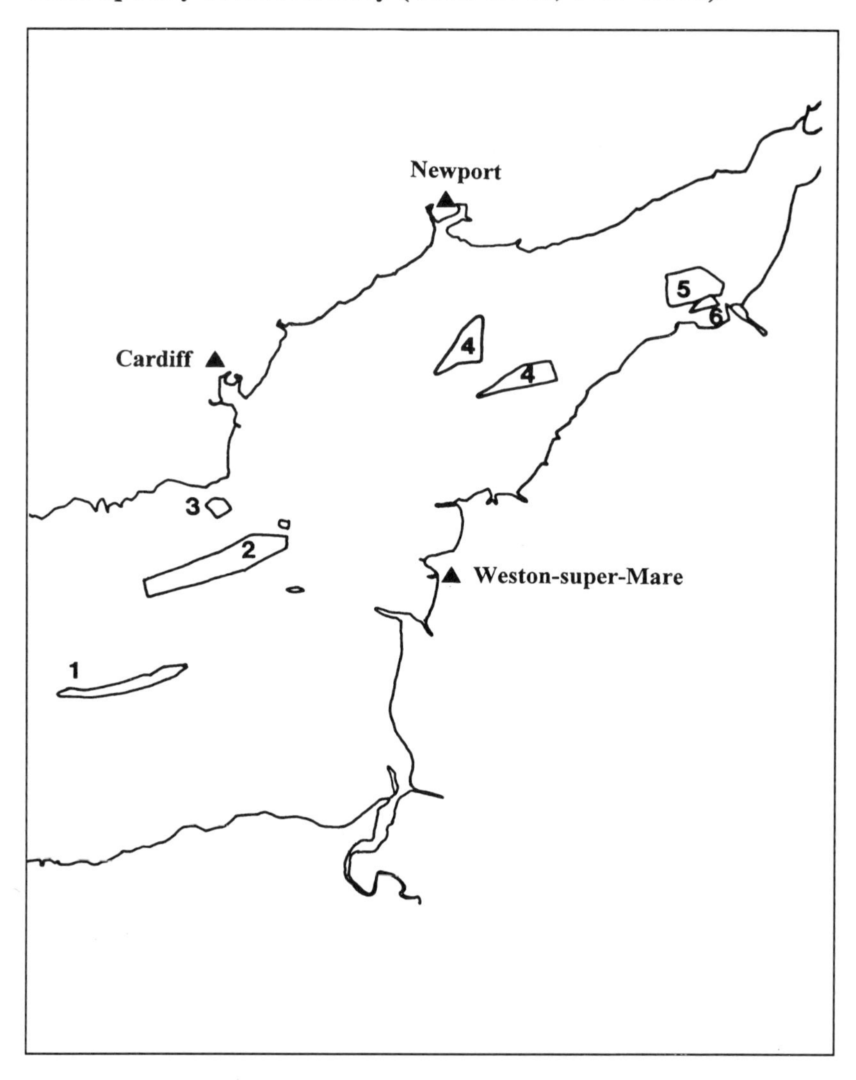

Fig. 16.5. Distribution of sand banks with licensed dredging areas in the inner Bristol Channel and outer Severn Estuary (with permission of the Crown Estate). 1=Culver Sand, 2=Holme Sand, 3=Lavernock Point, 4=Middle Grounds, 5=Denny Shoal, 6=Cockburn Shoal.

Dredging

The inner Bristol Channel and outer Severn Estuary supply an estimated 15% of United Kingdom sea-dredged aggregates (Fig. 16.5). In 1995 1.52×10^6 tonnes of sand and occasional gravel were landed at local ports of the region (Table 16.1). This means that the approximate equivalent of one year's fluvial sediment supply entering the system (i.e. $1.98–2.25\times10^6$ tonnes; Parker & Kirby 1982; Collins 1987; Allen 1990*a*) is at present yearly being dredged. It is impossible to establish how much sediment has been redistributed through over-spilling and screening, or how much has been introduced by severe wave action into the active sedimentary environments of the high intertidal zone. However, the volume of sandy beach pockets along the Wentlooge Levels in the outer estuary (Fig. 16.6), overlying saltmarsh deposits and contributing (by a sandblasting process) to the erosion of the marsh margins, appears to be steadily increasing.

Table 16.1. Summary of port statistics for marine dredged aggregates 1995 in the outer Severn Estuary and inner Bristol Channel (data supplied by The Crown Estate)

Landing ports	1995 Tonnages
Avonmouth	501 229
Barry	84 175
Bridgwater	52 928
Cardiff	447 814
Newport	436 331
Total	1 522 477

Tidal wetlands

Saltmarsh morphodynamics

In response to the post-glacial (Flandrian) rise in sea-level, some $3\,km^3$ (*c.* 4×10^9 tonnes) of mainly muddy sediment has accumulated over the last 9000 years along the Severn Estuary. The estuarine alluvium is up to 15–20 m thick (e.g. Allen & Rae 1987),

Fig. 16.6 Modern beach deposits overlying eroding Flandrian alluvium at Rumney Great Wharf [ST 250 786].

and has now been largely reclaimed, beginning in the Roman period (e.g. Allen 1986, 1988, 1990*b*, 1991; Allen & Fulford 1986, 1987, 1990*a,b*, 1992), and has given rise to the Wentlooge Surface, the oldest and lowest of the geomorphical surfaces (Fig. 16.7). The underlying Wentlooge Formation consists of alternating minerogenic and organogenic marsh deposits. The peats and organic-rich silts, which occur on an estuary-wide scale, were formed at several

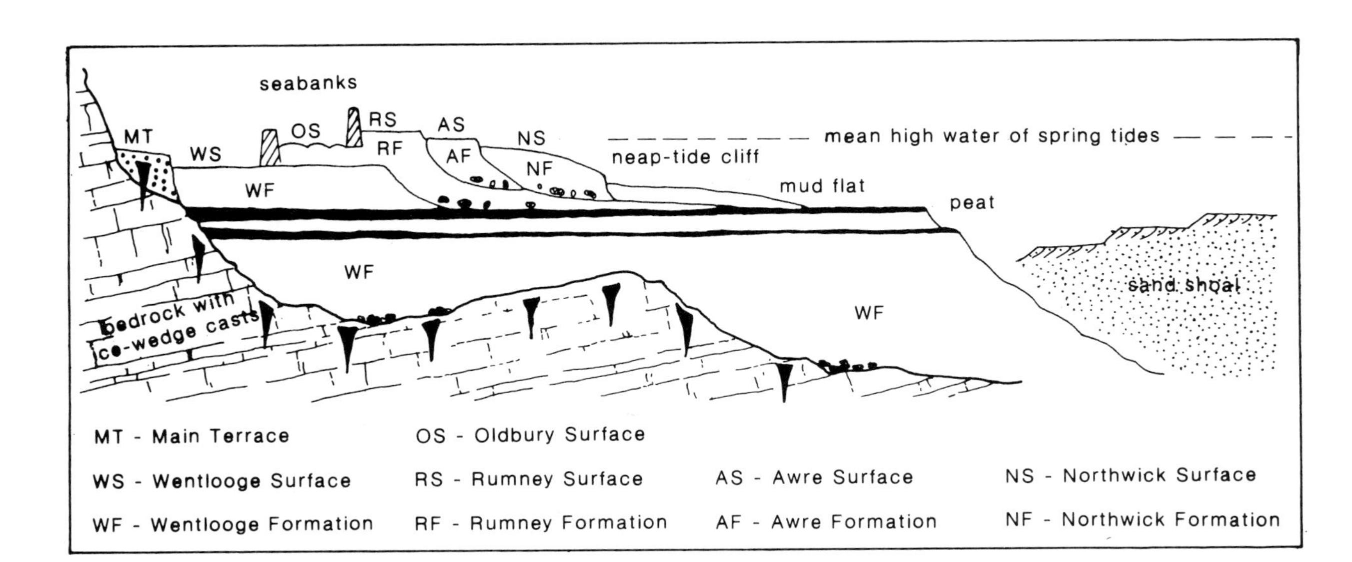

Fig. 16.7 Schematic representation of the main late glacial and post-glacial features of the margins of the Severn Estuary (modified from Allen & Rae 1987).

(a)

(b)

Fig. 16.8(a). Saltmarsh at Rumney Great Wharf [ST 250 785]. Strongly erosional *c.* 3 m high marsh cliff, with the (darker) Rumney Formation overlying erosively the (lighter) upper Wentlooge Formation. The cliff retreats inland over the stratigraphically highest peat horizon of mid-Flandrian age. (**b**) Saltmarsh at Peterstone Great Wharf [ST 270 797]. Northwick Surface fronting *c.* 0.6 m high clifflet cut into Rumney beds. This cliff was – prior to the deposition of the Northwick Formation – identical in height and morphology to the cliff shown in **(a)**. Note the recently raised sea bank in the background to accommodate the predicted sea level rise of 25 cm over the next 50 years.

periods during a mid-Flandrian stage dating from *c*. 6500 to 2500 conventional radiocarbon years ago. The outcrop of estuarine alluvium is completed by three younger, purely minerogenic morphostratigraphic saltmarsh units – the Rumney, Awre and Northwick formations – which began to form after the Little Ice Age (Fig. 16.7). The Rumney and Awre Formations started to accumulate at the end of the eighteenth and nineteenth century respectively, whereas the Northwick Formation commenced in the second quarter of this century.

These discrete units represent erosion and accretion cycles. The erosional phase, when the coastline retreats back into the former outcrop of estuarine alluvium, gives rise to a more or less bold cliff (Fig. 16.8a) with a wave-cut platform sloping gently down towards the axis of the estuary. The new mudflat, and subsequently the evolving marsh, build up and advance during the following accretionary phase, with the height of the cliff in between the units gradually diminishing (Fig. 16.8b). This is followed by another erosional cycle, though each morphostratigraphic unit represents in itself a regressive–transgressive cycle, as evidenced by the very common upward increase in grain size, and the apparently early commencement of cliff formation and subsequent retreat (Otto 1996*a*). The marsh surfaces therefore form an off-lapping, stair-like succession (Fig. 16.7), which accumulated during a period of continuous rise in sea-level. There is no evidence that sea-level in the area has varied in a stepwise fashion, at least during the past few decades (e.g. Woodworth 1987, 1990). Climatically controlled fluctuations in the discharge of suspended sediment load by rivers are also possible, but fail to provide an adequate explanation for the observed off-lapping marsh units in this case. However, an increase in suspended sediment in the tidal waters should make mud secretion more likely and faster in the areas least affected by wave and tidal action. The most plausible explanation for the alternation of erosion and accretion cycles was suggested by Allen (1992*b*), who investigated the historical wind record of the British Isles (Lamb 1972, 1977, 1982), in particular the frequency of southwesterlies and westerlies, within the context of marsh erosion and accretion periods in the Severn Estuary. There can be little doubt that the orientation and dimensions of the Severn Estuary determine the fetches which generate the most effective waves in the system. These waves arise when storms blow from the southwest and west, and less frequently, from the northeast. Consequently, erosional conditions are expected to prevail in the upper intertidal zone of the estuary during years when winds from the west, southwest and northeast are more frequent. Conversely, a decline in the frequency of winds from these directions should allow and encourage the deposition of estuarine alluvium.

Distribution of coastal morphologies

Figure 16.9 illustrates the distribution and extent of rocky and revetted coastal stretches, as well as the morphostratigraphic saltmarsh units which can be found at the contemporary mudflat–marsh boundary. Only the Welsh side of the outer estuary – and seaward from the present sea walls – will be considered here. Extensive stretches of the coastline (42%) are revetted, such as those north of Cardiff Bay, between Peterstone Great Wharf and the River Usk, and (almost) continuously between Goldcliff, where the saltmarsh was totally eroded, and Caldicot. In addition, the outer estuary is partly naturally rock bound (14%), on the Welsh side between Lavernock Point and Cardiff Bay and around the protruding coastline at Sudbrook. Neap-tide cliff exposures of the Rumney Formation (11%) are limited to Rumney Great Wharf – the type site of the Rumney Formation – with vertical cliffs of some 3 m high (Fig. 16.8a). The Awre Formation does not constitute any part of the mudflat–marsh boundary in the outer estuary, though Awre deposits occur inland from the Northwick Formation along the Caldicot Levels (Fig. 16.10). The Northwick Formation is the most common marsh unit (33%). Its deposits exhibit two distinct marsh edge morphologies: a clifflet ($\leq$1 m high), or a ramp of erosional mud mounds (Fig. 16.11), with spur and groove topography running at approximately right angles to the marsh edge. The former morphology (12%) is chiefly restricted to between Sudbrook and Beachley where incident wave energy is low but the main tidal channel relatively proximal, whereas the latter (21%) occurs along coastal stretches of considerably greater wave energy exposure, such as Peterstone

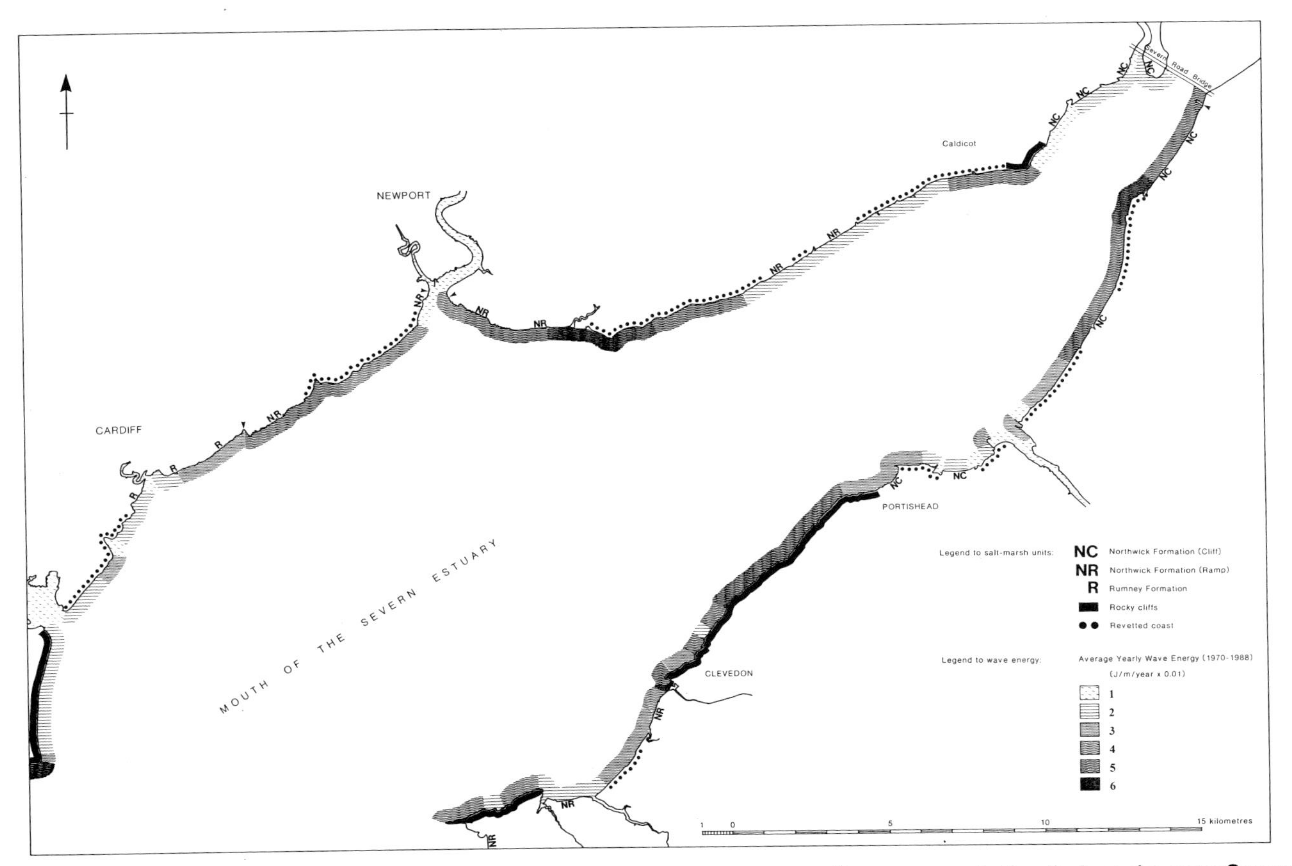

Fig. 16.9. Distribution of coastal morphologies and average yearly wave energy ('wave energy indices') along the outer Severn Estuary.

Fig. 16.10. Saltmarsh at Magor [ST 436 846]. The twentieth century Northwick Formation (left) abuts marsh deposits of the late nineteenth century Awre Formation (right), separated by an approximately 0.6 m high clifflet.

Fig. 16.11. Groove and spur morphology of the seaward margin of the Northwick Formation at Peterstone Great Wharf [ST 270 797]. The sub-parallel aligned elongated mud mounds face the prevailing winds (and waves).

Great Wharf – the largest continuous Northwick Surface – and as narrow fringing marshes between the mouth of the River Usk and west of Goldcliff Point, and fronting the Caldicot Levels around Magor (Fig. 16.10).

Saltmarsh erosion

The saltmarsh morphodynamics discussed above show that the shoreline of the Severn Estuary has a long history of instability, with the widespread estuarine alluvium being affected most. At present the marshes along the high-water line of the estuary have resumed an erosional regime (Fig. 16.8a), though with significant spatial differences and exceptions. The following will chiefly concentrate on Rumney and Peterstone Great Wharves, the seaward part of the Wentlooge Levels (Fig. 16.12), which constitute – with a combined area of almost 90 ha – the largest continuous area of still active saltmarsh outside the sea defences in the outer Severn Estuary. Figure 16.13 shows that up to 1956 Rumney Great Wharf was still considerably wider than Peterstone Great Wharf, owing to the difference in average yearly erosion rates in the nineteenth century. Between 1881 and 1900 the marsh cliffs at Rumney and Peterstone Great Wharves retreated on average by 1.79 and 3.17 $m\,a^{-1}$ respectively, thence their erosional rates slowed down for the next 50 years to an average of 0.48 and 0.37 $m\,a^{-1}$ respectively. From *c.* 1950 Rumney Great Wharf experienced a dramatic increase in the rate of erosion, which averaged (up to 1991) 3.09 $m\,a^{-1}$. At Peterstone Great Wharf, however, the width of the saltmarsh increased greatly between 1956 and 1965. By approximately 1960 both wharves had reached an average width of *c.* 140 m, one by retreat, the other by advance. Since 1964 the marsh edge at Peterstone Great Wharf has remained relatively stationary, though over recent years sediment removal has exceeded deposition (Otto 1996*a*). At Rumney Great Wharf, however, erosion of the cliff continues to increase the extent of the mudflat at the expense of the saltmarsh by between 2 and 3 $m\,a^{-1}$ (Otto 1996*a*).

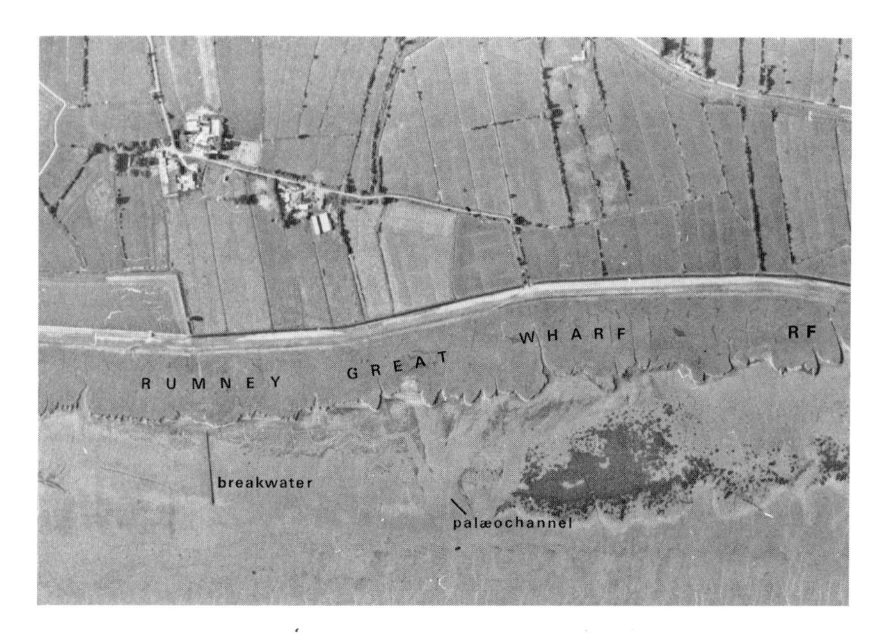

Fig. 16.12. Aerial photograph (7 September 1971; *c.* 2500×940 m; N towards
Photograph Crown Copyright reserved. RF = Rumney Formation, NF=Northwick

Table 16.2. Extent of active marshes outside the sea defences at the Wentlooge and Caldicot Levels in the outer Severn Estuary in 1830, 1909 and 1991 and intermittent rates of marsh loss

Active marsh area	Wentlooge Level (ha)	Caldicot Level (ha)
1830	381	252
1909	311	192
1991	198	79
Loss between 1830 and 1909	0.89 ha a^{-1}	0.76 ha a^{-1}
Loss between 1909 and 1991	1.38 ha a^{-1}	1.38 ha a^{-1}

The annual erosion rates for the Wentlooge and Caldicot Levels between 1830 and 1909 were 0.89 and 0.76 ha a^{-1} respectively (Table 16.2). Between 1909 and 1991 erosion appears to have accelerated

1 o'clock) of the boundary between Rumney and Peterstone Great Wharves.
Formation.

by approximately 40% to an average rate of 1.38 ha a^{-1} at both levels. This implies that marshes like that at Rumney which occur at both Levels must have actually eroded at a much faster rate, because in excess of one-quarter of the present active Wentlooge Levels is occupied by the post-1925 Peterstone Great Wharf (Otto 1996*a*), while Northwick and even Awre beds were deposited along the Caldicot Levels and have survived to the present day.

Methodology of wave climate modelling

Previous studies of sediment dynamics in the Severn Estuary did not include wave climate modelling along the morphologically very varied coastline nor did they shed light on the distribution of morphostratigraphic saltmarsh units within a hydrodynamic context. Due to the lack of any hydrodynamic data and model

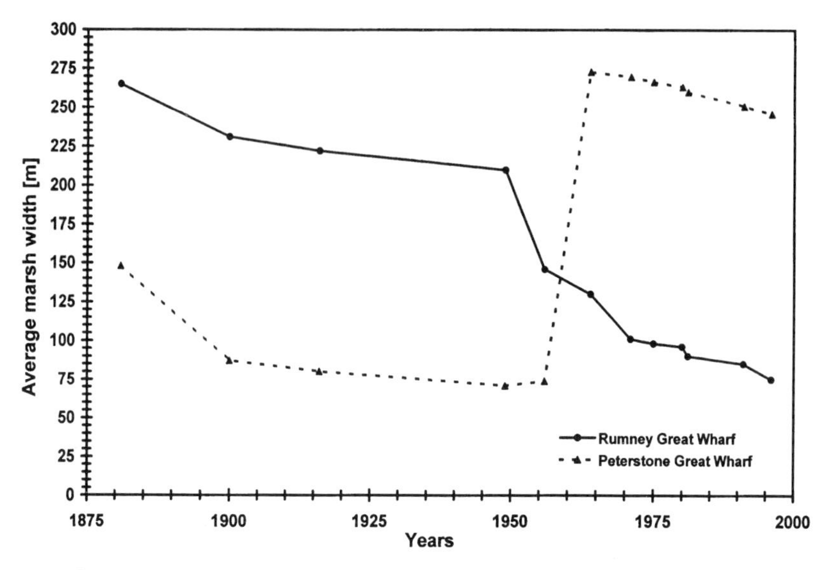

Fig. 16.13. Average marsh widths at Rumney and Peterstone Great Wharves between 1881 and 1996.

outputs suitable for investigating coastal morphodynamics in the study area, two separate approaches were chosen to establish: (1) the maximum significant wave heights and wave energy potentially incident along the coastline; and (2) how refraction by the offshore topography modifies these waves.

Significant wave heights and wave energy

The fetch-limited JONSWAP formula was chosen to hindcast average monthly and yearly wave conditions along the coastline of the outer Severn Estuary for the period 1970 to 1988 inclusive. This model (Hasselmann *et al.* 1973; Carter 1982; US Army Coastal Engineering Research Centre 1984) is: (1) depth-independent; (2) simple, and easily implemented; (3) empirical, based upon locally generated seas, and obtained in relatively shallow water; and (4) generates comparatively small wave heights for a given fetch and wind speed. The summary wind record for Avonmouth (1970 to 1988) was used for the wave climate modelling. For 415 points, spaced at 200 to 250 m along the banks of the outer estuary, 12 fetches at 30° intervals were established from Sheets 162, 171 and

172 of the Ordnance Survey's 1:50 000 maps. The effective fetches were subsequently calculated because fetch lengths are not constant within the 90° quadrant of the wind effect (method in HR Wallingford 1976). Following this the (JONSWAP) significant wave heights were computed for each standard wind speed class.

Each of the 415 wave height matrices was converted into wave energy, and subsequently multiplied by the matrices of yearly and monthly percentage frequencies of wind speed versus wind direction. The total average annual wave energy incident at anyone point is presented then as a unit-less index.

Wave refraction

To compute wave refraction for selected areas of the outer estuary a simple forward-tracking simulation model with graphic output was employed. The model (courtesy of R. Middleton, University of Hull) simulates the wave refraction and computes the shoaling transformation, using Snell's equation and Airy's theory respectively. The program does not incorporate wave reflection, and it is generally assumed that the wave energy contained between orthogonals remains constant as the wave front progresses, and this implies that there is no dispersion of energy laterally along the front, no reflection of energy from the rising bottom, and no loss by other processes. The model also takes no account of tidal currents and the surface slope due to coastal drag effects. Implementation of Galvin's coefficient decides at what water depth the waves break, following $GAL_{coeff}=h/d$, where h=wave height (corrected for the shoaling transformation), and d=water depth. The breaking condition is fulfilled if $GAL_{coeff}\leq0.78$.

Two adjacent areas of 125.44 km^2 each (i.e. 11.2×11.2 km) were digitized from Sheet 1176 (Hydrographer of the Navy 1988). A square grid of 4900 (70×70) nodes was generated, with water depth values (relative to local datum and LAT) for areas lacking sufficient density and coverage being interpolated from adjacent sites. This basic data grid was then run with additional wave and tidal parameters, (i.e. for wave heights and periods of between 1 m/4 s and 2.5 m/5.5 s), and also for MHWS and MHWN tide conditions.

The first area ('Lavernock Point to Peterstone Gout'; Figs 16.14 & 16.15) comprises the roughly southwest/northeast aligned coast-

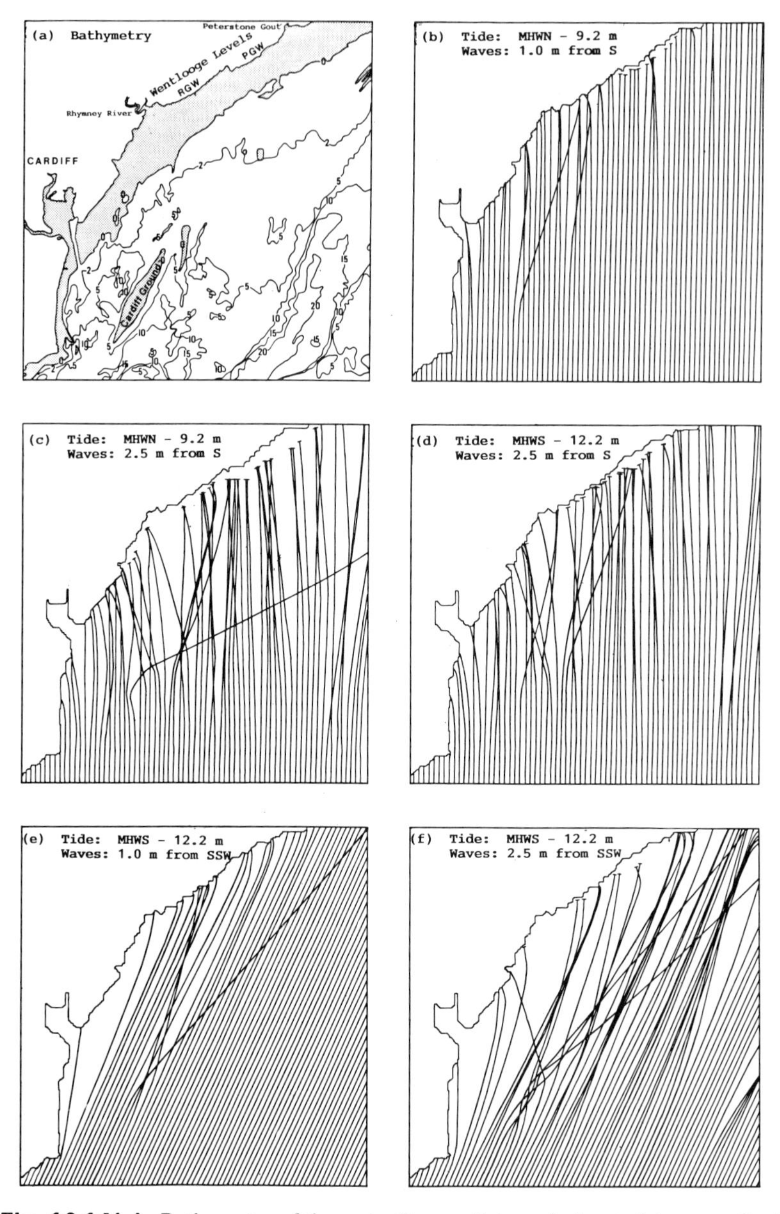

Fig. 16.14(a). Bathymetry of the outer Severn Estuary in front of the coastline between Lavernock Point and Peterstone Gout, and **(b)**–**(f)** wave refraction patterns for given tidal and wave characteristics (see text for discussion).

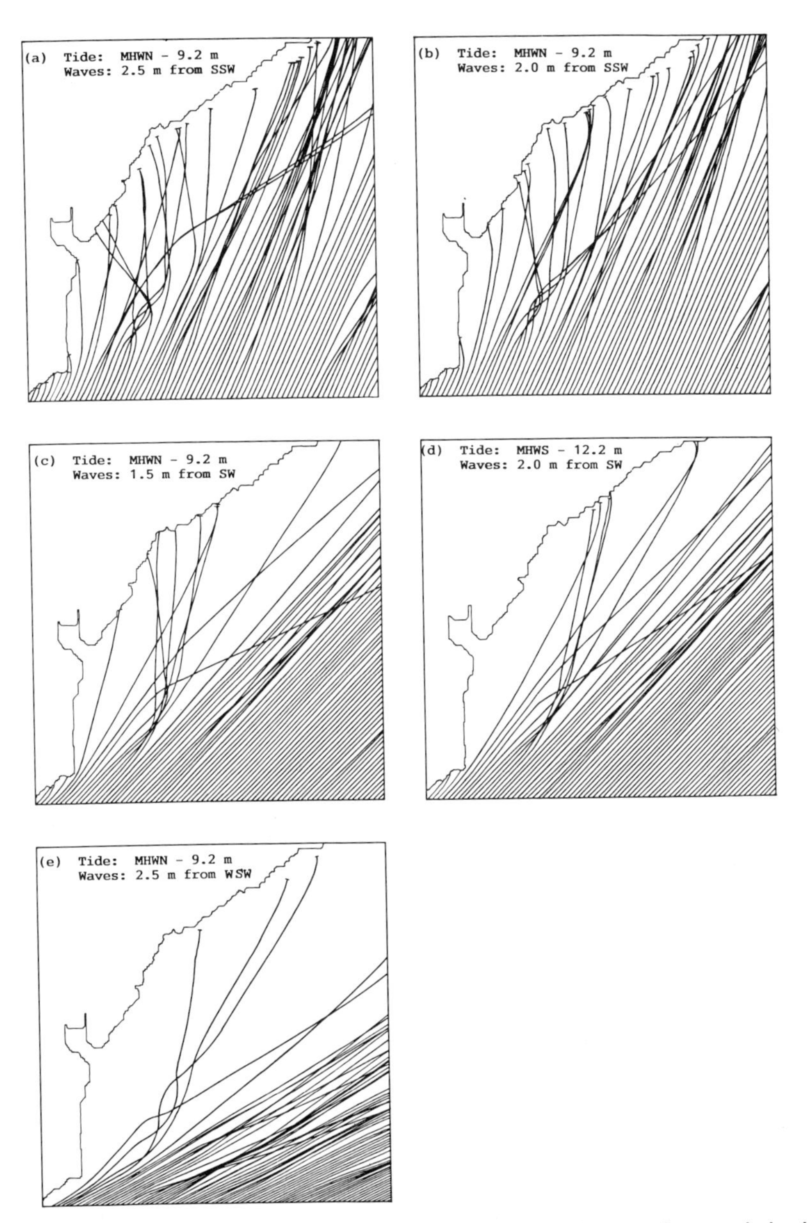

Fig. 16.15. Wave refraction patterns for given tidal and wave characteristics between Lavernock Point and Peterstone Gout (see text for discussion).

line between Lavernock Point and Peterstone Gout, including the mouth of the Rhymney River, Rumney Great Wharf and Peterstone Great Wharf. The bathymetry is characterized by: (1) a wide intertidal zone (above LAT) in front of the Wentlooge Levels, petering out towards Lavernock Point; (2) approximately 50% of the offshore area lying above the 5 m contour; (3) water depths of up to 10.7 m between Lavernock Point and the constantly changing sandbank (above LAT) to the north-east known as Cardiff Grounds; and (4) part of the main channel in the SE quadrant of the grid, with a maximum water depth of 23 m. It is noteworthy that the intertidal zone in front of Peterstone Great Wharf lies approximately 1 to 1.5 m above the one in front of Rumney Great Wharf. MHWS and MHWN tide levels between Cardiff and Newport are 12.2 m and 9.2 m respectively.

The second area ('Peterstone Gout to Magor'; Figs 16.16 & 16.17) is situated, with a very slight overlap, to the NE of the previous grid. The coastline runs from the mouth of the River Usk in the west, via Goldcliff Point (approximately in the centre) to the saltings lying outside the sea defences of Magor. The bathymetry of the area is characterized by a very extensive intertidal zone (above LAT) with many sand banks and flats (e.g. Usk Patch, Welsh Hook, Welsh Grounds and Middle Grounds), some parts (e.g. east of the Welsh Hook in the Welsh grounds) lying up to 5 m above the mudflats fronting the coastline. The most prominent sand banks of the Middle Grounds generally show greater elevations in the south than in the north. Usk Patch and Welsh Hook are separated from the Middle Grounds by the westward-shoaling Newport Deep (maximum depth of 10.2 m), whereas parts of the main channel (maximum depth of 18 m) can be seen in the south and south-east of the grid. The MHWS and MHWN tide levels used for the area are those for Newport (i.e. 12.1 m and 9.0 m respectively).

Results of wave climate modelling

Maximum significant wave heights and wave energy

The range of maximum significant wave heights incident along the right bank of the outer estuary is 1.06 to 2.51 m (Fig. 16.18). The

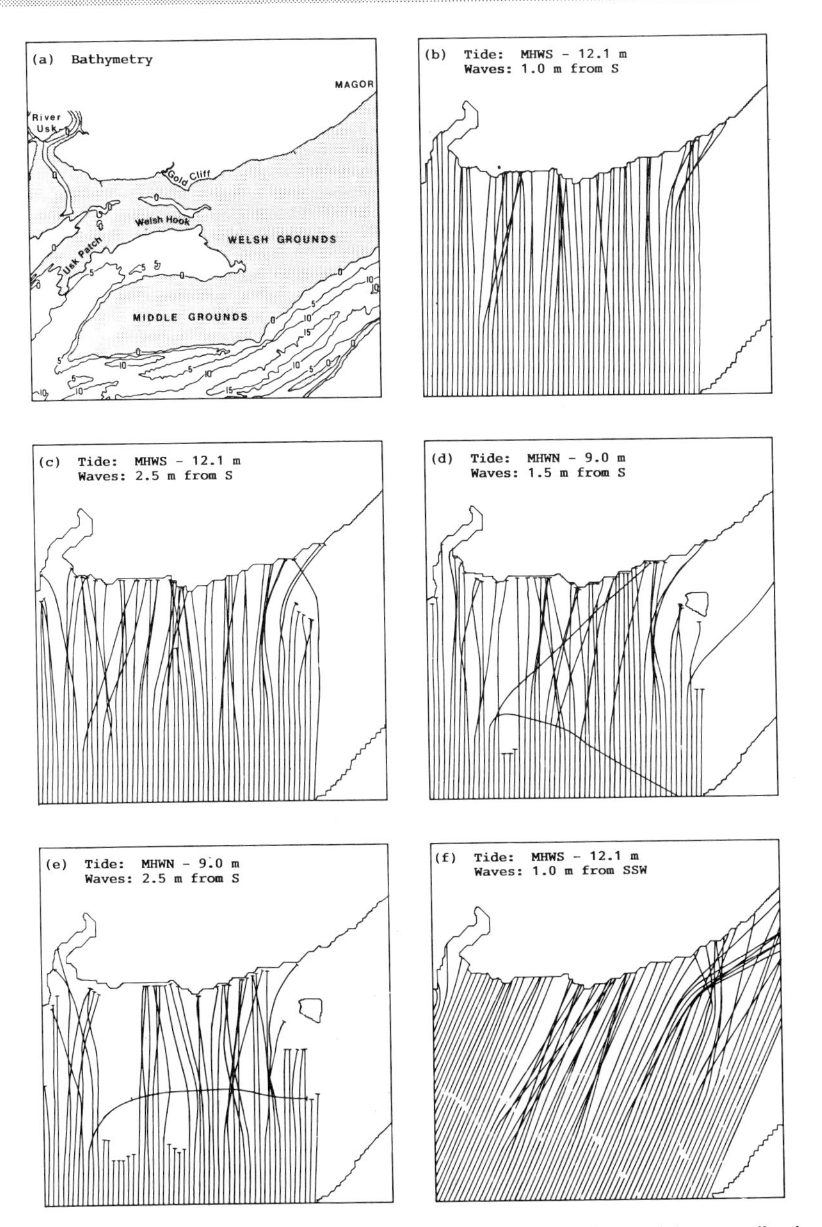

Fig. 16.16(a). Bathymetry of the outer Severn Estuary in front of the coastline between Peterstone Gout and Magor, and **(b)**–**(f)** wave refraction patterns for given tidal and wave characteristics (see text for discussion).

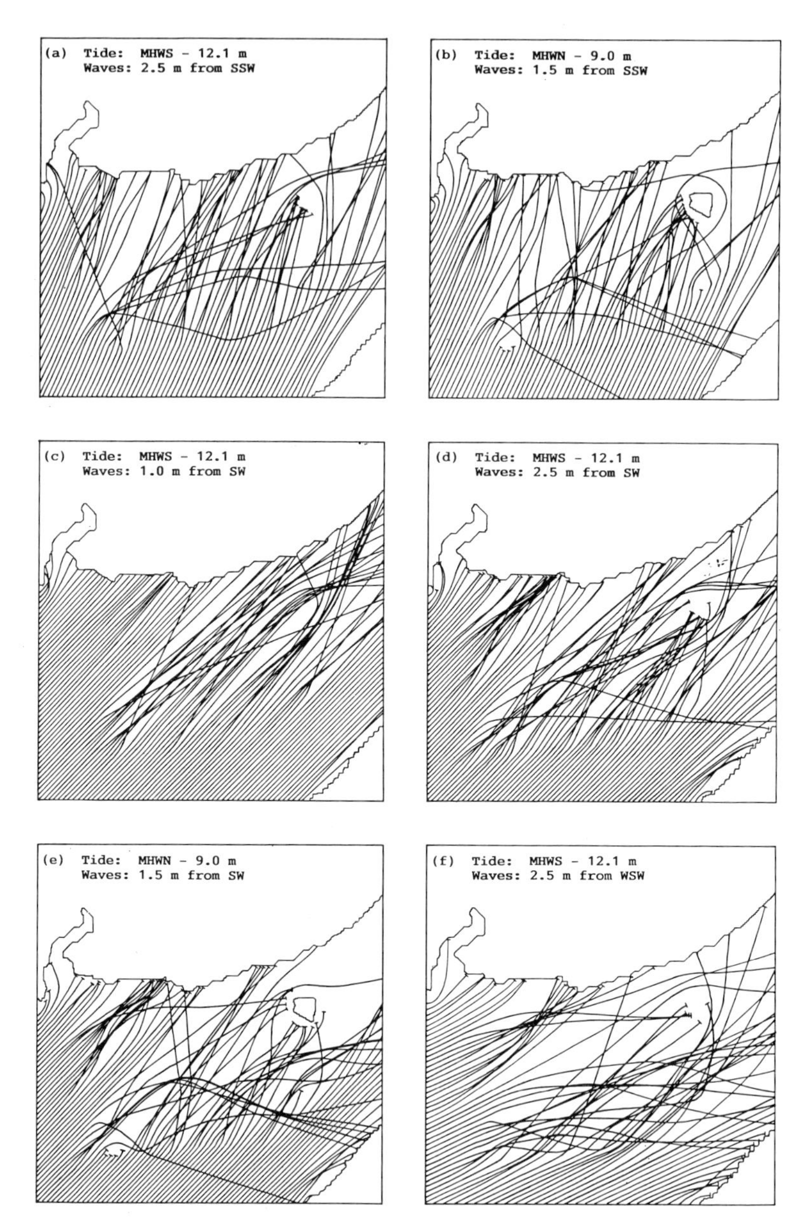

Fig. 16.17. Wave refraction patterns for given tidal and wave characteristics between Peterstone Gout and Magor (see text for discussion).

mean value is 1.99 m, and these waves are generally incident from a S to SSW direction, with a vector mean of 186°. Highest significant wave heights were computed for west and east (including Goldcliff) of the River Usk (i.e. 2.46–2.51 m), whereas the maximum significant wave heights generated at Rumney and Peterstone Great Wharves, and the saltings at Magor are 2.10, 2.39 and 1.69 m respectively. NE wind directions generating maximum wave heights are rare and occur only at Lavernock Point (2.4 m).

The coastal stretches on the Welsh side of the outer estuary (Fig. 16.9) which on average receive most wave energy per year (i.e. wave energy indices 4–6) are at Lavernock Point, between Peterstone Great Wharf and Redwick, though foremost around the projecting coast of Goldcliff, and at Caldicot. Low energy areas (i.e. wave energy indices 1–3) are found in sheltered embayments (e.g. SW of Beachley) and along coastal stretches which are protected from the SW by projecting headlands (e.g. NE of Lavernock Point). It is interesting to note that Rumney Great Wharf is placed in a lower wave energy index band than Peterstone Great Wharf, and that the division coincides with the boundary between the wharves.

Wave refraction

'Lavernock Point to Peterstone Gout'

Waves of 1.0 m height travelling from the south show very little evidence of refraction (Fig. 16.14b). With increasing height and particularly under MHWN tide conditions, waves are increasingly refracted over the shallow intertidal zone and converge over the Cardiff Grounds sandbank. This results in wave foci along the coast between Cardiff and south of the Rhymney River, and particularly at Rumney Great Wharf and the southwesterly end of Peterstone Great Wharf (Fig. 16.14c). At MHWS tide levels, waves up to 2.0 m break against the marsh cliff at Rumney Great Wharf, while they appear to break on top of the marsh surface at Peterstone Great Wharf. Under MHWS conditions, waves of 2.5 m height break at the seaward border of the two great wharves (Fig. 16.14d), while they do not appear to reach the marsh margins at MHWN tide levels (cf. Figs 16.14c & d).

Wave rays incident from SSW converge over Cardiff Grounds and refract over the wide intertidal zone fronting the two great

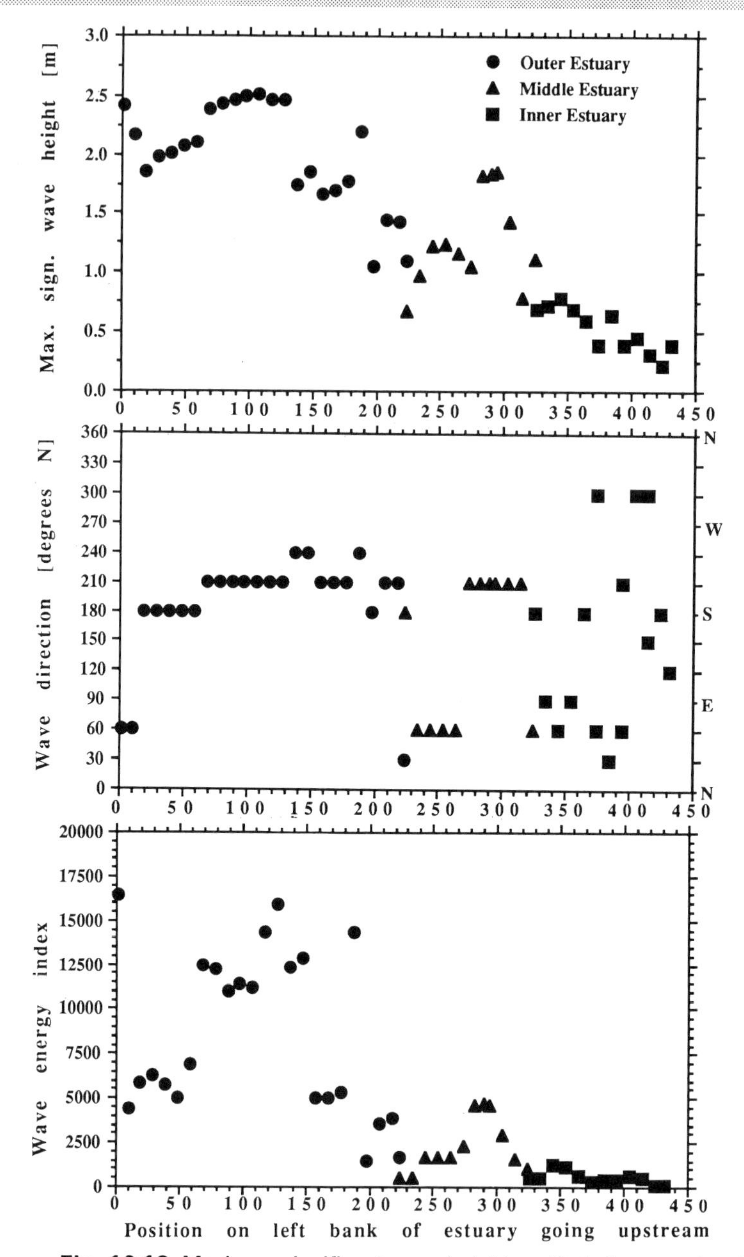

Fig. 16.18. Maximum significant wave heights, with their respective directions and average yearly wave energy ('wave energy index') incident along the right bank of the outer, middle and inner Severn Estuary respectively (calculated with the fetch-limited JONSWAP formula for the period 1970 to 1988 inclusive).

wharves. This results, with increasing wave height, in the concentration of waves incident at Rumney Great Wharf and Peterstone Gout respectively (Figs 16.14e & f). This occurs independently of tidal stage, although the wave foci are more pronounced at spring tide levels (cf. Figs 16.14f & 16.15a) and with large wave heights (cf. Figs 16.14e & f); and the wave focus at Rumney Great Wharf moves progressively further to the southwest with increasing wave height (cf. Figs 16.14e, 16.15a & b). Additionally, wave refraction over the intertidal zone in the northeast quadrant of the grid results in a focused path towards the mouth of the River Usk and the coastline east of it. Peterstone Great Wharf, however, appears to be in general the most sheltered part of the Wentlooge Levels.

Wave rays from the SW experience at MHWN tide levels, and independent of wave height, refraction by Cardiff Grounds towards the area of the Rhymney River and Rumney Great Wharf, and to a lesser extent towards Peterstone Gout (Fig. 16.15c). The same general pattern is generated by 2.0 and 2.5 m high waves at MHWS tides, though more wave orthogonals find their way to Rumney Great Wharf and their points of incidence are more focused on its centre (cf. Figs 16.15c & d).

No waves incident from WSW which parallel the orientation of the shoreline, are refracted towards it. The only exception is shown in Fig. 16.15e where only three orthogonals of 2.5 m wave height find their way to Rumney Great Wharf, Peterstone Great Wharf and Peterstone Gout respectively, and all break offshore from the marsh margins due to MHWN tide levels.

To summarize, Cardiff Grounds sandbank and the wide and shallow intertidal zone control the pattern of wave refraction in front of the Wentlooge Levels. While the Cardiff Grounds sand bank is chiefly responsible for wave ray refraction towards Rumney Great Wharf, convergence over the intertidal zone results in another wave focus forming at Peterstone Gout. These wave foci bracket Peterstone Great Wharf, which appears to be more sheltered than Rumney Great Wharf – although lying further towards the east than the latter, and therefore being less protected by the rocky coast extending north from Lavernock Point – and the coast approaching the mouth of the River Usk.

'Peterstone Gout to Magor'
Refraction of waves from the S (180° true) caused by the elevated Middle Grounds results, at MHWS tide levels, in a concentration of wave rays incident at and just west of Goldcliff Point (Fig. 16.16b). With increasing wave heights this focus becomes more pronounced, and waves are also progressively more refracted against the sides of the mouth of the River Usk (cf. Figs 16.16b & c). The most elevated sand bank on the Welsh Grounds causes convergence of wave orthogonals resulting in a coastal wave focus north of it, while some wave rays are being diverted up-estuary towards Magor. At MHWN tide levels part of the Welsh Grounds is emergent and waves of 1.5 to 2.5 m height break along elevated sand banks of the South Middle Grounds (Figs 16.16d & e). Waves focus more at Goldcliff Point itself up to a wave height of 1.5 m (Fig. 16.16d), whereas a further increase in wave height results in an almost symmetrical diverging of wave rays resulting in the majority impinging on either flank of Goldcliff Point (Fig. 16.16e).

Wave rays from SSW give rise to wave foci in the area of Goldcliff and at the headland west of it at times of MHWS tides (Fig. 16.16f), while the most elevated sand bank on the Welsh Grounds causes converging refraction towards the coast and in an up-estuary wave pathway – though away from the shoreline. With increasing wave height the focus at Goldcliff moves slightly eastward, while the number of wave rays incident at the headland west of it increases dramatically (Fig. 16.17a). At MHWN tide levels the wave focus east of Goldcliff is more diffuse, whereas the wave ray concentration west of it persists (Fig. 16.17b). With increasing wave height, however, the waves break progressively further off shore without reaching the coastline. Under MHWN tide conditions waves are unable to overcome the Welsh Grounds, giving rise to a very protected estuary bank north and northeast of it. With increasing wave height progressively more wave rays enter the mouth of the River Usk, owing to wave divergence chiefly caused by the relatively deep water of the navigation channel.

Under MHWS tide conditions wave rays from the SW converge over the Welsh Grounds. This results in a focused wave path towards the coastline at Magor, while simultaneously less waves are being refracted towards mid-channel. At low wave heights only an embryonic wave focus forms just west of Goldcliff (Fig. 16.17). With increasing wave height this focus becomes stronger, though the number of wave rays being able to 'circumnavigate' the Welsh Grounds decreases (Fig. 16.17d). At MHWN tide levels wave foci occur at Goldcliff Point and on either side of it, while the Welsh Grounds block (again) the up-estuary pathway of waves (Fig. 16.17e).

AT MHWS tides wave rays from the WSW concentrate on Goldcliff (Fig. 16.17f). This focus becomes stronger with increasing wave heights, while progressively less waves are able to pass the Welsh Grounds in an up-estuary direction. They either break at the most elevated sand bank, are refracted towards the northeastern shore, or continue up-estuary in a focused group, but in a mid-estuary position. At MHWN tide levels the wave focus concentrating on the Goldcliff headland becomes stronger with increasing wave heights, though it moves slightly to the west. Very few wave rays are able to negotiate the Welsh Grounds, giving rise to approximately the same wave ray pattern as observed for southwesterly waves under MHWN conditions (Fig. 16.17e).

To summarize, the complex and extensive system of very elevated sand shoals determines the wave refraction pattern between the River Usk and Magor. The area around Goldcliff is the most strongly affected by wave focusing for most tidal and wave conditions, owing to the position and elevation of the Middle Grounds. Conversely, the extreme elevation of an area of sand banks on the Welsh Grounds protects the coastline extending northeast towards Magor from the most extreme wave conditions independent of tidal stage. Needless to say, at more elevated tidal stages (e.g. HAT) the effectiveness of the sand bank in providing shoreline protection will certainly be reduced. The pattern of wave refraction in this part of the outer estuary described above explains well the totally eroded saltmarsh at Goldcliff and the residual mud mound morphology of the Northwick Formation to the west of Goldcliff. At the same time it provides an explanation for the still relatively healthy state of the Northwick and Awre developments at Magor.

Discussion

Along the Welsh coast in the outer Severn Estuary the pattern of incident wave energy in general and wave refraction in particular appears to be in good agreement with the present day distribution and morphologies of saltmarshes and their erosive behaviour. Although lateral marsh growth was in the past attributed to the planting and subsequent vigorous natural spreading of *Spartina* in the early 1950s up to about 1956/7 (Martin 1990; HR Wallingford 1992), differences in the level of incident wave energy are probably the overriding control as *Spartina* appears to be sensitive to changes in wind/wave climate with increased wave action and exposure adversely affecting the vigour of its growth and the size of its niche (Gray 1992; Clarke *et al.* 1993; Otto 1996*e*).

Although the waxing and waning fortunes of the Severn marshes can principally be explained by the low and high frequencies of southwesterly winds blowing along the longest fetches in the estuarine system, this pattern is locally modified by the presence of extensive sand banks and flats. These are chiefly responsible for controlling wave refraction, giving rise to areas of hydrodynamic quiescence along coastal stretches where saltmarshes are able to survive in spite of the more severe wave climate prevailing over the last two decades. Wave refraction gives rise simultaneously to localized wave foci resulting here in enhanced coastal erosion, though such coastal stretches are limited in the area investigated and the overall benefit of the sand shoals is apparent. It should be pointed out that in the long term the sand shoal complexes will only be able to retard coastal erosion but not wholly prevent it, as the estuary continues to be forced inland by the continued rise in sea-level.

Although the sediment transport paths suggest a contemporary constructive process with sand banks and flats either in equilibrium or in a state of net-accretion, intense dredging activity over several decades has probably removed more sediment than has entered the system. There is no modern source of sand, and if we accept the concept of bedload parting to be active (somewhere) in the Bristol Channel, the sand resources up-estuary from this zone must be considered relatively finite. Sediment is undoubtedly being redistributed by the vigorous hydrodynamic regime and pushed further into the estuary by a continuous rise in sea level, but a net gain of sand is not possible by these means. Parker & Kirby's (1982) suggestion that the volume of sand banks has increased significantly over the last 100 years has, as yet, not been investigated or quantified, and is indeed less likely. Intense dredging in the inner Bristol Channel and outer Severn Estuary must therefore be very carefully considered, as sand banks and flats may become sediment starved by dredging of the sources which would otherwise feed them.

It is noteworthy that the dredging areas shown in Fig.16.5 fall, as part of The Crown Estate, under the jurisdiction of the Department of the Environment, and therefore require an environmental assessment prior to dredging approval in order to establish 'whether the dredging will cause, or modify, refraction of waves and thus lead to significant changes in the wave pattern' (Department of the Environment 1989). The author is not aware of any wave refraction studies undertaken to date to assess the impact of aggregate dredging on coastal sedimentary processes in the outer Severn Estuary. Additionally, the area up-estuary from Avonmouth is not owned by the Crown, and although dredging is taking place the above regulations do not, as yet, apply.

Any proposed aggregate extraction and spoil dumping operations, or building of coastal protection measures should be scrutinized in the light of comprehensive environmental impact studies. Ideally this should include the type of wave climate modelling approach adopted here for a relatively small area of an otherwise large estuary. It explains successfully the distribution of coastal morphologies and the causes for their erosional or accretional behaviour, and will also allow predictions to be made regarding bathymetric changes as a consequence, for example, of dredging, spoil dumping, sea-level rise, and changes in the local wind regime. However, prerequisite for a full appreciation of the results of any type of coastal hydrodynamic modelling is a thorough understanding of the temporal and spatial variations of coastal change, whether naturally caused or by anthropogenic interference.

Concluding on a more general note, erosional problems in the coastal zone cannot be viewed without consideration of shallow offshore belts which are still affected by wave action, in particular during storms and/or low tide conditions, and which in turn modify heights and directions of waves subsequently incident along the coast. Poor understanding of the causes and effects of sedimentary processes in the wider coastal zone may result in inadequate planning to counteract saltmarsh and beach erosion and flooding of the hinterland. Within the context of coastal zone management and flood defence issues, offshore sand banks, intertidal mud and sandflats, saltmarshes and beaches, must all be appreciated as a tiered defence system. Any large-scale modifications in the offshore zone by natural events (e.g. extreme storms) or anthropogenic interference (e.g. dredging or dumping of spoil) will inevitably have a significant impact on the shoreline. It is in this light that dredging and sand/gravel extraction policies should be reviewed, and options considered to modify, and possibly incorporate, these sand banks into a system of submerged shoals, as large scale hydraulic modifications of the system are most likely to yield best results for the retardation of saltmarsh erosion.

Acknowledgements

University of Reading PRIS Contribution No. 540

References

Al Mansi, A. M. A. 1990. Wave refraction patterns and sediment transport in Monifieth Bay, Tay Estuary, Scotland. *Marine Geology*, **91**, 299–312.

Allen, J. R. L. 1986. A short history of salt-marsh reclamation at Slimbridge Warth and neighbouring areas, Gloucestershire. *Transactions of the Bristol and Gloucestershire Archaeological Society*, **104**, 139–155.

——— 1987. Dimlington Stadial (late Devensian) ice-wedge casts and involutions in the Severn Estuary, south-west Britain. *Geological Journal*, **22**, 109–118.

——— 1988. Reclamation and sea defence in Rumney parish

(Monmouthshire). *Archaeologia Cambrensis*, **137**, 135–140.
——— 1990*a*. The Severn Estuary in south-west Britain: its retreat under marine transgression, and fine-sediment regime. *Sedimentary Geology*, **66**, 13–28.
——— 1990*b*. Late Flandrian shoreline oscillations in the Severn Estuary: change and reclamation at Arlingham, Gloucestershire. *Philosophical Transactions of the Royal Society of London*, **A330**, 315–334.
——— 1991. Salt-marsh accretion and sea level movement in the inner Severn Estuary: the archaeological and historical contribution. *Journal of the Geological Society, London*, **148**, 485–494.
——— 1992*a*. Trees and their response to wind: mid Flandrian strong winds, Severn Estuary and inner Bristol Channel, south-west Britain. *Philosophical Transactions of the Royal Society, London*, **B338**, 335–364.
——— 1992*b*. Tidally influenced marshes in the Severn Estuary, south-west Britain. *In*: Allen, J. R. L. & Pye, K. (eds) *Saltmarshes: morphodynamics, conservation and engineering significance*. Cambridge University Press, Cambridge, 123–147.
——— & Fulford, M. G. 1986. The Wentlooge Level: a Romano-British salt-marsh reclamation in south-east Wales. *Britannia*, **17**, 91–117.
——— & ——— 1987. Romano-British settlement and industry on the wetlands of the Severn Estuary. *Antiquaries Journal*, **62**, 237–289.
——— & ——— 1990*a*. Romano-British and later reclamations on the Severn salt-marshes in the Elmore area, Gloucestershire. *Transactions of the Bristol and Gloucestershire Archaeological Society*, **108**, 17–32.
——— & ——— 1990*b*. Romano-British wetland reclamations at Longney, Gloucestershire, and evidence for the early settlement of the inner Severn Estuary. *Antiquaries Journal*, **70**, 288–326.
——— & ——— 1992. Romano-British and later geoarchaeology at Oldbury Flats: Reclamation and settlement on the changeable coast of the Severn Estuary. *Archaeology Journal*, **149**, 82–123.
——— & Rae J. E. 1987. Late Flandrian shoreline oscillations in the Severn Estuary: a geomorphological and stratigraphical reconnaissance. *Philosophical Transactions of the Royal Society*,

B315, 185–230.
——— & ——— 1988. Vertical salt-marsh accretion since the Roman period in the Severn Estuary, south-west Britain. *Marine Geology*, **83**, 225–235.
Andrews, J. T., Gilbertson, D. D. & Hawkins, A. B. 1984. The Pleistocene succession of the Severn Estuary: a revised model based upon amino acid recemization studies. *Journal of the Geological Society, London*, **141**, 967–974.
Barrie, J. V. 1980*a*. Heavy mineral distribution in bottom sediment of the Bristol Channel, U.K. Estuarine. *Coastal and Marine Science*, **11**, 369–381.
——— 1980*b*. Mineralogy of non-cohesive sedimentary deposits. *In*: Collins, M. B., Banner, F. T., Tyler, P. A., Wakefield, S. J. & James, A. E. (eds) *Industrialised Embayments and their Environmental Problems*. Pergamon Press, Oxford.
——— 1981. Hydrodynamic factors controlling the distribution of heavy minerals (Bristol Channel). *Estuarine, Coastal and Shelf Science*, **12**, 609–619.
Bascom, W. N. 1954. The control of stream outlets by wave refraction. *Journal of Geology*, **62**, 600–605.
Bowen, D. Q. 1970. South-east and central Wales. *In*: Lewis C. A. (ed.) *The Glaciations of Wales and Adjoining Regions*. Longmans, London.
Brownlie, W. R. & Calkin, P. E. 1981. *Great Lakes coastal geology; effects of jetties, Sodus Bay, New York*. New York Sea Grant Institute, Albany, N.Y., United States.
Carter, D. J. T. 1982. Prediction of wave height and period for a constant wind velocity using the JONSWAP results. *Ocean Engineering*, **9**, 17–33.
Caston, V. N. D. 1967. A note on cliff erosion in southern Lleyn. *Geological Magazine*, **104**, 393–395.
Clarke, R. T., Gray, A. J., Warman, E. A. & Moy, I. L. 1993. *Niche modelling of saltmarsh plant species*. Institute of Terrestrial Ecology (NERC), Project Report No. T08059P1 (ETSU/NERC Contract).
Clayton, K. M. 1990. Sea-level rise and coastal defences in the UK. *Quarterly Journal of Engineering Geology*, **23**, 283–287.
Collins, M. B. 1987. Sediment transport in the Bristol Channel: a

review. *Proceedings Geological Association*, **98**, 367–383.
Cornish, V. 1901. On sand waves in tidal currents. *Geographical Journal*, **18**, 170–200.
Crickmore, M. J. 1982. Data collection – tides, tidal currents, suspended sediment. *In*: Institution of Civil Engineers (ed.) *Severn Barrage*. Thomas Telford, London, 19–26.
Culver, S. J. & Banner, F. T. 1979. The significance of derived pre-Quaternary foraminifera in Holocene sediments of the north-central Bristol Channel. *Marine Geology*, **29**, 187–207.
Davies, C. M. 1980. Evidence for the formation and age of a commercial sand deposit in the Bristol Channel. *Estuarine and Coastal Marine Science*, **11**, 83–99.
Department of the Environment, 1989. Offshore dredging for minerals: review of the procedure for determining production licence application. *Department of the Environment News Release*, **265**.
Dodd, N., Bowers, E. C. & Hampton, A. H. 1995. *Non-linear modelling of surf zone processes: a project definition study*. Report SR 398, HR Wallingford.
Evans, C.D.R. 1982. The geology and superficial sediments of the inner Bristol Channel and Severn Estuary. *In*: Institution of Civil Engineers (eds). *Severn Barrage*. Thomas Telford, London, 35–42.
Ferentinos, G. & Collins, M. 1985. Sediment response to hydraulic regime of the coastal zone: a case study from the northern Bristol Channel, U.K. *In*: *Proceedings of European Workshop on Coastal Zones, as related to physical processes and coastal structures*. Council of Europe, Loutraki (Greece), 2.1–2.14.
French, P. W. 1990. Coal dust – a marker pollutant in the Severn Estuary and Bristol Channel, UK. *Marine Pollution Bulletin*, **26**, 692–697.
Garrard, R. A. & Dobson, M. R. 1974. The nature and maximum extent of glacial sediments off the west coast of Wales. *Marine Geology*, **16**, 31–34.
Gornitz, V., Lebedeff, S. & Hansen, J. 1982. Global sea level trend in the past century. *Science*, **215**, 1611–1614.
Gray, A. J. 1992. *Saltmarsh plant ecology: zonation and succession revisited. In*: Allen, J. R. L. & Pye, K. (eds) *Saltmarshes:*

Morphodynamics, Conservation and Engineering Significance. Cambridge University Press, Cambridge, 63–79.

Griffiths, E. C. 1974. *Sedimentary response to the tidal regime in the Upper Severn Estuary.* PhD thesis, University of Bristol, Bristol.

Hamilton, D. 1979. *The high energy, sand and mud regime of the Severn Estuary, south west Britain. In*: Severn, R. T., Dineley, D. L. & Haeker, L. E. (eds) *Tidal Power and Estuary Management.* Proceedings Symposium Colston Research Society, 30th Scientechnica, Bristol, 162–172.

——— 1982. Sand size populations, transport paths and currents, South West Britain. *International Congress on Sedimentology*, **11**, 94.

———, Sommerville, J. H. & Griffiths, E. C. 1975. Sediment movement in the south western approaches to Britain, with a contrast in tidal energy levels of three sands. *International Congress on Sedimentology*, **9**, Theme 6, 67–72.

Hamm, L., Madsen, P. A. & Peregrine, D. H. 1993. Wave transformation in the nearshore zone: a review. *Coastal Engineering*, **21**, 5–39.

Harris, P. T. 1982. *The distribution and dynamics of sedimentary bedforms in the central and inner Bristol Channel.* MSc dissertation, University College of Swansea, Swansea, Wales.

——— 1988. Large-scale bedforms as indicators of mutually evasive sand transport and the sequential infilling of wide-mouthed estuaries. *Sedimentary Geology*, **57**, 273–298.

——— & Collins, M. B. 1985. Bedform distributions and sediment transport paths in the Bristol Channel and Severn Estuary, U.K. *Marine Geology*, **62**, 153–166.

——— & ——— 1988. Estimation of annual bedload flux in a macrotidal estuary: Bristol Channel, U.K. *Marine Geology*, **83**, 237–252.

——— & ——— 1991. Sand transport in the Bristol Channel; bedload parting zone or mutually evasive transport pathways? *Marine Geology*, **101**, 209–216.

———, Pattiaratchi, C. B., Collins, M. B. & Dalrymple, R. W. 1995. What is bedload parting? *Special Publication, International Association of Sedimentologists*, **24**, 3–18

Hasselmann, K., Barnett, T. P., Bouws, E., Carlson, H., Cartwright, D. E., *et al.*, 1973. Measurements of wind-wave growth and swell decay during the Joint North Sea Wave Project (JONSWAP). *Deutsche Hydrographische Zeitschrift, Supplement A*, **8**, No. 12.

Hawkins, A. B. 1971. *Sea-level changes around south-west England. In*: Colston Papers No. 23: Marine Archaeology, 67–88. Butterworths, London.

——— 1990. Geology of the Avon coast. *Proceedings of the Bristol Naturalists' Society*, **50**, 3–27.

——— & Sebbage, M. J. 1972. The reversal of sand waves in the Bristol Channel. *Marine Geology*, **12**, M7–M9.

Hicks, D. M. & Inman, D. L. 1987. Sand dispersion from an ephemeral river delta on the Central California coast. *Marine Geology*, **77**, 305–318.

Hodgson, W. A. 1966. Coastal processes around the Otago peninsula. *New Zealand Journal of Geology and Geophysics*, **9**, 76–90.

HR Wallingford 1976. *Lower Severn Basin Study – Wave heights in the Severn Estuary*. Report No. EX738.

——— 1992. *Wentloog Levels*. Project Study report: Drainage – Seawall overtopping – Seawall stability. 2 volumes, Report No. EX2511.

Hydrographer of the Navy 1988. International Chart Series - Sheet 1176: Bristol Channel, Severn Estuary (Steep Holm to Avonmouth), 1:40,000; Newport, 1:20,000.

——— 1992. *Admirality Tide Tables. Vol. 1. 1993. European Waters including the Mediterranean Sea*. Hydrographer of the Navy, Taunton, England.

Jackson, W. H. & Norman, D. R. 1979. Port Talbot – accretion and dredging in the harbour and entrance channel. *In*: Collins, M. B., Banner, F. T., Tyler, P. A., Wakefield, S. J. & James, A. E. (eds) *Industrialised Embayments and their Environmental Problems*. Pergamon Press, Oxford, 573–582.

Johnson, M. A., Kenyon, N. H., Belderson, R. H. & Stride, A. H. 1982. Sand Transport. *In*: Stride, A. H. (ed.) *Offshore Tidal Sands; processes and deposits,*. Chapman and Hall, London, 58–94.

Kenyon, N. H. & Stride, A. H. 1970. The tide-swept continental shelf sediment between the Shetland Isles and France. *Sedimentology*, **14**, 159–173.

King, C. A. M. 1953. The relationship between wave incidence, wind direction and beach changes at Marsden bay, county Durham. *Institution of British Geographers, Publication*, 13–23.

Kirby, R. & Parker, W. R. 1982*a*. A suspended sediment front in the Severn Estuary. *Nature*, **295**, 396–399.

——— & ——— 1982*b*. The distribution and behaviour of fine sediment in the Severn Estuary and inner Bristol Channel, U.K. *Canadian Journal of Fisheries and Aquatic Sciences*, **40**, 83–95.

Laitananda, P. A., Hanumanta, R. K., Veenadevi, Y. & Lakshman, R. G. R. 1981. Wave refraction and sediment transport along Kunavaram, East coast of Indian. *Indian Journal of Marine Sciences*, **10**, 10–15.

Lamb, H. H. 1972. *British Isles types and a register of the daily sequence of circulation patterns 1861–1971*. HMSO, London (Meteorological Office, Geophysical Memoirs No. 116).

——— 1977. *Climate: Present, Past and Future*, Vol. 2: *Climatic History and the Future*. Methuen, London.

——— 1982. *Climate, History and the Modern World*. Methuen, London.

Lamb, G. M. 1987. Erosion downdrift from tidal passes in Alabama and the Florida Panhandle. *Bulletin of the Association of Engineering Geologists*, **24**, 359–362.

Lennon, G. W. 1963*a*. A frequency investigation of abnormally high tidal levels at certain west coast ports. *Proceedings of the Institution of Civil Engineers*, **25**, 451–484.

——— 1963*b*. The identification of weather conditions associated with the generation of major storm sturges along the west coast of the British Isles. *Quarterly Journal of the Royal Meteorological Society*, **89**, 381–394.

Martin, M. H. 1990. A history of *Spartina* on the Avon coast. *Proceedings of the Bristol Naturalists' Society*, **50**, 83–94.

McLaren, P., Collins, M. B., Gao, S. & Powys, R. I. L. 1993. Sediment dynamics of the Severn Estuary and inner Bristol Channel. *Journal of the Geological Society, London*, **150**, 589–603.

Mii, H. 1956. Cuspate deposits within wave shadows along the coast of the Natsudomari peninsula, Aomori prefecture. *Saito Ho-On Kai Museum Research Bulletin*, **25**, 27–33.

Mitchel, G. F., Penny, L. F., Shotton, W. F. & West, R. G. 1973. *A correlation of Quaternary deposits in the British Isles.* Geological Society, London, Special Report, **4**, 99.

Murray, J. W. & Hawkins, A. B. 1976. Sediment transport in the Severn Estuary during the past 8000–9000 years. *Journal of the Geological Society, London*, **132**, 385–398.

Oliver, J. 1960. Wind and vegetation in the Dale Peninsula. *Fluid Studies*, **1**, 37–48.

Onishi, Y. 1987. Simulation of sediment transport in coastal water. *In*: Ragan, R.M. (ed.) *Proceedings of the 1987 national conference on hydraulic engineering, 291–296.* American Society of Civil Engineers, New York, N.Y., United States.

Otto, S. 1996*a*. *The erosion of saltmarshes along the Severn Estuary, SW Britain.* PhD thesis, University of Reading, Reading, England.

——— 1996*b*. *Sea-level rise and saltmarsh erosion with special reference to the Severn Estuary, SW Britain.* Publication No. 96/1, Kimberley Services, Reading.

——— 1996*c*. The Wentlooge Levels: past, present and future. *In*: Otto, S. (ed.) *Coastal Zone Management in the Severn Estuary: Contributions towards a concerted strategy plan. Field Guide and Proceedings, Saltmarsh Field Workshop – 26 September 1996.* Publication No. 96/4, Kimberley Services, Reading, 2–25.

——— 1996*d*. Sea-level rise in the Severn Estuary: considerations for managed retreat. *In*: Otto, S. (ed.) *Coastal Zone Management in the Severn Estuary: Contributions towards a concerted strategy plan. Field Guide and Proceedings, Saltmarsh Field Workshop – 26 September 1996.* Publication No. 96/4, Kimberley Services, Reading.

——— 1996*e*. Some observations on saltmarsh vegetation along the Severn Estuary: implications for the re-evaluation of grazing practices and marsh restoration schemes. *In*: Otto, S. (ed.) *Coastal Zone Management in the Severn Estuary: Contributions towards a concerted strategy plan. Field Guide and Proceedings, Saltmarsh Field Workshop – 26 September 1996.* Publication No. 96/4, Kimberley Services, Reading.

Parker, W. R. & Kirby, R. 1982. Sources and transport patterns of sediment in the inner Bristol Channel and Severn Estuary. *In*: Institution of Civil Engineers (eds.) *Severn Barrage*. Thomas Telford, London, England, 181–194.

Pattiaratchi, C. & Collins, M. 1988. Wave influence on coastal sand transport paths in a tidally dominated environment. *Ocean & Shoreline Management*, **11**, 449–465.

Peterson, C. D., Pettit, D. J., Darienzo, M. E., Jackson, P. L., Rosenfeld, C. L. & Kimberling, A. J. 1991. Regional beach sand volumes of the Pacific Northwest, USA. *In*: Kraus, N. C., Gingerich, K. J. & Kriebel, D. L. (eds) *Coastal Sediments '91*. American Society of Civil Engineers, New York, N.Y., United States. 1503–1517.

Pierce, J. W., So, C. L., Roth, H. D. & Colquhoun, D. J. 1970. Wave refraction and coastal erosion, southern Virginia and northern North Carolina. Abstracts with Programs. *Geological Society of America*, **2**, 237.

Pingree, R. D. & Griffiths, D. K. 1979. Sand transport paths around the British Isles resulting from M_2 and M_4 tidal interactions. *Journal Marine Biology Association U.K.*, **49**, 497–513.

Pirazzoli, P. A. 1989. Present and near future global sea level changes. *Palaegeography, Palaeoclimatology and Palaeoecology*, **75**, 241–258.

Pye, K. & Neal, A. 1993. *The Sefton Coast and Ribble Estuary: Holocene evolution, modern sedimentary environments and coastal zone management*. British Sedimentological Research Group Field Excursion – 18 December 1993. Cambridge Environmental Research Consultants Ltd, Cambridge.

Ramanadham, R. & Sastry, J. S. R. 1959. *A study of the beach cycle at Visakhapatnam on the basis of wave refraction*. 1st International Oceanography Congress, 807–808.

Reddy, M. P. M. & Varadachari, V. V. R. 1972. Sediment movement in relation to wave refraction along the west coast of India. *Journal of the Indian Geophysical Union* **10**, Special number, 169–91.

Roberts, J. A. & Chien, C. W. 1965. The effects of bottom topography on the refraction of the tsunami of 27–28 March

1964; The Crescent City case. *In*: *Ocean Science and Ocean Engineering 1965*, Joint Conference of the Marine Technology Society and the American Society of Limnology and Oceanography, Washington, DC, United States. Transactions, **2**, 707–716.
Shennan, I. 1989. Holocene sea level changes and crustal movements in the North Sea region: an experiment with regional eustasy. *In*: Scott, D. B., Pirazzoli, P. A. & Honig, C. A. (eds) *Late Quaternary Sea Level: Correlation and Applications*. Kluwer, Dordrecht, 1–25.
Shrivastava, P. C. 1979. Wave refraction study off Paradip, Orissa coast during the month of February. *Indian Minerals*, **33**, 56–59.
Shuttler, R. M. 1982. The wave climate in the Severn Estuary. *In*: Institution of Civil Engineers (eds) *Severn Barrage*. Thomas Telford, London, 27–34.
Stride, A. H. 1963. North-east trending ridges of the Celtic sea. *Proceedings of the Ussher Society*, **1**, 62–63.
——— & Belderson, R. H. 1990. A reassessment of sand transport paths in the Bristol Channel and their regional significance. *Marine Geology*, **92**, 227–236.
——— & ——— 1991. Sand transport in the Bristol Channel east of Bull Point and Worms Head; a bed-load parting model with some indications of mutually evasive sand transport paths. *Marine Geology*, **101**, 203–207.
Trupin, A. & Wahr, J. 1990. Spectroscopic analysis of global tide gauge sea level data. *Geophysical Journal International*, **100**, 441–453.
US Army Coastal Engineering Research Centre 1984. *Shore Protection Manual* (4th edn). US Government Printing Office, Washington.
Valentin, H. 1953. Present vertical movements of the British Isles. *Geographical Journal*, **119**, 299–305.
Warrick, R. A. & Oerlemans, H. 1990. Sea-level rise. *In*: Houghton, J. T., Jenkins, G. J. & Ephraums, J. J. (eds) *Climate Change*. Cambridge University Press, Cambridge.
Wigley, R. M. L. & Raper, S. C. B. 1987. Thermal expansion of sea water associated with global warming, *Nature*, **330**, 127–131.
Williams, D. J. 1970. *The Draining of the Somerset Levels*.

Cambridge University Press, Cambridge.

Woodworth, P. L. 1987. Trends in U.K. mean sea level. *Marine Geology*, **11**, 57–87.

Woodworth, P. L. 1990. A search for accelerations in records of European mean sea level. *International Journal of Climatology*, **10**, 129–143.

Worsley, P. 1985. Pleistocene history of the Cheshire-Shropshire Plain. *In*: Johnson, R. H. (ed.) *The Geomorphology of North-West England*, Manchester University Press, Manchester, 201–221.

17 'Sense and sustainability' – achieving geological conservation objectives as part of the present Shoreline Management Plan process

Jonathan McCue

Summary

- There are many examples where coastal defence structures have impeded the natural erosion of cliffs, reduced the level of geological exposure and, therefore, the importance of a particular site.
- The recent introduction of Shoreline Management Plans (SMPs) around the country has attempted to address these problems brought about by piecemeal defence schemes.
- SMPs have exposed another 'inadvertent' imbalance – that, compared to the conservation of 'living' (biotic) features, 'non-living' (abiotic) conservation measures are commonly not granted the protection emphasis they equally deserve.
- The development of shoreline management policies is dependent on achieving a successful balance between the requirements of all coastal interests in an area. Whether achieving geological conservation objectives falls within the responsibility of the SMP or not is currently open to debate.
- This chapter assesses the problems involved in achieving 'sustainability' of geological assets and judges whether SMPs are really the correct vehicles to 'manage' valuable sites.

The conservation of natural geological assets is becoming a popular issue. Not only does it aim to preserve Britain's earth heritage, but it is also a potentially saleable commodity. The developing paradigm of 'geo-tourism' which provides an appreciation of the perceptions, knowledge and understanding of the earth sciences to site visitors, particularly at coastal locations, is presently being promoted by conservation bodies around the country. Hand in hand with this growth of geological awareness is the production of Shoreline Management Plans (SMP) around the English coastline which seeks to put forward 'sustainable' policies for future coastal defence. The two aim to 'dovetail' together, though in reality, is this really the case? At present, there is a divide in what practitioners/coastal managers believe to be the best approach. Some advocate that this is best achieved through the effective integration of earth and natural science conservation interests, coastal defence requirements and socio-economic constraints within a wider Integrated Coastal Zone Management framework (ICZM). Others would argue that earth science conservation objectives should initially be achieved within a Shoreline Management Plan, even if the aim of the plan is to outline strategies for future coastal defence. Does it really matter? The fact is that the conservation of natural assets (biotic and abiotic) along the coast is vitally important for the future. However, when situations arise within the Shoreline Management Plan framework that put into the balance the priorities of protection or conservation, it is common that conservation objectives play a secondary role. This is a problem that is experienced in many locations around Britain's coastline but is one that needs to be tackled in a sensible and sustainable manner depending on the requirements of each specific site.

Engineering versus conservation

There are many examples where coastal defence structures have impeded the natural erosion of cliffs, thereby reducing the level of geological exposure and decreasing the importance of the site for education and research. Cliff exposure sites in particular are often threatened by engineering schemes designed to prevent further erosion. Commonly, there is intense local pressure to find engineering solutions to the problem of coastal recession. In the extreme, such schemes can lead to the obliteration of geological exposures and consequently the scientific interest of a site. Problems are particularly severe where piecemeal defence schemes are adopted as remedial measures to protect property or infrastructure from erosion. Quaternary sections pose a particular problem as unconsolidated deposits are readily eroded by the sea. Many such sites have been designated as Sites of Specific Scientific Interest (SSSI) over the years, though where these sites are backed by developments or high grade agricultural land, an immediate conflict arises between maintaining the quality of the geological exposure and safeguarding the landward asset. Coastal erosion is also important in providing the sediment flux necessary to help maintain natural beach defensives. For example, shingle and sand features, such as Hurst Spit in Hampshire, are often the beneficiaries of erosion from elsewhere along the coast whilst the erosion of rock features provides material which may protect land downdrift from erosion. Proposed protection works updrift may therefore result in disbenefits through loss to the downdrift feature where this is of shingle or sand. Similarly, protection of a non-cohesive cliff may also have disbenefits through loss of sediment elsewhere (Penning-Rowsell *et al.* 1992).

The recent onset of SMP around the country has attempted to redress this issue by reducing 'myopic' planning and instead, seeks to promote more strategic thinking towards the impacts of carrying out such work on the natural environment. Nevertheless, compared to the conservation of 'living' (biotic) features, 'non-living' (abiotic) conservation measures are commonly not granted the protection 'emphasis' that they equally deserve. This imbalance needs to be rectified.

SMP seek to propose coastal defence strategies for 'lengths' of shoreline. Nevertheless, this approach may overlook the larger scale issues of coastal geological conservation which would benefit from having different strategies assigned not only along the coast, but also 'across' the profile of a location. It is therefore the intention of this chapter to review the existing SMP process, assess whether geological conservation objectives are sensibly being incorporated within this non-statutory planning system and finally, to put forward possible suggestions for a more appropriate framework for future earth heritage conservation. It is not the purpose of this chapter to critically appraise the relevance of coastal geological conservation within the SMP, nor the quality of present SMP production, but instead to make coastal engineers aware of the importance of this topic within the SMP process.

Types of coastal geological exposures

Coastal geological sites fall along a continuum between two end points which are defined according to their conservation management needs (Nature Conservancy Council 1990; Doyle & Bennett 1998). 'Exposure Sites' such as cliffs and foreshores are those whose value is in providing accessible exposures of a particular deposit. 'Integrity Sites' are of value because they contain landforms or restricted deposits which are irreplaceable if destroyed, for example, shingle bars or spits. For the purpose of this chapter, earth heritage features are divided into two aspects; 'dynamic' landforms and 'static' features. Those geological assets or landforms that are 'dynamic' (i.e. mobile) are important components of the coastal landscape and in themselves pose special conservation problems. However, where conservation demands that the dynamic nature of the coast is preserved, this may threaten coastal infrastructure (i.e. continued cliff erosion to ensure clean geological exposures or the continued natural roll-back of a 'dynamic' landform, such as a shingle bank), measures will need to be pursued to reduce such risks. Where geological conservation assets are 'static' (i.e. rock structures/mineral assemblages), their preservation is determined by the maintenance of a characteristic set of forms which has evolved over much longer geological timescales to the aforementioned features. This is more of an issue where hard geological features are found.

Three major groups of coastal geological conservation site have been identified by English Nature (Nature Conservancy Council 1990). They are: (1) coastal cliffs; (2) foreshore exposures; and (3) geomorphological sites.

Coastal cliff sites are widely threatened by the construction of coast protection works, particularly mass concrete structures, and by uncontrolled dumping at the base of cliffs. This obscures rock sequences directly and more importantly, prevents the slow erosion necessary to maintain exposure in softer deposits. Consequently, tendency towards a precautionary approach is preferred as an option for such sites. The precautionary principle means that action is not to be postponed because of scientific uncertainty. It is a valid concept if a decision-maker definitely wants to avoid a certain consequence, irrespective of how costly this may be (Tol *et al.* 1996). Consequently, in this situation, the precautionary principle is no more and no less than a handy short cut in decision-making.

Second, foreshore exposures, which are more characteristic of hard rock areas, are of considerable importance in many coastal sites. They are subject to periodic pressures arising from larger coastal schemes or proposals to extend schemes seaward thereby jeopardizing the integrity of the feature. In the longer term, a rise in sea-level is a potential threat although it seems unlikely that this will lead to a substantial loss of foreshore exposure.

Third, geomorphological sites, often associated with Quaternary deposits, span a wide range of interests and include active coastal process sites and fossil shorelines. Again, these sites are sensitive to interference from activities such as coast protection or stabilization of sand dunes and certain other uses, such as sand extraction.

The management problems associated with hard and soft geologically important sites

The majority of geomorphological assessments on this topic have been concerned with soft geological assets. Bray & Hooke (1995) have discussed in great detail the need for special approaches to be adopted to conserve dynamic coastal landforms and their associated habitats along with soft geological features. It is important to understand that it is rare for geological conservation objectives (relevant to both hard and soft coastal sites) to be isolated from other issues of relevance to the coast, for example, archaeological preservation or natural habitat protection. This is why the SMP must seek to address all issues when assigning policies and strategies for 'lengths' of coast. In terms of geological conservation, a remedy to the problem in certain locations may be assisted by setting up a '3-dimensional' strategy that deals with longitudinal and lateral management of the shoreline depending upon the assets that occur in a particular location.

Interesting examples of important geological conservation sites on the south coast include, amongst many others, those at Barton-on-Sea and at Hengistbury Head. The cliffs around Barton-on-Sea in Hampshire are renowned for their Tertiary fossils and are still visited by many tourists and amateur fossil hunters each year. The site gives its name to a division of the Palaeogene sub-Period, the Bartonian Stage. Fossils are still abundant and are revealed by regular marine erosion events, however, because of this erosion, the site has been under threat and damaged by coastal protection schemes on numerous occasions in the past. Sections of natural cliff do remain immediately west and east of the Barton frontage, although these could be obscured by the expansion of existing protection schemes. The problem here is the need to protect land which has experienced an accelerated erosion rate as a consequence of piecemeal coastal engineering schemes downdrift. The SMP for this stretch of coast recognizes the problem, though debate is currently taking place to establish the most 'sustainable' option for the future.

A similar classic 'multi-user' management problem occurs further west at Hengistbury Head. This is an exposed site which experiences erosion and cliff retreat. It affords valuable protection to Christchurch Harbour and the surrounding lowlands as well as being important for its environmental, archaeological and geological qualities. It represents a stratigraphically important Tertiary site since it facilitates correlation of the more westerly exposures of strata adjacent to the Bracklesham/Barton Groups junction with those further to the east (Daley 1996). A recently produced Management Plan for Hengistbury Head has highlighted the problems associated with balancing the needs of conservation and the protection of property in low lying areas. It was clear from the earliest formulation of this plan that the earth science needs were not an easy interest to satisfy, given the other multi-sectoral qualities of the area. A method of dynamic beach management has been proposed to help resolve the problems at Hengistbury Head. This will involve maintaining the beach at a critical width such that marine erosion of the cliff toe is controlled, but not completely halted, ensuring that the stratigraphical exposures are kept fresh by natural processes. The significance of this in terms of SMP production will be addressed during Stage 2 of the SMP for Subcell 5f, planned to be completed by 1999. Here the option of assigning two strategies 'across' the shoreline is a possibility (i.e. one for the cliff itself and one for the foreshore).

A similar problem occurs at Brean Down, in Somerset, where there is a classic Pleistocene SSSI showing a fine sequence of breccias and aeolian sands exposed in a sand cliff. There is considerable palaenotological interest in the site which includes terrestrial molluscs, mammals and re-worked forams. Rapid erosion of the sandcliff caused considerable alarm to the National Trust who own the site and also to archaeologists involved in the excavation of the nationally important Bronze Age site located nearby. A scheme was designed here to stop erosion immediately in front of the sensitive archaeological area whilst allowing erosion to continue along the rest of the Pleistocene section. This foreshore berm defence was constructed prior to the establishment of the SMP which is due for completion during 1998, however, the problem of assigning strategies to longitudinal stretches of coast in locations like these that have distinct changes in 'preferred' policy along short geographic limits is an issue that needs to be addressed during SMP production and is the main point of this chapter. It would appear that this issue appears to become of greater relevance along crenulate hard rock coastlines that possess pocket beaches and rocky headlands experiencing limited littoral drift.

The above, which provide examples primarily concerned with softer non-cohesive sediments (Palaeogene and younger in age), are also influenced by increased storm frequency, combined with rising sea-levels and a trend towards increased coastal development. This may lead to increased threats in the future but does not at present pose a significant threat to the more harder rock shorelines. Conservation problems encountered at hard rock cliff sites are usually site specific and commonly relate to safety considerations. By its very nature, a hard geological site commonly does not contribute significant amounts of sediment to the local sediment budget and so is not as important in terms of being a supplier of sediment (i.e. a source) to adjacent shorelines. For this reason, SMPs having to deal with similar situations would seek to adopt a precautionary strategic approach involving minimal intervention techniques. Nevertheless, this is not universally the case and a framework should be established that allows change. At present, this is believed to be too restrictive within the existing management system.

The Shoreline Management Plan process

It is now widely understood that the goal of developing sustainable shoreline management policies through close dialogue and consultation is dependent on achieving a successful balance between the requirements of all coastal interests in an area. The advent of SMPs marks an important step in achieving this and aims to help eliminate conflicts over the way in which the coastline is protected.

SMPs are 'non-statutory', but the need to develop strategic management planning for the coastlines of England and Wales, is now widely recognized. More importantly, they are being increasingly linked to the availability of Ministry of Agriculture, Fisheries & Food (MAFF) grant aid for coastal defences. For the purpose of any non-statutory document, such as an SMP, suggestions and policy strategies for the future must consider the existing statutory planning system. It is also vital that the SMP focuses on all the issues relative to the shoreline within the geographical area concerned.

For the purpose of assigning and implementing shoreline strategies or 'policies', the MAFF guidelines, produced in 1995, suggest that the coastline should be divided into 'management units' (MAFF 1995). These units are sections of coastline that are sensibly consistent in terms of the coastal processes acting upon them and in terms of land use and existing features. The MAFF guidelines define a management unit as 'a length of shoreline with coherent characteristics in terms of both natural coastal processes and land use'.

Each management unit should also have a coastal defence policy and implementation guidelines. For each unit a range of generic options are appraised. These comprise the following choices: (1) do nothing; (2) advance the line; (3) retreat the line; and (4) hold the line.

Before being accepted, each option is considered in relation to its impacts (both positive and negative) upon the various factors which are influenced by, or influential upon, the condition of the coastline. It is reviewed on the basis of its compatibility with natural processes, the implications for the human environment, natural environment acceptability, technical soundness and sustainability, economic viability and its wider impacts.

In terms of earth heritage conservation, it is commonly the case that where geologically important areas exist along the coast, it may not be appropriate to assign one specific generic policy to a particular 'length of coast'. This is essentially dependent upon the type of asset that is to be protected (foreshore or cliff feature) along with the preference for future 'management' techniques at the site (i.e. allow natural erosion to continue at the present day rate or at a more reduced rate). An issue along the Dorset coastline, for example, is the problem of ensuring that the concept of sustainable fossil collection is adhered to. Whether this is achieved through some kind of 'anthropogenic – intervention' management approach or not is pertinent to this SMP debate. Similarly, whether this issue should be addressed at all within the scope of an SMP is likewise arguable.

Figure 17.1 has been prepared to outline a possible update to the present SMP system by introducing a 'lateral foreshore strategy' comprising four new generic options that may prove of use where shoreface/defence line strategies do not reflect the requirement of the foreshore asset (i.e. important natural habitat or geological exposure site). This new approach is not, however, without its pitfalls. There is a danger that by introducing additional 'lateral' units, each with specific strategies, units may be seen to be managed independently of each other. With a potentially large number of units established along any one stretch of coast, it may be easy to lose sight of the overall strategic requirements by becoming inadvertently focused on small sections of coast, as has been the case with the numerous authorities responsible for short lengths of coast in the past.

It is not disputed that subdivision into defined units is important, but the management approach needs to be considered in a wider context than purely a linear 'longitudinal' unit. It is recommended that the issue should be spatially widened to include cross-sectional strategies. In terms of geological conservation, some coastal sites would benefit by having, for example, separate strategies that are principally concerned with cliff or foreshore exposures. This approach would also be advantageous to other coastal 'uses', particularly other conservation aspects such as for archaeology or

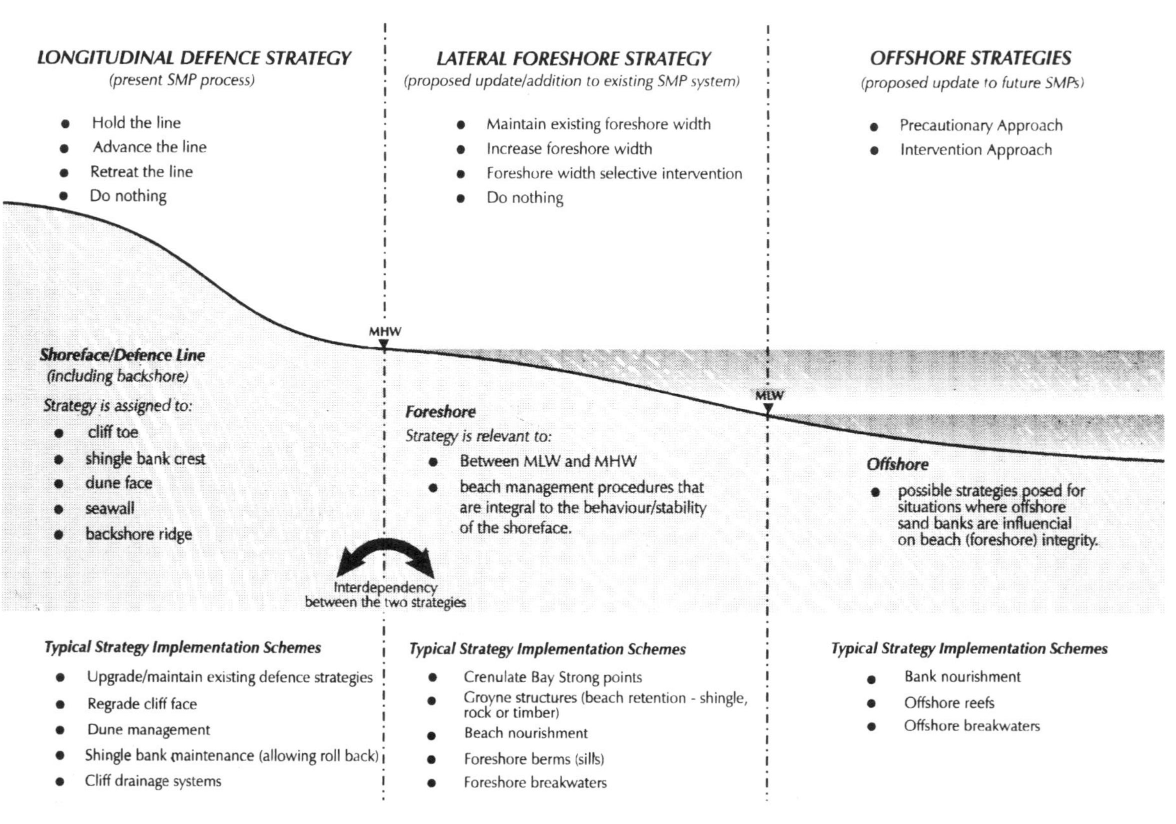

Fig. 17.1 The lateral approach to shoreline management decision making.

nature conservation where future strategies for the shoreline would possibly benefit from two separate strategies being assigned 'across' a management unit. Figures 17.2 and 17.3 display some hypothetical examples of how this 'lateral' approach to strategy implementation may benefit such situations as well as highlighting the inefficiencies of the present system. Examples do occur where cliff and foreshore geological assets exist together in one location. Ideally, in these situations, two separate strategies for the cliff (protection) and the foreshore (do–nothing) would prove to be beneficial. In reality, to preserve the hypothetical situation presented in Fig. 17.2, some sort of intervention would need to be carried out (supposing favourable cost/benefit ratios to protect the cliff top property were forthcoming). Achieving the geological objectives of both these sites would be a particularly difficult and expensive task without sacrificing the integrity of one of the exposures.

Figure 17.3, presents a more realistic example of the benefits of adopting a 'lateral' strategy approach in an area of geological importance. Here, under the present SMP process, a policy to allow erosion to continue, if at a somewhat reduced rate, would most probably be proposed for the shoreface/defence line (i.e. retreat the line). However, the necessity behind ensuring that 'managed retreat' of the geologically important cliff site is achieved is interlinked with 'holding' the foreshore width in front of it. Therefore, we have a situation that would benefit from having two separate strategies assigned to a site.

The key issue here boils down to the protection of a non-tangible coastal asset. Similar situations arise where other 'living' natural assets occur along the shoreline (habitats etc.) especially when their protection or 'sustainability' in the long term is of international importance. Consequently, it is believed that this 'lateral' approach to shoreline management may potentially be of wide ranging significance to all aspects of conservation.

Are coastal geological sites a sustainable resource?

The 1992 Earth Summit in Rio de Janeiro, Brazil, placed the concepts of sustainable development and biodiversity firmly on the

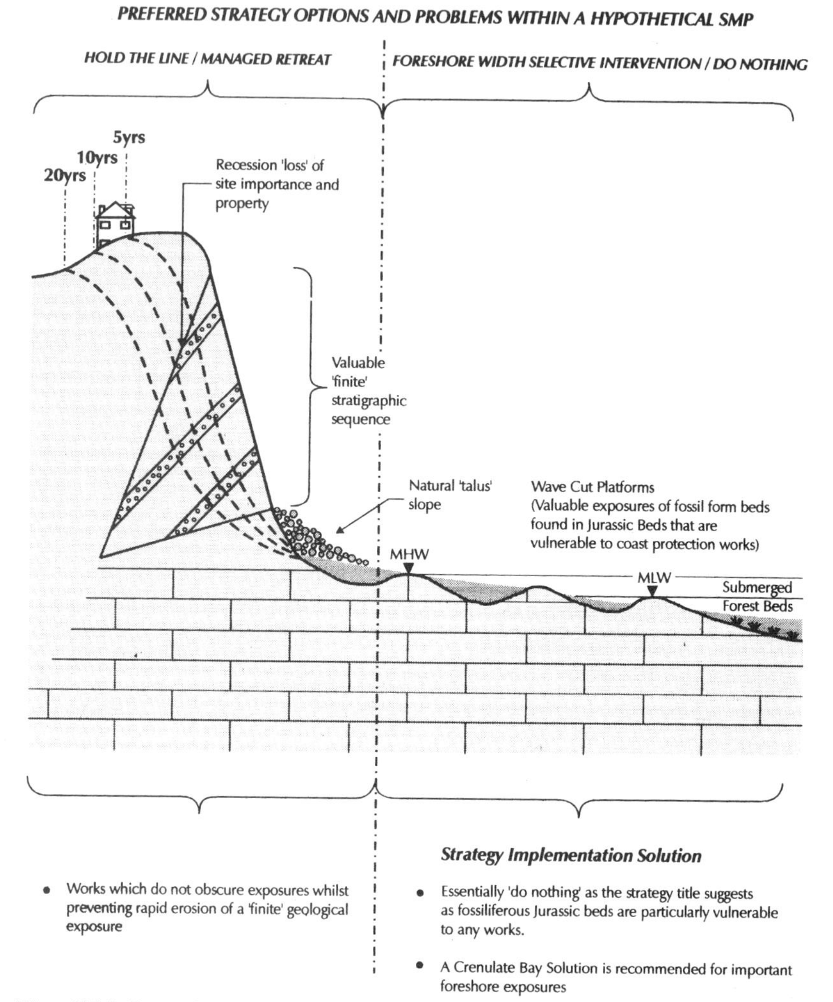

Fig. 17.2 Example of the problems associated with a lateral strategy approach where cliff and foreshore geological sites occur together.

agendas of many governments. Agenda 21 places emphasis on government, business, the voluntary sector and individual citizens to participate in the formation of national and local strategies for sustainable development. It also provides tremendous opportunities for earth heritage conservation. The 'Malvern International Conference on Geological and Landscape Conservation' held in

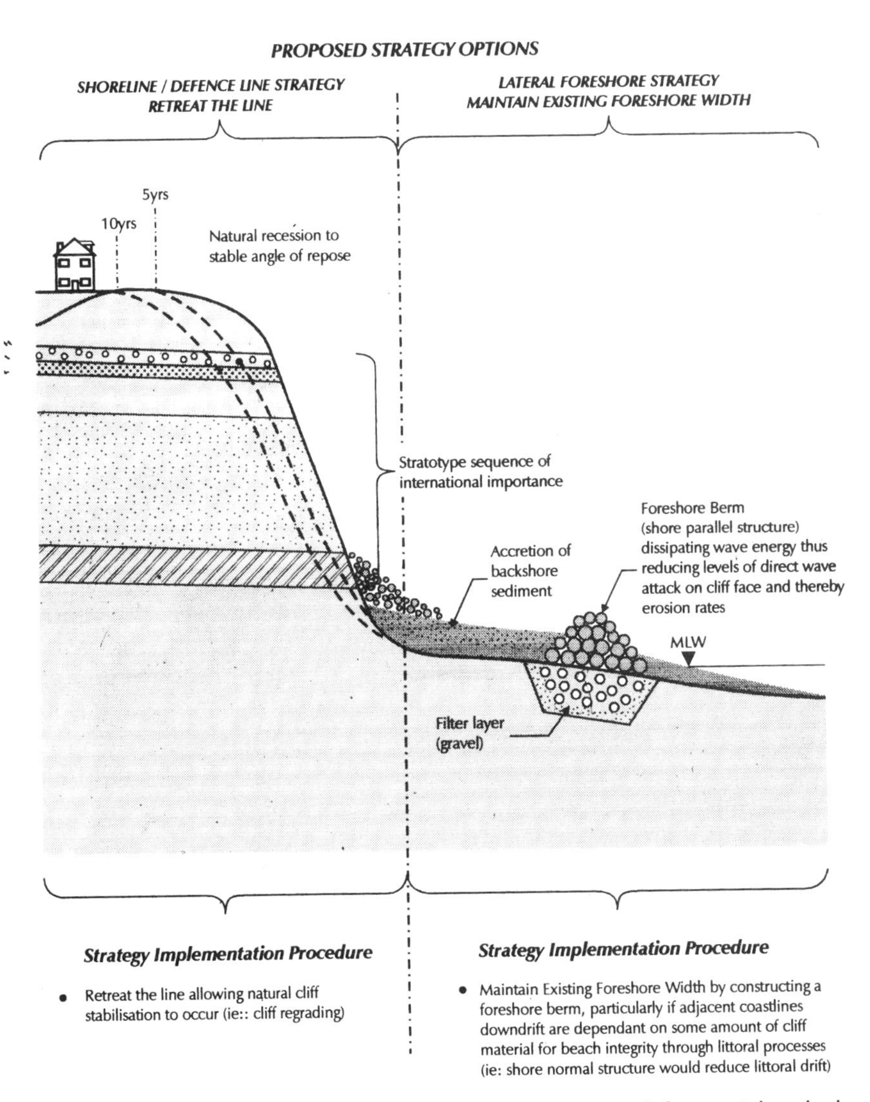

Fig. 17.3 Example of the benefits of a lateral strategy approach for coastal geological conservation.

July 1993, strengthened the links between nations and promoted the appreciation of earth resources and processes and their sustainable use. These welcome advances in future planning, however, are tailored towards the wider holistic management approach, and it is subject to interpretation how such strategies are implemented from the global level down to the local level. The recent onset of SMP has provided great opportunities for such holistic management strategies to be adopted within the SMP process providing conservationists with the opportunity to achieve these wider goals.

The definition of the new buzz word 'sustainable' has been discussed in some detail over recent years. Whilst sustainable development is now not thought of as a new paradigm, actually how to achieve sustainability is becoming a more prominent issue. In the context of this chapter, one needs to qualify how sustainable geological conservation features are. Geological resources, such as oil, gas and fossil fuels are finite and their continued use in the future will be dependent upon world demand. Coastal stratigraphical type sites or highly fossiliferous strata that require international protection (i.e. geological assets) themselves are commonly finite for particular situations, as they cannot be reproduced elsewhere. Consequently, achieving their integrity as scientifically important features needs to entail a continuous process of decision-making. The concept of sustainability is important here. The working definition of 'ecological sustainable development', derived from the Organization for Economic Co-operation and Development report, is abbreviated as meeting '... the needs of the present generation of humans ... without jeopardising the ability of the biosphere to support ... the reasonably foreseeable needs of humans and other species'. In terms of 'ecologically sustainable development', it is believed that the concept is well understood. Adapting this to sustainable earth heritage conservation practices is essentially charting new territory as in general, sustainable outputs are dependent upon:

1. the interaction of various natural morphodynamic systems;
2. the complementary use of important geological sites;
3. scientific evidence on uniqueness (e.g. stratotype site) and its tolerance to natural/anthropogenic change;

4. public consensus and participation in reaching decisions concerning geological conservation.

No studies actually exist that attempt to document individual cases of sustainable geological conservation. In part, this can be attributed to the difficulty in defining sustainability in this context. Undoubtedly, there are similarities between establishing sustainable practices for ecological development, 'eco-tourism' (or 'geo-tourism' for that matter) or management of a geological World Heritage Site (e.g. proposed site being the Dorset Jurassic Coast). In any discussion on sustainability, one commonly focuses on four key aspects, these include: (1) visits to the site in question; (2) financial implications of protecting it; (3) science in this case, earth science information; and (4) economics tangible and intangible costs associated with the location. Without doubt, decisions about whether a management practice adopted for a length of geologically important coast is sustainable over the long term requires a complex analysis and evaluation involving the co-ordination of many disciplines (including economics), a number of administrative decision-making levels (local, regional, national) and a high degree of public participation. Whether all these components are achievable within a SMP is, however, very debatable. To this end, a sensible framework must be set up to ensure that those points raised above are at the very least, brought into discussion early on within the SMP process. Opportunities to study the Earth's history through its rocks, fossils and minerals need to be safeguarded as many sites are being threatened by future development proposals. The author believes that the SMP should be the document to initiate this debate, though there are limitations within the present system that inhibit progress. The SMP process will undoubtedly evolve over the next few years to ensure that geological conservation objectives are fully realized and achieved.

Management and implementation

The nature of SMP are that strategies are produced for the future management of lengths of coast. These strategies are commonly assigned to 'linear' longitudinal lengths of coast. This chapter has developed the argument that there is a need for strategies to look beyond the immediate area in order to negate the impact of a certain coastal strategy upon adjacent stretches of shoreline downdrift. Whilst this approach is a necessity, and is to some extent presently being carried out in SMP around the country, the three-dimensional importance of shoreline management is, however, somewhat taken for granted when assigning appropriate defence strategies. Perhaps this oversight, which does not benefit geological conservation in coastal areas (i.e. cliff and/or foreshore exposures of importance) should be rectified by a more three-dimensional approach.

Precisely how a length of coast is to be defended is not the role of the SMP. Only the strategy for a stretch of coast is requested as part of the plan. Consequently, it does not need to suggest how the future strategy should be implemented. The selection of suitable defences is dependent upon the nature of the geological interest (stratigraphic, palaeontological, coastal geomorphological or mass movement), the location of the interest (cliff or intertidal) and the cause of erosion (marine or groundwater).

For geological conservation sites subject to marine erosion, appropriate coastal defence techniques are likely to be governed by whether adjacent shorelines are dependent on littoral drift of material derived from erosion of the cliff site. Examples of shore parallel structures which reduce the risk of littoral starvation downdrift include the construction of wave attenuation structures such as breakwaters or sills, revetments (both less appropriate where the geological interest extends into the intertidal zone), cliff strongpoints, cliff strengthening and sediment management structures. Where groundwater induced slumping is also occurring (commonly where there are Head or alluvial deposits present), consideration is often given to installing drainage systems which would reduce water pressure. Such systems include horizontal drainage systems, drainage galleries, tunnels or adits, wells or vertical drains and trench drains.

The key point that needs addressing here is the relationship between the various cross-sectional components of a section of shoreline and the impact that one lateral strategy may have on the integrity or sustainability of a geological resource elsewhere. By adopting a framework that allows for lateral shoreline strategies to be introduced, this will allow the flexibility for appropriate strategies to be assigned avoiding any clash of objectives that may arise in certain situations. The benefit to earth heritage conservation is equally applicable to other aspects of nature conservation that require the protection of habitats, notably within the intertidal area. Examples of this include habitat which has been granted internationally important status under recent EU legislation. Such land should seek to be maintained and not lost, however, the strategy for coastal defence may appear to be detrimental to this cause. Undoubtedly, there will be situations where the allocation of lateral shoreline strategies are not required, though by establishing the framework to include this where necessary, a sensible approach to future sustainable management and the implementation of strategies will surely be achieved.

Conclusion

This chapter has attempted to see whether SMP are really the correct vehicles to manage valuable geological assets.

It is concluded that SMP are the most suitable means of achieving geological conservation objectives, but that, one should not feel constrained by the four generic policy options presently being used within the SMP planning framework. A possible enhancement of the existing system has been outlined in this chapter. The important issue here undoubtedly is the strategy. By introducing a new lateral strategy concept into the existing generic option selection criteria, then the SMP should be able to provide the necessary framework to enable the successful management of non-tangible assets (i.e. conservation interests).

In the context of the first round of SMP that are currently being completed for the English coastline, it is essential that one recognizes the SMP for what it is – the foundation for shoreline management planning. It is not the definitive solution. In reality, the SMP is a document which uses existing information that seeks to set out strategies for future coastal defence 'taking account of natural coastal processes and human and other environmental influences and needs'. This point is unquestioned, although as the

SMP process evolves, there is a need for it to pay greater attention to the importance of earth heritage conservation issues where these situations arise. Policies to conserve dynamic landforms or unstable cliff exposures can be formulated within the shoreline management process by including specific objectives towards 'maintaining and enhancing the natural coastal environment' (MAFF 1995) whilst advocating that where applicable, more than one strategy may be adopted laterally across a unit.

It is strongly recommended that achievable site specific earth heritage conservation objectives are coherently established for individual locations early on in the SMP process. These should be focused on whether they need to be concerned with dynamic landforms (mobile spits) or static assets (petrology or structural interests). Raising awareness of the issues by involving the public through consultation is believed to be the best mechanism for achieving success. Only by influencing the coastal decision-makers in authority, by forging working partnerships, disseminating information and promoting positive management of sites will the objectives of sustainable earth heritage be successfully implemented and attained. However, common sense must prevail along strategically important lengths of coastline and by updating the present SMP process by implementing a 'lateral' strategy approach, it is believed that a more competent understanding of earth heritage conservation will be attained. This will then help to ensure that the sustainability of finite geological/geomorphological resources is achieved along our coasts.

References

Bray, M. J. & Hooke, J. M. 1995. Strategies for Conserving Dynamic Coastal Landforms. *In*: Healy, M. & Doody, P. (eds) *Directions in European Coastal Management*. Samara Publishing Limited, Cardigan, 275–290.

Daley, B. 1996. The Tertiary geological succession in Christchurch Bay. *In*: Bray, M. (ed). *Coastal Defence and Earth Science Conservation Field Excursion Guide*. University of Portsmouth, Portsmouth, 6–10.

Doyle, P. & Bennett, M. R. 1998. Earth heritage conversation: past,

present and future agendas. *This volume*
MAFF. 1995. *Shoreline Management Plans – a guide for Coast Defence Authorities*. HMSO, London.
Nature Conservancy Council. 1990. *Earth Science Conservation in Great Britain: a Strategy. Appendices – a handbook of earth science conservation techniques*. Nature Conservancy Council, Peterborough.
Penning-Rowsell, E., Green, C. H., Thompson, P. M., Coker, A. M., Turnstall, S. M., Richards, C., Parker, D. J. 1992. *The Economics of Coastal Management – a manual of benefit assessment techniques*. Belhaven Press, London.
Tol, R. S. J., Klein, R. J. T., Jansen, H. M. A. & Verbruggen, H. 1996. Some economic considerations on the importance of proactive integrated coastal zone management. *Ocean and Coastal Management*, **32,** 1.

Part Four

Environmental geology and education

Environmental geology is increasingly considered by students as a subject suitable for a first degree. Most geology departments of British universities now offer degrees in this subject and this growth of interest in the importance of geology in the human environment has breathed new life into traditional applied geology. Part Four examines the issues of the provision of environmental geology to the new generation of university sector students.

18 Environmental geology in the urban environment: a comment on the benefits of collaboration between universities and local authorities

Nick R. G. Walton

Summary

- As new courses and new developments occur within the rapidly expanding environmental geology sector, there is a growing opportunity to focus some of the environmental geoscience expertise into the local, mainly urban, environment.
- The majority of university geoscience departments are situated within urban environments, most of which have suffered industrial contamination, dereliction or decline in recent years leading to a host of local problems which could be addressed by environmental geologists getting involved in their local environment.
- The urban environment offers excellent training opportunities for undergraduate environmental geologists and a combination of student final-year projects and local environmental action programmes, under the aegis of the local authority, can assist in creating a better, healthier urban environment for the benefit of the whole community.
- This chapter shows how both the BSc Applied Environmental Geology and the new MSc Contaminated Land courses at the University of Portsmouth have integrated local environmental problems into the curriculum resulting in better trained, more experienced and confident graduates, whilst accomplishing tangibly useful results within the local urban environment for the benefit of all.

Most university geology departments contain research groups which focus on the traditional areas of geoscience and much of this work is conducted in remote regions. Environmental geology has the ability to be truly global in nature, addressing, as it does, earth systems within the current context of global change and their potential impact on the future course of human development. However, these subjects overlap with physical geography and environmental science, and so this chapter argues that a 'modern frontier' for environmental geologists actually exists on our own doorsteps, in the nature of our urban environment. This change of focus was recently addressed both by the initiation of the NERC URGENT research programme and in a widely publicized speech by the Director of the British Geological Survey who urged geologists to spend less time and resources looking at the theories of deep geological processes and focus more on the everyday problems of shallow crustal and surface processes which directly affect the world we live in and the quality of our lives. These include a diverse array of everyday problems such as: land fill and waste disposal; contaminated and derelict land; polluted rivers and waterways; rising and/or falling water tables; contaminated groundwaters; underground void development; ground movements; and the decay of building stones. All of these essentially urban problems fall within the scope of the environmental geologist.

This chapter focuses on Portsmouth as being typical of an urban environment which suffers from many of the above problems, particularly those relating to land fill and soil contamination. It shows how local environmental geologists, in this case from the University of Portsmouth, have played a significant role in defining and studying these problems in the local urban environment in collaboration with the City Council. In addition, the development of new BSc and MSc courses at Portsmouth utilizes this urban backyard as an economical training ground enabling both class mini-projects and final-year undergraduate and graduate projects to investigate real environmental problems within the local urban environment. The results of these projects ultimately bring tangible improvements to the local environment, whilst students gain experience, skills, confidence and satisfaction in tackling 'real-world' problems.

Environmental geology in Portsmouth

Portsmouth, although fairly typical of many British cities in terms

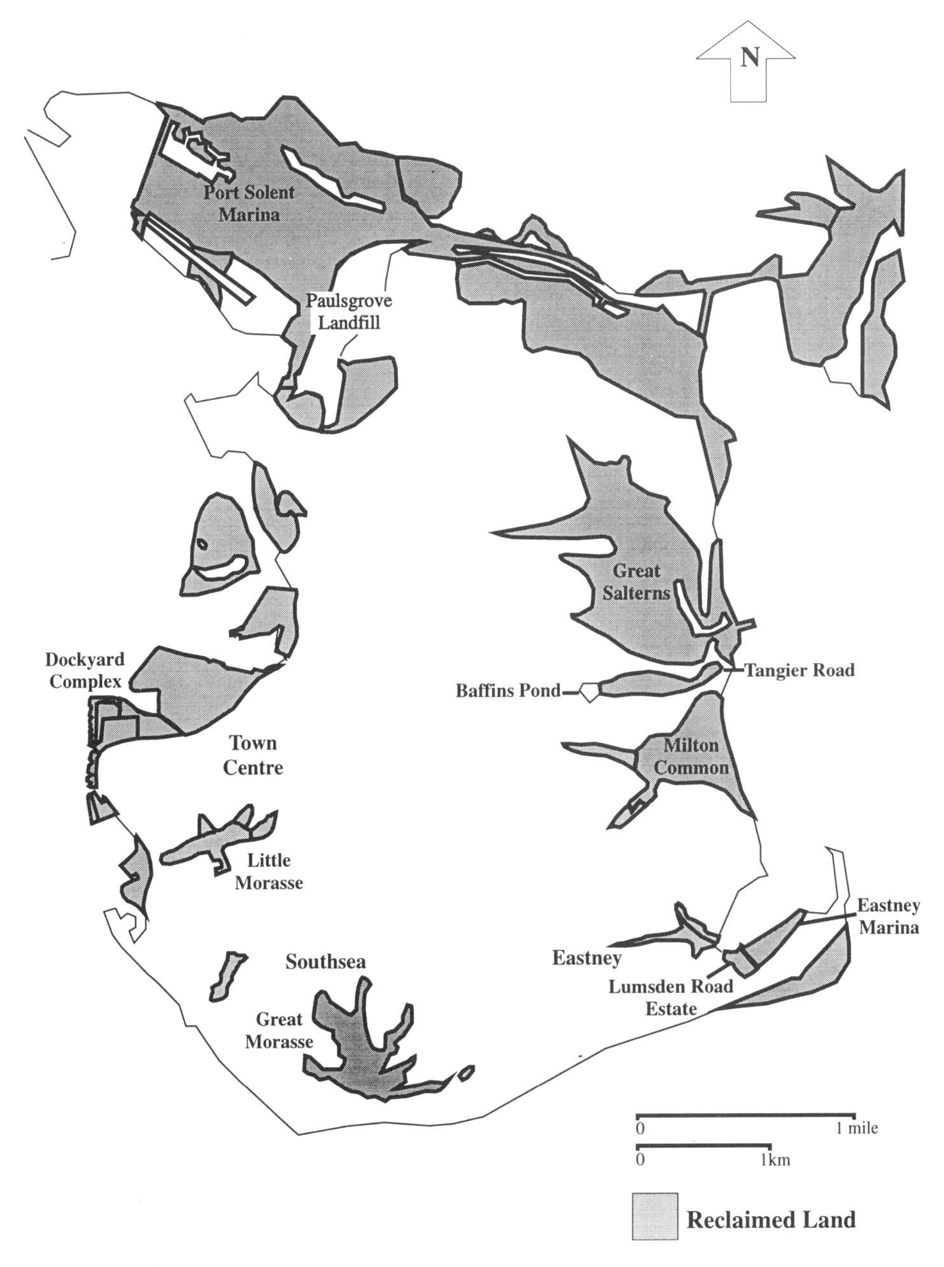

Fig.18.1. Location of reclaimed land areas in Portsmouth.

of its industrial past and consequent legacy of contamination, has a number of special problems because of its flat, island topography and its important naval harbour and military history.

The development of Portsmouth as Britain's premiere Naval Port and Royal Dockyard over the last few hundred years, and its subsequent decline as the Royal Navy has contracted, has had a profound effect on the island City of Portsmouth. During phases of its historical expansion, the dockyard developed into what was, at one time, the largest industrial complex in the world. This led to the creation of two pressing problems: (1) the need for more land within the finite space of an island; and (2) the need to dispose of increasing volumes of waste materials.

In common with other island and coastal cities, Portsmouth chose the most obvious and practical solution to these two problems by filling the low-lying marshy coastal fringes and creeks of Portsea Island with a wide variety of dockyard, industrial and municipal wastes (Fig.18.1). Although some land filling took place in the sixteenth and seventeenth centuries, major land fill reclamation did not really commence until mid-Victorian times, when the last major expansion of the dockyard and city itself took place.

The legacy of this contaminated land and the pressures to find more land on which to build resulted in problems at Lumsden Road in 1991, when families had to be re-housed rapidly because of health fears from a 'toxic cocktail' of contaminants contained within an old military dump under their housing estate (Walton & Higgins 1998).

A key factor which brought this situation to a head was legislation contained within the new Environmental Protection Act (1990) which, under Section 143, proposed a register of all 'Potentially Contaminated Land' and placed a legal responsibility on local authorities to inspect these areas for contamination and then respond appropriately by investigating and subsequently remediating the situation (Cuckson 1998). Although Section 143 was eventually abandoned under pressure from developers, the situation with regards to contaminated land had become increasingly public and responsible local authorities felt obliged to continue their investigations into searching out areas of contamination, investigating its extent and then prioritizing the worst cases for subsequent remediation. This provided an increasingly important role for local environmental geologists in Portsmouth, and for the development of student involvement in local problems.

Student project involvement

Desk studies and site investigations

Portsmouth City Council has been very pro-active in addressing the problems of local land contamination, and has managed to secure a significant slice of the annually available Government fund for investigating and remediating the worst areas of urban pollution. However, since there are so many potentially contaminated sites, with unquantified and probably variable levels of public risk, it is necessary to undertake a large number of preliminary desk study projects to delineate all the sites and categorize them by degree of risk. This is done so that the worst sites can be prioritized for early site investigations and typical examples are old gas works, metal manufacturers, shipyards and industrial waste sites. It is the preliminary desk study phase that is ideal for student project work, as it involves extensive research in libraries, museums, records offices and council archives to find old maps and trade directories which indicate the former uses of any particular site going back to early Victorian (pre-industrial) times. Such work is time consuming and it is unlikely that councils would be able to undertake the task themselves given their tight budgets and limited personnel, and it would be too expensive for the routine employment of external consultants to undertake. Students can fill this role very ably, thereby providing the background data to a substantial number of sites that the Council can then prioritize in terms of initiating a full intrusive site investigation to discover the actual degree of contamination present.

Due to the specialist training given to students at Portsmouth, they are not only capable of conducting the background research, but also of carrying out a preliminary site investigation themselves. This usually involves a walk-over observational survey of the site, spike testing for soil gases such as methane and carbon dioxide as well as hand augering of some soil samples down to about one to two metres depth for subsequent laboratory processing to determine the heavy metal content. This goes some way to determining the overall contamination status of the near-surface environment, although further study of the organic compounds is necessary. A combination of the results of the desk study, which highlights the potential for contamination, and the initial site investigation results enables a site prioritization ranking by the Council, and gives the necessary supporting background information required by central government for Supplementary Credit Approval (SCA) funds. This has enabled Portsmouth City Council to be one of the major recipients of government SCA money, which helps pay for full site investigations undertaken by engineering consultancy companies, and which subsequently helps in funding the necessary site remediation proposals.

Due to the Health and Safety implications of students working on potentially contaminated sites, student field projects are very carefully vetted to ensure that they are only involved in site investigations where there is a low contamination potential. In fact, a number of recent projects were conducted in public parks, common lands and cemeteries in an effort to define local background levels of urban contaminants in order to provide practical data against which to assess the growing data bank of contaminated soils. This has proved very valuable, since urban soils can contain high levels of contaminants derived from airborne deposition, such as lead from urban traffic and arsenic fall-out from earlier coal burning industries, which can be significantly above current government (ICRCL) guidelines for uncontaminated soils.

Environmental monitoring

Environmental field projects usually require a large amount of monitoring data, often spanning several seasons. Such intensive data collection requires either a lot of time or commitment of expensive capital equipment with dedicated sensors and data-loggers, or the employment of expensive consultants who will use either or both of the above methods. Whichever method is chosen, it will be expensive and therefore difficult to fund or justify, especially for any extended and diffused periods of time. However, final-year student projects, which traditionally span several weeks of the summer for intensive fieldwork, can be extended and diffused for routine monitoring purposes to allow for weekly local data gathering over a period of up to nine months.

This is required for water pollution projects which, in contrast to soil pollution, tend to require much longer time periods to monitor seasonal changes in both water quality and levels. Portsmouth University students have investigated a number of urban lakes and ponds, as well as the local harbour, for water pollution and hydrological investigations. These have given some surprising results which have often resulted in considerable environmental benefits and direct financial savings to all concerned.

Examples of recent collaborative projects

Hilsea Moat

The study of this urban water body, which the Council wanted to clean up for increased leisure and boating activities, produced a number of important results. These defined the local hydrological regime and tidal influences, found leaking storm sewers, defined both the depth of water and sediment in the lake and identified a perennially leaking water main inflowing at a rate of some $2 \mathrm{l\,s^{-1}}$ which equates to *c.* £3.60 per hour of potable water loss. The quality of the water body was improved after the identified sewer inflows were all stopped, and quantification of the total volume and quality of the sediment allowed the Council to draw up a contract which enabled the sediment to be dredged and disposed of locally, for very reasonable costs.

Baffins Pond

A study of this urban lake, which was suffering from eutrophication, lack of water and periodic fish deaths, has provided a wealth of background data with which the Council has been able to draw up an action plan for remedying the complex ecological problems. This has included the planting of reed beds, the use of barley straw bales, the introduction of biological controls on the aquatic micro fauna and flora, the identification of a potentially new source of water and the evaluation of the quantity and quality of the lake sediment. This evaluation enabled a contract to be drawn up for both the dredging of the lake and the beneficial disposal of the spoil over the adjoining contaminated lands, thereby resolving both environmental problems simultaneously. This project has continued over four successive years, with each new student taking over the weekly monitoring schedule from the last, but with each new project containing a specifically different emphasis. This has given excellent continuity to a Council-sponsored environmental improvement scheme, and the project student has become an integral member of the regular neighbourhood forum meetings which discuss the progress of the lake rejuvenation scheme.

Horse Sands Fort

This project was the chance to assist English Heritage and the Naval Property Trust in their restoration works on one of the old Solent gun forts and had as its goal the understanding of the local hydrogeology. This was undertaken by studying the old, deep water wells which are present on each of the Solent forts, and relating this to the known geology and hydrogeology on either side of the Solent. The practical work has involved test pumping of the deep water wells to define their sustainable quantity and water quality, and has resulted in the ability to use the well water for jet scouring and other cleaning and sanitary uses on the old fort. The academic benefit has enabled the change in groundwater quality and aquifer characteristics to be defined under the Solent by being able to access previously inaccessible sampling points in the middle of the sea.

Project requirements and benefits

The foregoing project case histories illustrate very real examples of how the goals of local environmental improvements and academic project requirements can be met simultaneously, with a range of beneficial spin-offs in terms of both student experience and local Council financial savings, as summarized in Table 18.1.

The overriding requirement for conducting these projects is excellent relations and communications with the local Council at project leader level. In our case, this has involved three separate Council departments: Parks and Leisure; Engineering; and Environmental Health.

Table 18.1. Summary of benefits deriving from local student urban environmental geoscience projects

Benefits to students	Benefits to University and Department	Benefits to local Council
Involvement in a real project breeds enthusiasm	Students enthused with real projects study better in the final year as they realize the relevance of specific course material. This translates into better grades and ultimately a better overall rating of the department	Allows a large number of possible environmental projects to be screened and ranked in importance from the student project appraisals
Involvement in the local community engenders self respect and builds oral communication skills	Departmental involvement in the field investigations and analytical laboratory work is usually re-imbursed on a cost-plus basis for the equipment and chemicals used	Enables the targeting of scarce resources to the most urgent sites
Involvement with local democracy, Council politics, committees, budgets and financing is good overall experience	Additional analytical and/or consultancy work for the staff results in a significant income for the department which enables the purchase of more equipment to extend the analytical capabilities of the department	Enables additional SCA funding to be claimed from central government to investigate and cleanup more project areas than would otherwise be possible
Background (desk study) research is good training and practice	Closer ties with the local Council and neighbourhood communities result in further projects and requests for speakers at neighbourhood meetings etc.	By limiting Council involvement to providing advice, permissions and basic resources, allows more efficient use of limited Council resources in bringing about environmental improvements
Organizing and scheduling personnel and equipment for sitework is excellent practice and training	University and/or Departmental involvement is often noted in press reporting of the more serious public projects, thus raising the university's profile within the wider community	Supplies large amounts of background data which can be used in a wide variety of ways for evaluating local development projects
Use of field equipment in real project situations is good training.		Promotes closer ties with the University and its departments which lead to mutually beneficial future collaboration
Use of laboratory equipment to generate real results facilitates good practice and cultivates responsibility for quality work		
Realizing an environmental improvement, as the end product of the work reaches fruition, builds morale and satisfaction		
The accomplishment of the project is good for the CV		

Continuing contacts with the Council enable future student projects to be identified. Project recognition is then followed by initial meetings to lay down the overall guidelines and areas of responsibility which leads to discussion of details, such as site access, security, health and safety precautions and responsibilities. In this way, a series of projects can be offered to the students. Once a student has chosen a project and been accepted, it is then necessary to have further meetings to introduce the student to the Council, work up a schedule of site visits/fieldwork, and to clarify the overall project objectives in conjunction with the student. This can be rather time consuming for the supervisor, who also needs to keep a watch over the student's work during the summer period to ensure that good quality data are being collected. Such close contacts require that the project area, the university department and the Council and/or Non-Governmental Organization (NGO) are all physically close to each other so that meetings can be arranged and attended at short notice and good overall control can be kept. Additionally, some quality assurance needs to be exercised, especially on the laboratory results, to ensure that good data are produced, since important decisions are likely to be taken on the basis of the data.

References

Cuckson, D. 1998. Contaminated land: problems of liability. *This volume*.

Walton, N. R. G. & Higgins, A. 1998. The legacy of contaminated land in Portsmouth: its identification and remediation within a socio-political context. *In*: Lerner, D. N. & Walton, N. R. G. (eds) *Contaminated Land and Groundwater: Future Directions*. Geological Society, London, Engineering Geology Special Publications, **14**, 29–36.

19 Developing a degree in environmental geology

Michael Rosenbaum

Summary

- The concept of environmental geology encompasses a wide range of natural science subjects and, while a third of the earth science departments in British universities are currently offering some kind of environmental option within their degree courses, developing a separate degree in environmental geology with an effective curriculum therefore requires a clear statement of aim.
- An understanding of how geology interacts with the activities of humans, emphasizing recognition of problems and site assessment, is one such aim.
- Practical applications of the subject can be developed, particularly in the context of environmental site assessment, by students gaining an understanding of the theoretical, technical and legislative issues concerning environmental geology and how these influence the work of the professional environmental geologist dealing with environmental impact assessment and resource management.
- Is environmental geology really appropriate for an undergraduate (BSc) course, or would a postgraduate (MSc) course based on a relevant first degree suit both student and employer better? A feasible alternative may be the new Master in Science (MSci) degree.

What is meant by environmental geology?

According to the 1994 Careers Research and Advisory Centre (CRAC) guide discussed by Woodcock (1995), one-third of the earth science departments in British universities are currently offering some kind of environmental option within their degree courses (Careers Research & Advisory Centre 1994). However, a specific meaning for the term 'environmental geology' has yet to find wide acceptance; it can range from global concepts of environmental science to specific issues such as the problems of pollution and contamination. For example, Nathanail (1995) has reviewed the definitions for this discipline, but in fact subsequent correspondence with a range of university departments reveals a lack of agreement.

Most definitions seem to accept environmental geology as being the science devoted to the investigation, study and solution of problems which arise as a result of the interaction between geology and human activities. This includes, the prediction, prevention and mitigation of geological hazards, which are also studied by other applied earth scientists, for example earthquakes by seismologists, volcanoes by volcanologists, groundwater supply by hydrogeologists, and landslides by engineering geologists or geotechnical engineers.

The environmental geologist provides the scientific framework within which the other specialists operate and is particularly concerned with the interactions between the various different earth surface processes and in providing a risk assessment of potential hazards to humans and their infrastructure. The discipline appears at first sight to be similar to that covered by engineering geology, and some would consider the two fields to be essentially the same (Oliveira 1993). Nevertheless, the emphasis for the engineering geologist tends to be with the problems associated with civil engineering design and construction, where a detailed consideration of geotechnical engineering is necessary (notably hydrogeology, soil and rock mechanics). This contrasts with the role of the environmental geologist which tends to be concerned with broader issues of ground behaviour and places greater emphasis on problems associated with pollution and resource use notably concerned with hydrogeology and geochemistry.

The starting point for developing any degree course in environmental geology must be a clear statement of the aim and objectives, responding to the perceived future needs and demands of potential students and their future employers. The key objectives for the environmental geology student will be to gain an understanding of the theoretical, technical and legislative issues of

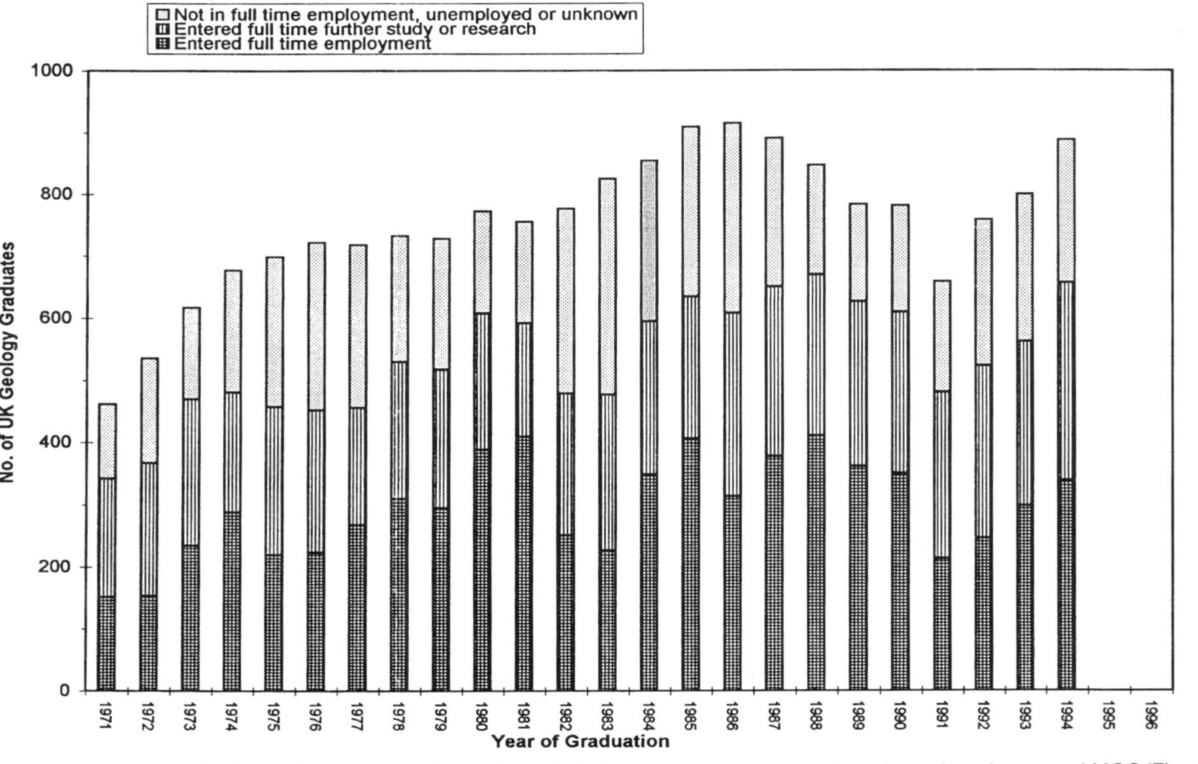

Fig. 19.1. First destinations of geology graduates from UK Higher Education institutions, based on the annual UGC 'First Destinations of University Graduates', succeeded by the annual Universities Statistical Record.

the subject and how these influence the work of professional environmental geologists concerned with environmental impact assessment and resource management.

In brief, given the wide range of topics involved, to work as an environmental geologist requires a broad knowledge of science and an awareness of engineering practice. It is necessary to work alongside other specialists whose background (university training and industrial experience) is unlikely to be the same, but with whom it is necessary to establish respect and be able to communicate effectively.

The need for environmental geologists

Planning and environmental divisions within local government and civil engineering consultants are currently the main employers of environmental geology undergraduates. The environmental industry increasingly requires new employees to have a broad environmental knowledge, to be proficient in the basic sciences (mathematics, physics, chemistry and biology) and to have wider skills in communication than achieved hitherto. The work is primarily concerned with defining ground conditions in the context of pollution, as part of the environmental site assessment process, and as an integral part of the team concerned with ground remediation policy, design and execution. Scientific and economic investigations are also increasingly conducted by environmental geologists for national geological surveys, environmental protection agencies and national government departments.

Ideally, environmental geology graduates should be trained to be articulate and logical, have a sound grasp of scientific principles, be practical in their approach and yet sympathetic to the needs of society. During their undergraduate studies, students should acquire organizational skills, self-reliance and practise working in teams, either during mapping projects or during industrial placements.

Based on figures for the past 20 years the overall demand for geology graduates in Britain has remained fairly steady. Wider recognition of environmental problems and enhanced trade with our partners in Europe should broaden employment prospects further. The demand for postgraduate geologists remains high, especially where individuals also have relevant work experience.

The components of environmental geology

Environmental geology is usually concerned with defining the ground conditions in and around a site of concern. This requires a knowledge of the geomorphology, lithology, stratigraphy, structure and groundwater conditions in three dimensions along with the environmental state (e.g. stress, pore fluid pressure, temperature and chemical equilibrium) and how these variables are likely to change with time. The subjects which need to comprise a degree with the title 'Environmental Geology' are therefore determined by the aim of the course, whether this be academic – involved in broad exploration of scientific principles – or vocational – focused towards employment.

In the definition of scope of environmental geology outlined in this chapter the following attributes would be required in a degree course to enable practice as an environmental geologist.

1. A thorough understanding of geology is required. This can be provided by the first and second years of an earth science degree.
2. An awareness of how humans interact with the geological environment. This can be provided in specialist courses such as engineering geology, waste management and water pollution.
3. Knowledge as to which interactions are likely to be adverse, typically provided by courses in geohazards and environmental assessment.
4. Ability to put this knowledge into practice, requiring a knowledge of techniques comprising the investigation and analysis of results. This can be provided by courses in environmental geochemistry and geophysics, Geographical Information Systems (GIS), and backed up by individual project work, course work and presentations.

The broadening of the school curriculum with the arrival of the National Curriculum is leading to a lower information content in each subject at GCSE level, reflected also in A-level courses, but a greater reasoning power. More pupils are becoming aware of the importance of understanding environmental issues and interest in studying environmental geology is increasing. However, the greater range of A-level subjects available is leading to a reduction in the number of university candidates able to offer mathematics and two fundamental sciences at A-level.

Nevertheless, a sound foundation of physical science and mathematics is required to pursue environmental geology at research level or within industry, implicitly including proficiency at data acquisition, compilation and analysis. Chemistry plays a principal role, particularly with regard to pollution. Mathematics is increasingly important, notably in data handling and hydrogeology, and is essential in geophysics. Physics is useful in understanding many geological concepts, and of course is essential in geophysics. Nuffield physical science is welcomed, since it allows a wide spectrum of A-level study. Biology is a most useful background for palaeontology, and for the understanding of geological environments and their evolution, as well as giving wider recognition to the role of the biota in geological processes. Geology is useful but not essential; about half the students admitted to current courses have geology A-level. Students registering for a four-year Master in Science (MSci) undergraduate programme in environmental geology would be expected to have similar qualifications to those entering for a BSc degree.

Numbers of students applying for admission to the environmental geology degree courses in Britain remain buoyant, largely due to the diversity of courses available within the framework of the undergraduate programme. However, their preparation for an honours course is uneven. Many candidates do not offer more than two of mathematics, physics and chemistry, some only one, and just half offer geology at A-level. Good candidates are nevertheless sought who, through no fault of their own, lack the broad range of foundation science required for studying geology.

Course provision

BSc degrees

Most of the BSc degree programmes offering environmental

Table 19.1. Selected list of Environmental Geology and related courses in the British Isles

Institution	Type of course
Birkbeck College, University of London	BSc Environmental Geology
The University of Birmingham	BSc Environmental Geoscience
Camborne School of Mines	BSc Environmental Science and Technology
University of Wales Cardiff	(1) BSc Environmental Geoscience
	(2) MSc Applied Environmental Geology
University of Derby	BSc Applied Environmental Earth Science
The University of Edinburgh	BSc Environmental Geoscience
The University of Greenwich	(1) BSc Environmental Geology
	(2) MSc Environmental Assessment
	(3) MSc Environmental Risk Assessment
	(4) MSc Geomaterials
Imperial College, University of London	(1) MSci Environmental Geology
	(2) MSc Environmental Technology
	(3) MSc Soil Mechanics and Environmental Geotechnics
Keele University	BSc Jt Hons in Geology and Environment
Kingston University	(1) BSc Environmental Geology
	(2) MSc Earth Sciences and the Environment
Lancaster University	BSc Environmental Science
University of Leeds	BSc Environmental Science: Geology Option
University of Newcastle	MSc Environmental Biogeochemistry
	MSc Environmental Geotechnology
Nottingham Trent University	MSc Contaminated Land Management
Open University	BSc module entitled Physical Resources and Environment
University of Portsmouth	(1) BSc Applied Environmental Geology
	(2) MSc Contaminated Land
	(3) MSc Geohazard Assessment
Royal Holloway, University of London	(1) BSc Environmental Geology
	(2) MSci Environmental Geoscience
Staffordshire University	BSc Environmental Geology
The University of St Andrews	BSc Environmental Geology
Trinity College, Dublin	BSc Geology (Environmental Geology module)
University College, University of London	MSc Hydrogeology

geology (Table 19.1) within the three-year model do so by enabling students to select from a number of options in their final year, sometimes guiding their choice through the second year as well, closely linked to the traditional BSc geology degree rather than providing a stand-alone course. The foundation of geology obtained in the first two years of the standard geology/earth science degree needs to be developed into an understanding of its environmental applications and to integrate all the taught elements in the third year.

The patchy coverage of physical sciences and mathematics by the first year intake discussed above generally necessitates, at Imperial College at least, the early introduction of foundation courses in mathematics, physics and chemistry. The variable knowledge possessed by incoming students is requiring progressively longer courses in these subjects, eroding the time available for the geological sciences and their applications. Since less than half of the first year intake have studied geology at A-level, the majority of geological courses start with the assumption of no previous formal training in the subject.

In their second year, students generally continue to study the main branches of geology as in the first year. This subject core is then usually augmented by courses designed to: (1) teach the necessary statistical and computing skills relevant to the earth sciences; (2) to introduce the economic and industrial aspects of geology; and (3) prepare students for the specialized courses of their third year and for industrial or independent work in the summer vacation between the second and third year.

During the summer vacation students generally undertake an independent field project. This vacation work occupies from three to eight weeks and is usually followed in the third year by an equivalent period of laboratory and office work necessary to complete maps and prepare a research report.

The workload on students has been progressively increased over recent years in an attempt to redress the loss in foundation teaching of geological subjects and to incorporate the developments, particularly in techniques such as computing and data analysis, which are seen as being of paramount importance to successful career development in research or industry. However, in many degree programmes, the upper limit has been reached for the student load, in particular for the number of lectures which students can absorb each week and for the number of hours of private study required to support the practicals. Compulsory attendance of field courses further erodes the available time and energy for students to cope with the work demands.

MSci degrees

The increased workload on students has led some departments to investigate the possibility of a four-year structure to their undergraduate degree, awarding the Master in Science (MSci) degree on successful completion. This mirrors the developments in other physical sciences (e.g. chemistry and physics) and in engineering (e.g. civil engineering and environmental engineering). This enables students to develop their skills of scientific analysis and their understanding of the practical and industrial applications of their discipline beyond the somewhat restricted exposure which can be provided through short option courses in a typical BSc programme. This has the potential to allow the research strengths and industrial links of a department to be used in a direct way within the teaching programme, particularly in connection with project work and industrial placements.

Discussion

The course structure adopted by most departments provides a common set of geological courses attended by all geoscience students for the first two years, enabling a sound foundation to be laid in geological and physical sciences. In later years, many of the topics are directed towards specific applications, or to the tools and methodologies needed for those applications. In the degree course modules listed in the Careers Research and Advisory Centre (CRAC) guide for undergraduate degrees covering the geological sciences (Careers Research & Advisory Centre 1994), a number have clearly been accepted as forming an essential basis for any geoscience degree. However, out of the 24 modules listed, it can be argued that just seven are of direct relevance to environmental geology: geochemistry, geophysics, earth resources, geomorphology, engineering geology/hydrogeology, computer geology and Quaternary geology. Nevertheless, on inspecting the contents of the undergraduate degrees concerned with environmental geology, only 20% of such degrees have all seven courses compulsory within their curriculum. A further 10% would allow all seven to be taken if optional courses are included in the analysis.

Therefore, more than two-thirds of the environmental geology undergraduate degrees lack one or more of the seven environmentally-orientated courses listed in the CRAC tabulation. Most significant amongst these omissions appear to be geophysics, geomorphology and engineering geology/hydrogeology, compulsory in just half of the environmental geology degree courses, and Quaternary geology which is compulsory in only 40% of such degree courses. One environmental geology degree within the CRAC listing would appear to allow the students to opt out of all seven of these courses, yet still manage to acquire a degree entitled BSc Environmental Geology! Courses commonly regarded as being essential within this field are geochemistry and earth resources, and these are compulsory in 80% of the environmental geology degree courses. There is evidently a very wide range of degree syllabi which carry the title BSc Environmental Geology and a significant number of these lack at least one major environmental topic.

The common course structure also enables the geological applications to be brought to the attention of students pursuing a more academic course of study, widening their appreciation of the applications of their chosen discipline. During the first two years, courses of a suitably broad nature include titles such as Applied Geology, Elements of Environmental Geology, Foundation Geology, Science of the Global Environment, and Surface Processes. At the present time, the proportion of students pursuing an environmental geology degree course usually falls between 15% and 25% of the total number studying geological sciences, a figure which, with approximately one-third of British university geology departments offering this course of study, gives an annual graduation of between 100 and 150 environmental geologists.

The breadth of options available in Year 3 is seen by many students as a great attraction to pursuing a particular department's undergraduate programme, but for the department this leads to a dilemma between teaching combined classes, inevitably with students of varying background knowledge, and teaching similar subject material several times over to small classes. Nevertheless, the options add breadth to a course and help students develop specific interests during their final year.

The specific courses which can be offered have, of necessity, to reflect the experience of the staff, but should not be at the expense of providing a broad overview of the most relevant areas. This means that external lecturers may need to participate, although opportunities to involve practitioners from industry is seen by many as being of positive benefit. The list of courses required at this level is by no means firmly established. Table 19.2 lists those commonly offered as part of environmental geology degrees in the British Isles.

The length of the list in Table 19.2 alone makes the task of timetabling, let alone staffing, an environmental geology degree far from trivial. Yet few broad-based environmental consultants would be likely to agree on which of these aspects could be readily omitted. There would seem to be little prospect of undergraduate training on its own being capable of providing a sufficiently in-depth education for individuals to enter industry with a strong technical expertise; yet to raise their awareness of the important issues and to develop their skills of research and communication should be well within grasp. A practical goal should therefore inspire and encourage rather than swamp with technical detail.

Independent work by students normally becomes increasingly important towards the end of their degree programme, and individuals need to be encouraged to develop their own scientific studies, starting with data handling and analysis within a well-defined project controlled by academic staff, but then going on to problem recognition, hypothesis creation and rigorous testing using well-planned data acquisition strategies. Presentation includes traditional reports and vivas, but needs to be extended to increase awareness of other important issues such as legislation, social and political constraints, and professional practice. Group work supported by tutorials, followed up by open seminars and peer-reviewed presentations can effectively develop both technical skills and professional awareness.

Table 19.2. Options typically offered as part of Third Year Environmental Geology courses in Britain. Note that they are arranged alphabetically and not in order of importance.

Construction geology
Contaminated land
Data analysis (statistics, geostatistics)
Environmental auditing
Environmental geochemistry
Environmental geotechnics
Environmental impact assessment
Environmental impact of quarrying and mining
Environmental policy and legislation
Environmental site assessment
Geohazards
Geomaterials
Ground remediation
Hydrogeology
Palaeoclimatology
Personal transferable skills
Professional role of the environmental geologist
Quaternary environments
Risk analysis
Rock mechanics
Seismology
Site investigation
Soil mechanics
Soil pollution
Surficial geochemistry
Surficial geophysics
Techniques (sampling, surveying, GIS, remote sensing)
Urban geology
Waste management and disposal
Water pollution

Several methods of assessment can be employed, including: fieldwork reports, tutorial presentations, poster presentations, seminars and practical work, reports, and dissertations on individual projects. To learn to form part of a team and to develop skills of leadership and active participation are seen as important personal skills to acquire. Indeed, a course specifically devoted to personal transferable skills forms part of the curriculum in some institutions, often focused towards the requirements of industrial employers and career development, so making the applicability to students' needs all the more apparent.

The pressures on departments to optimize the use of their staff has inevitably led to financial criteria being used as a measure of efficiency; it has therefore become important for duplicate and additional teaching to be minimized. Nevertheless, environmental geology by its very nature encompasses a broad spectrum of the geosciences and requires cross-discipline frameworks to be established. It is not feasible for the majority of students to bridge the discipline gaps on their own, and therefore considerable synthesis tuition is required. This is not covered by existing courses, and so has significant implications for the management of staff time. This means that unless new staff are appointed with specific teaching responsibilities for the new degrees in environmental geology, the teaching for any new courses must take advantage of existing ones, either within the host department or elsewhere within the university. Even so, additional staff time will inevitably be required to supervise independent project work in the final year, and additional staff time is also required to set up tutorials. Allowances of the order of £1000 per student will probably need to be budgeted to cover the costs of field work, analytical support and consumables.

The issue of staffing remains a problem for many departments. Lack of investment in both staff numbers and training in the new field of environmental geology is still significantly restricting module development, and preventing students from realizing their full potential from the courses on offer. The size of most geology departments is too small to have more than one or two academic members of staff specializing in environmental geology, although others may have a significant interest in related subjects, perhaps in geochemistry or hydrogeology. This is still a new area and the relative importance of the component disciplines has not yet become established. Development is being driven from three directions: (1) training students for work in the environmental industry; (2) satisfying the inquisitive undergraduate; and (3) boosting the student population entering the geological sciences.

The question remains unanswered as to what should actually be covered within a degree entitled BSc Environmental Geology. Some authorities would see this as a geological extension of environmental science, providing a broad view of human influence on the environment and *vice versa*, while others would see environmental geology as a study of the ways in which sustainable development of natural resources could be achieved. Yet others would take a much more focused view, towards the investigation of specific sites with regard to the potential risks arising from geohazards and previous land use, involving legal and technical aspects of the management of contaminated land. For the potential student wishing to select a course appropriate to their interests, these different approaches provide a significant obstacle in the selection process. Degree title alone cannot give the full flavour of the course and even at interview this may not easily come across. Flexibility for course transfer therefore remains an important aspect in degree planning, not only to help students develop a course to their satisfaction but also to let others join in who may not have realized hitherto the opportunities which exist in this new field at the time of admission!

Should environmental geology be an undergraduate or a postgraduate course?

The concepts of environmental geology developed so far tend towards the vocational, but it is recognized that considerable differences lie between the expectations of an undergraduate class and those of a postgraduate class. Undergraduates need to acquire a firm knowledge of the basics of their subject and gain confidence in the techniques used for description and analysis. Postgraduates have already acquired the basic knowledge and have often experienced the requirements and expectations of industry, so need to develop their skills of problem recognition and increase their confidence for analysis and decision-making.

As recognized by De Freitas (1994) in the context of engineering geology courses, the teaching of an environmental geologist essentially has three principal components.

1. The acquisition of knowledge in order to be able to identify problems, so requiring an understanding of scientific theory.

An intellectual framework needs to be established in order for this knowledge to be used.

2. An ability to investigate a site where a potential problem exists, including the acquisition, management and analysis of the information, so requiring a knowledge of technological methods. This requires a logical and methodological approach together with an awareness of acceptable practice and the constraints of cost and time.
3. An awareness of possible preventative and remedial measures which could be brought to bear, so requiring an understanding of engineering practice, particularly civil engineering and environmental engineering.

The training of an environmental geologist can be seen to have three similar principal components:

1. gaining an awareness of the nature of real problems and the social, commercial and political pressures which need to be considered;
2. appreciating the standards of working practice, technical knowledge, legislation and the expectations of society at the company, national and international level;
3. learning a professional approach for coping with unknown processes, inadequate information and extreme ground conditions and being given limited resources with which to tackle the problems.

Such teaching and training will take time to acquire and will require industrial experience as well as academic study. In the field of engineering geology, a similar education is required and the individual's development is guided by the accreditation procedures leading to Chartered status organized through the Geological Society and the Institutions of Civil Engineers and Mining & Metallurgy. Similar training is required for those wishing to pursue professional careers in environmental geology, for which the Institution of Water and Environmental Management provides an alternative route to Corporate Membership and thus Chartered Engineer status.

No comparable continental European undergraduate course is as short as three years; most students take at least five years after leaving school before they graduate. The validity of the British degree in Europe is therefore in question, a situation which the professional institutions are trying to address. The BSc is already unacceptable as an entry qualification for pursuing a PhD in Germany, and an extra (DEA) year of preparation is required in France. Full professional registration by European bodies such as FEANI generally requires at least four years of university education as well as some years training in industry, for which the three-year BSc is clearly inadequate, putting pressure on the British education system to extend their course duration.

Should environmental geology therefore be an undergraduate or a postgraduate course, or can either be acceptable? The ability to recognize problems requires an extensive knowledge of the subject and an understanding as to the significance of the various possible situations which can arise. This is difficult to acquire from lectures and reading alone, judging from personal experience of teaching courses in applied geology, and needs exposure to industrial practice to be effective. The time within the traditional British three-year BSc undergraduate degree is too limited to do more than superficially introduce the student to the basic operations of industrial practice. The 'sandwich' concept with a year in industry prior to the final year of study is a successful approach applied to engineering degrees, but is rare in the geosciences. However, prior to commencing a one-year MSc postgraduate degree there is opportunity for acquiring direct industrial experience. The recent moves towards developing the Master in Science (MSci) degree as an undergraduate course goes some way towards providing those students who wish, the opportunity of gaining experience from industry, primarily through project work with attachments to industrial organizations supplemented by literature reviews of case histories, but personal experience tends still to be academically orientated.

In the author's experience, a junior professional environmental geologist's work is primarily concerned with defining ground conditions concerning pollution, as part of the environmental site assessment process, and as an integral part of the team concerned with ground remediation policy, design and execution. The main employment opportunities for environmental geologists are provided by government organizations and engineering consultancies. Industry requires new employees to have a broad knowledge of the environmental field but also to be proficient in the basic sciences and to have wider skills in communication than frequently achieved by three-year BSc courses. Indeed, it is becoming increasingly the case that employers are looking for a competent individual to work as part of a team within their organization, to solve a particular problem or undertake a specific task. Employers are now often looking for skills such as team work, willingness to learn, flexibility and enthusiasm, rather than technical knowledge. The undergraduate programme therefore, which focuses on comprehensive coverage of the subject at the expense of personal transferable skills, is increasingly likely to fail, both at providing a technical coverage to the standards required for industrial applications, and at developing the skills of the individual for performing within the professional team.

Some might argue that only a period spent in industry will equip the individual with a sufficiently broad experience to cope with the demands of a professional job as an environmental geologist, but there remains the problem for the individual of getting accepted even for that first post. Once in post, graduate environmental geologists can begin to develop their skills and experience to the extent that their ambition allows. This suggests that the BSc + industrial experience + MSc route still provides the best foundation for a career in the environmental geology field. However, an appropriate balance of training may be difficult to achieve. Indeed, it is not clear from where the individual obtains the advice and support necessary to guide them through this process. There is, therefore an argument for the concept of an extended undergraduate programme within which experienced staff will be able to guide the student's development. The four-year MSci programmes are beginning to fulfil this role, although it is doubtful whether the direct experience of industrial practice can reach the level of achievement possible by working directly within industry, an approach facilitated by the sandwich structure of a traditional engineering degree.

Most of the new MSci courses are designed to equip their students for such roles by providing a thorough foundation of physical science and providing a deeper coverage of the applications of geology to environmental problems, paying particular attention to data acquisition and analysis. Substantial research projects form an important element of the fourth year, and can benefit greatly from being undertaken in collaboration with industry or conducted in an approved European institution.

Conclusions

Proposals to launch four-year MSci degrees in environmental geology have arisen in response to changing demands in professional employment and to changes in school education. Departments which have been running geology degrees with environmental options have been faced with increasing difficulties in providing adequate coverage of theory within the traditional three-year structure. Increasingly the practical applications have had to give way to make space for increased attention to foundation courses in physical science, and there is little time for students to pursue their own interests in the subject. The existing BSc courses continue to run, but are increasingly being designed for students not intending to commit themselves to a career in environmental geology or to pursuing research at postgraduate level. The geological science foundations will achieve a similar standard of training in both the BSc and MSci programmes, but less attention will be paid to the techniques and analysis required for identifying and solving environmental geology problems during the BSc courses and there will be comparatively limited opportunities for individual research projects. Since it is felt unlikely that most students will be in a position to realistically choose their career path before they commence their undergraduate studies at university, departments need to maintain a policy of providing courses which enable students to defer as long as possible their decision as to whether to pursue the MSci or BSc route.

The aim of the new four-year Master in Science (MSci) Environmental Geology degree is to prepare students for entering the environmental industry or pursuing an industrial or academic career after taking their doctorate, providing them with an education which adequately prepares them for their chosen professional life at a standard equivalent to that achieved in first degrees at leading universities in continental Europe. It has been argued that there would seem to be little prospect in the future of undergraduate training on its own being capable of providing an in-depth training for an individual wishing to enter directly the environmental industry, yet to provide an education able to raise the students' awareness of the important issues and to develop their skills of research and communication should be well within grasp. A practical goal is therefore to inspire and to encourage rather than to swamp with technical detail.

Acknowledgements

The author would particularly like to thank the following individuals and their institutions for having furnished details of their environmental geology courses, enabling the trends identified in this chapter to be substantiated: Hilary Downes (Birkbeck College); Bill Gaskarth (Birmingham); David Watkins (Camborne School of Mines); Simon Wakefield; Charles Harris (Cardiff); Peter Regan (Derby); Nicola McEwan (Edinburgh); Peter Doyle (Greenwich); Peter Floyd (Keele); Bob Stokes (Kingston); Harry Pinkerton (Lancaster); Jared West (Leeds); Matthew Collins (Newcastle); Paul Nathanail (Nottingham Trent); Gill Foulkes (Open University); Andrew Poulsom, Nick Walton and Bill Murphy (Portsmouth); Robert Hall (Royal Holloway); David Roberts (Staffordshire); Grahame Oliver (St Andrews); Chris Stillman (Trinity College); and William Burgess (University College London). The author would also like to thank former colleagues at Imperial College for their interest and support in developing the environmental geology courses which now form part of the College curriculum.

References

Careers Research & Advisory Centre (CRAC). 1994. *Degree course guide: Geology and Environmental Science 1994/95*. Hobsons, Cambridge.

De Freitas, M. H. 1994. Keynote lecture: teaching and training in engineering geology: professional practice and registration.

Proceedings of the 7th Congress of the International Association of Engineering Geology, Lisbon, September 5–9, 6: lvii–lxxv.
Nathanail, C. P. 1995. What is environmental geology? *Geoscientist*, **5**, 14–15.
Oliveira, R. 1994. Engineering geology and the environment. *Newsletter of the International Association of Engineering Geology*, **20**, 3.
Woodcock, N. 1995. Environmental geology: educational threat or opportunity? *Geoscientist*, **5**, 11–13.

Index

Page numbers in *italics* refer to Figures or Tables